An Introduction to Modern Nonparametric Statistics

Daniel, *Applied Nonparametric Statistics*, 2nd ed.

Derr, *Statistical Consulting: A Guide to Effective Communication*

Durrett, *Probability: Theory and Examples*, 2nd ed.

Graybill, *Theory and Application of the Linear Model*

Johnson, *Applied Multivariate Methods for Data Analysts*

Kuehl, *Design of Experiments: Statistical Principles of Research Design and Analysis*, 2nd ed.

Larsen, Marx, & Cooil, *Statistics for Applied Problem Solving and Decision Making*

Lohr, *Sampling: Design and Analysis*

Lunneborg, *Data Analysis by Resampling: Concepts and Applications*

Minh, *Applied Probability Models*

Minitab, Inc., *MINITAB™ Student Version 12 for Windows*

Myers, *Classical and Modern Regression with Applications*, 2nd ed.

Newton & Harvill, *StatConcepts: A Visual Tour of Statistical Ideas*

Ramsey & Schafer, *The Statistical Sleuth*, 2nd ed.

SAS Institute Inc., *JMP-IN: Statistical Discovery Software*

Savage, *INSIGHT: Business Analysis Software for Microsoft® Excel*

Scheaffer, Mendenhall, & Ott, *Elementary Survey Sampling*, 5th ed.

Shapiro, *Modeling the Supply Chain*

Winston, *Simulation Modeling Using @RISK*

To order copies, contact your local bookstore or call 1-800-354-9706. For more information, contact at 20 Davis Drive, Belmont, CA 94002-3098, USA, or go to www.cengage.com/statistics.

An Introduction to Modern Nonparametric Statistics

James J. Higgins

Kansas State University

BROOKS/COLE
CENGAGE Learning

Australia • Brazil • Japan • Korea • Mexico • Singapore • Spain • United Kingdom • United States

An Introduction to Modern Nonparametric Statistics
James J. Higgins

Editor: Carolyn Crockett

Assistant Editor: Ann Day

Editorial Assistant: Julie Bliss

Technology Project Manager: Burke Taft

Marketing Manager: Joseph Rogrove

Project Manager, Editorial Production: Ellen Brownstein

Print/Media Buyer: Barbara Britton

Permissions Editor: Sue Ewing

Production Service: Scratchgravel Publishing Services

Copy Editor: Carol Reitz

Cover Designer: Vernon Boes

Compositor: Scratchgravel Publishing Services

© 2004 Brooks/Cole, Cengage Learning

For product information and technology assistance, contact us at
Cengage Learning Customer & Sales Support, 1-800-354-9706

For permission to use material from this text or product,
submit all requests online at **www.cengage.com/permissions**
Further permissions questions can be e-mailed to
permissionrequest@cengage.com

Library of Congress Control Number: 2002117217

ISBN-13: 978-0-534-38775-4

ISBN-10: 0-534-38775-6

Brooks/Cole Cengage Learning
20 Davis Drive
Belmont, CA 94002-3098
USA

Cengage Learning is a leading provider of customized learning solutions with office locations around the globe, including Singapore, the United Kingdom, Australia, Mexico, Brazil, and Japan. Locate your local office at **www.cengage.com/global**

Cengage Learning products are represented in Canada by Nelson Education, Ltd.

To learn more about Brooks/Cole, visit **www.cengage.com/brookscole**

Purchase any of our products at your local college store or at our preferred online store **www.cengagebrain.com**

Printed in the United States of America
3 4 5 6 14 13 12 11

Contents

3 *K*-Sample Methods   79

4 Paired Comparisons and Blocked Designs 109

5 Tests for Trends and Association 145

8 Nonparametric Bootstrap Methods   249

9 Multifactor Experiments   301

10 Smoothing Methods and Robust Model Fitting   323

Preface

Modern nonparametric statistics comprises a broad range of methods for data analysis. Rank-based methods, permutation tests, bootstrap methods, and curve smoothing fall under this heading. Most normal-theory methods have nonparametric counterparts that may be used when the normal-theory assumptions are violated. Nonparametric methods may also solve problems for which conventional methods do not apply or are difficult to implement. These problems include the analysis of sparse contingency tables, tests for ordered alternatives, one-sided multivariate tests, and the analysis of censored data.

This book presents a wide array of nonparametric methods that researchers will find useful in analyzing their data. It is appropriate for advanced undergraduates and beginning graduate students in fields such as the life sciences, engineering, medicine, social sciences, and statistics. Students are expected to be familiar with topics typically covered in an introductory methods course such as t tests, inferences for proportions, simple linear regression, and one-way analysis of variance. A few topics in the latter part of the book require a more advanced background, including multiple regression and two-way analysis of variance.

At one time nonparametric methods were thought of as quick, hand-calculation methods, suitable only for simple designs and small data sets. However, the availability of high-speed computing to carry out time-consuming or difficult computations has enabled nonparametric statistics to come into its own as a field of statistics. Computationally intensive methods such as permutation tests, bootstrap estimation, curve smoothing, and robust methods, along with the classic methods based on ranks, have become part of the statistician's tool kit.

The selection of material for this book was guided by data analysis problems that commonly arise in practice. In addition to the familiar one-sample, two-sample, and k-sample procedures, nonparametric counterparts to analysis of variance, multiple regression, and multivariate analysis are considered. Categorical data methods include exact permutation tests for contingency tables and tests for tables with ordinal classifications. A chapter on censored data covers common nonparametric techniques for the analysis of survival data. Bootstrap methods, nonparametric density estimation, and curve smoothing are also considered.

Where possible, the connection among methods is stressed. For instance, rank tests are introduced as special cases of permutation tests applied to ranks, and methods for censored data are treated as special cases of permutation tests applied to general scores. Some statistical theory is sketched where it is felt that it would help illuminate the methods. In most cases, the theory is set apart from the methods and may be omitted at the option of the instructor. Selected computer output and code show how methods are implemented. Although generally not needed in light of the availability of statistical packages, statistical tables are given for some of the more common tests.

Chapter 1 treats one-sample methods based on the binomial distribution. The issue of the power of statistical tests is discussed, and an example is given to show how a simple nonparametric test for a location parameter can have substantially greater power than the one-sample t test.

Chapter 2 sets a pattern that is used for the next six chapters. A permutation test is introduced to solve a problem, and then rank tests or other tests are developed as special cases or modifications of the permutation test. Chapter 2 introduces two-sample methods. K-sample methods, including multiple comparisons, follow in Chapter 3. Then come paired comparisons and blocked designs in Chapter 4, tests for trends and association including contingency table analysis in Chapter 5, multivariate tests in Chapter 6, and tests for censored data in Chapter 7. Asymptotic approximations are introduced in Chapter 2 and appear throughout. However, the need for such approximations is not as compelling as it once was as a result of the availability of computer software to carry out exact tests.

Chapter 8 introduces bootstrap methods for one-sample, two-sample, k-sample, and regression problems. In Chapter 9, both bootstrap and aligned-rank methods are given for two-factor analysis of variance with the extension to the multifactor setting being straightforward. A section on lattice-ordered alternatives extends tests for ordered alternatives to the multifactor setting. In Chapter 10, nonparametric density estimation, curve smoothing, and robust model fitting are introduced. The availability of computer software to carry out the computations makes this material accessible to the practitioner.

The material in this text is sufficient for a one-semester course. Problems at the end of each chapter are included to provide practice with the methods and, in some cases, to extend theory or methods not covered in the text.

Statistical software is indispensable for the implementation of modern statistical methods, and the area of nonparametrics is no exception. We feature four statistical packages in this book: Resampling Stats, StatXact, S-Plus, and MINITAB.

Resampling Stats

Resampling Stats comes in three versions: a stand-alone package, an add-in to Microsoft Excel, and an add-in to Matlab. The stand-alone version is a simple-to-use programming language that is specifically designed to do the type of resampling of data needed for permutation tests and bootstrap sampling. The programming language has functions that enable programmers to construct a variety of statistical procedures. With intuitive commands such as "shuffle," "sample," "repeat," and "score," Resampling Stats code may also serve as pseudocode for describing the steps needed to program nonparametric procedures in other languages. The add-ins are designed to work in conjunction with the features of the host software. Most of the methods in Chapters 2–7 can be done with this software. If students have the option of only one statistical software package to use with this book, Resampling Stats is the one to choose. The web address is www.resample.com.

StatXact

StatXact is a powerful, simple-to-use, menu-driven program that has more than 80 nonparametric procedures for continuous and categorical data. The software also has data manipulation capabilities. Power and sample size computations are available for tests for one and two binomial populations. Most of the procedures in Chapters 2–5 and Chapter 7 are included in this software package, as well as a number of procedures that we do not cover. Those who make use of exact nonparametric procedures in their statistical consulting and research will find this program extremely useful. The web address is www.cytel.com.

S-Plus

S-Plus is a powerful programming language that can be used to carry out the procedures in this book. We limit our use to those procedures that can be accessed through the S-Plus menu options. This program is especially useful for the smoothing techniques and robust methods discussed in Chapter 10. The web address is www.insightful.com.

MINITAB

MINITAB has long been a favorite of statisticians for both instruction and consulting. Some of the standard statistical procedures in Chapters 1–4 and Chapter 7 are included in the MINITAB menu. Data manipulation capabilities are included. MINITAB also has commands for carrying out rank-based regression as discussed in Section 10.3, although these are not presently documented in the help menu. The web address is www.minitab.com.

Other

The SAS® programming language, a standard for applied statisticians, has several capabilities for nonparametric statistics. PROC NPAR1WAY has a variety of two-sample and k-sample tests, and there is an option for doing exact tests. PROC FREQ has an option to do exact tests for contingency tables. PROC MULTTEST allows bootstrap and permutation sampling and may be applied in the multivariate setting. The web address is www.sas.com.

One other source of software that we have used is at a web site created by J. W. McKean. Here one can carry out rank-based regression and analysis of variance on-line. The web address is www.stat.wmich.edu/slab/RGLM.

Most sections of the book have a subsection entitled "Computer Analysis" that outlines the use of selected statistical packages. We do not intend this to be a comprehensive discussion, but rather it is to give an indication of the capabilities of various software packages.

Acknowledgments

I would like to thank the reviewers of this book for their valuable suggestions and comments. They are: R. Clifford Blair, University of South Florida; Joshua D. Naranjo, Western Michigan University; and Thomas H. Short, Villanova University. I would also like to thank Scott J. Richter, University of North Carolina at Greensboro, who taught from the original manuscript and provided valuable input.

James J. Higgins

0

Preliminaries

0.1
Cumulative Distributions and Probability Density Functions

We begin by reviewing some terminology. Random sampling can be pictured as the process of drawing numbers at random from a large basket of numbers, as in a lottery. The hypothetical basket of numbers is called a *statistical population*, and the set of randomly selected numbers is called a *random sample*.

Suppose X is a *random variable* that denotes an observation selected randomly from the population. The *cumulative distribution function (cdf)* of X is the probability that the random variable X takes on a value less than or equal to x. We denote this as $F(x) = P(X \leq x)$. For instance, if the heights of 40% of a population are less than or equal to 70 inches, then the probability that a randomly selected height X is less than or equal to 70 is .4, or $F(70) = .4$. If we are measuring continuous variables, such as weight, height, or temperature, then probabilities can be expressed as areas under a curve $f(x)$ called the *probability density function*. A probability density function $f(x)$ and its cumulative distribution function $F(x)$ are depicted in Figure 0.1.1 (see page 2). For a precise mathematical discussion of random variables and probability distributions, see Hogg and Craig (1995).

0.2
Common Continuous Probability Distributions

The most widely used statistical procedures are based on the normal distribution. The normal distribution, or more precisely the normal probability density function, is the famous bell-shaped curve with the form

$$f(x) = \frac{e^{-(x-\mu)^2/2\sigma^2}}{\sigma\sqrt{2\pi}}, \quad -\infty < x < \infty$$

1

FIGURE 0.1.1

Cumulative Distribution Function $F(x)$ and Probability Density Function $f(x)$

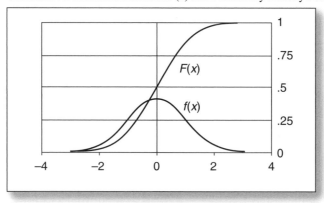

The quantities μ and σ are the mean and standard deviation, respectively, of the distribution. The special case of $\mu = 0$ and $\sigma = 1$ is called the *standard normal distribution*. We will denote the ($100p$)th percentile of the standard normal distribution as z_p. That is, if Z is a standard normal random variable, $P(Z \le z_p) = p$. Common values for z_p are $z_{.90} = 1.282$, $z_{.95} = 1.645$, and $z_{.975} = 1.96$. At times it will be convenient to index the percentiles by their upper-tail probabilities. If we do so, we will use the notation $z(\alpha)$. For instance, $z(.05) = z_{.95} = 1.645$. The usage should be clear from the context. Table A2 in the Appendix contains standard normal probabilities for selected values of z.

The normal distribution is important not only as a model for the distribution of a population, but also as a *sampling distribution*. If $\overline{X}$ is the sample mean of a random sample of size n from any population with mean μ and standard deviation σ, then as a consequence of the *central limit theorem*, $\overline{X}$ has an approximate normal distribution with mean μ and standard deviation $\sigma/\sqrt{n}$ for sufficiently large n. Intuitively, averages have approximate normal distributions regardless of the distribution of the population, provided σ is not infinite.

We will have occasion to deal with families of distributions of the form

$$f(x) = \frac{1}{b} h\!\left(\frac{x-a}{b} \right)$$

where $h(z)$ is a standard form of the distribution. The normal distribution is like this, where $a = \mu$, $b = \sigma$, and $h(z)$ is the standard normal distribution probability density function. The parameters a and b are termed *location* and *scale* parameters, respectively. They have the effect of shifting and scaling the standard distribution.

Figure 0.1.1 plots the standard normal probability density function and its distribution function. The functional forms and plots of the *uniform, exponential, Laplace,* and *Cauchy* probability density functions, which we have occasion to use in the

TABLE 0.2.1

Selected Probability Density Functions with Mean and Variance

Type	Functional Form	Mean	Variance
Uniform	$h_1(z) = 1, \quad 0 < z < 1$	1/2	1/12
Exponential	$h_2(z) = e^{-z}, \quad z > 0$	1	1
Double exponential or Laplace	$h_3(z) = \dfrac{e^{-\lvert z\sqrt{2}\rvert}}{\sqrt{2}}, \quad -\infty < z < \infty$	0	1
Cauchy	$h_4(z) = \dfrac{1}{\pi\left(1 - z^2\right)}, \quad -\infty < z < \infty$	Does not exist	Does not exist

book, are shown in Table 0.2.1 and Figure 0.2.1. A discussion of other distributions, such as the Weibull, gamma, and lognormal, is found in Bain and Engelhardt (1992).

Two features of such distributions are important in nonparametric statistics. One is the *tail weight* of the distribution. A *heavy-tailed* distribution is one that will occasionally produce observations that are much more extreme than the others.

FIGURE 0.2.1

Four Probability Density Functions

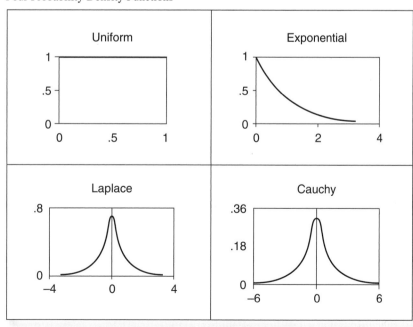

Light-tailed distributions do not produce these extreme observations. The uniform and normal are lighter-tailed distributions. The exponential and Laplace distributions have heavier tails, and the Cauchy distribution has very heavy tails, so much so that the mean and standard deviation of the distribution do not exist. The other important feature is the *skewness* or asymmetry of the distribution. The exponential distribution is skewed to the right, whereas the uniform, normal, Laplace, and Cauchy distributions are symmetric.

0.3
The Binomial Distribution

The binomial distribution plays an important role in applied statistics. It arises when random samples are taken from populations with elements that can be classified into exactly one of two categories, such as success-failure, yes-no, good-defective, and heads-tails. We may label each element of the population either 0 or 1 depending on the category to which the element belongs. We will let p denote the fraction of 1's in the population. For instance, suppose we have a population of voters and 40% of this population favors Candidate Smith. We may associate 1 with those who favor Smith, and 0 with those who do not, so $p = .4$.

Suppose we select n elements randomly from the population. Let X denote the number of 1's. The probability distribution of X is called the *binomial distribution*. It has the mathematical form

$$\binom{n}{x} p^x (1-p)^{(n-x)}, \quad x = 0, 1, \ldots, n$$

Table A1 in the Appendix lists selected probabilities for the binomial distribution. The expected value and variance of the binomial random variable are given by

$$E(X) = np, \quad \text{var}(X) = np(1-p)$$

For large n, the binomial distribution can be approximated by the normal distribution with the same mean and variance. For derivations related to the binomial distribution, see Higgins and Keller-McNulty (1995).

EXAMPLE 0.3.1 Suppose X denotes the number of heads in 100 tosses of a coin. Here $E(X) = 100(.5) = 50$ and $\text{var}(X) = 100(.5)(1 - .5) = 25$. The standard deviation is $\sqrt{\text{var}(X)} = 5$. Based on the normal approximation, there is approximately a 95% chance that X will fall within 2 standard deviations of the mean—that is, between 40 and 60. ■

0.4
Confidence Intervals and Tests of Hypotheses

The purpose of this section is to give a brief review of the notions of confidence intervals and tests of hypotheses. Most introductory textbooks have a treatment of these concepts—for instance, Moore and McCabe (2002). It is expected that students have been introduced to these topics in the context of sampling from a normally distributed population.

Suppose we wish to estimate the mean of a population. The sample mean based on a random sample of size n is called a *point estimate* of the population mean. What we want is an interval that likely contains the population mean. Let us consider an interval that may be computed when the population has a normal distribution with unknown mean but known standard deviation σ.

Since the population has a normal distribution, the standardized quantity

$$Z = \frac{\overline{X} - \mu}{\sigma/\sqrt{n}}$$

has a standard normal distribution for any sample size. If the population does not have a normal distribution, it has an approximate normal distribution for large sample sizes. Since a standard normal random variable Z has the property $P(-1.96 < Z < 1.96) = .95$, we have

$$P\left(-1.96 < \frac{\overline{X} - \mu}{\sigma/\sqrt{n}} < 1.96\right) = .95$$

A *95% confidence interval* for the mean of the population is obtained by solving this inequality for μ. We get the interval

$$\overline{X} - 1.96\frac{\sigma}{\sqrt{n}} < \mu < \overline{X} + 1.96\frac{\sigma}{\sqrt{n}}$$

It is a 95% confidence interval in the sense that among all intervals of this type, 95% actually contain the population mean and 5% do not. For other levels of confidence, the 1.96 is replaced by other values. For instance, 1.645 leads to a 90% confidence interval and 2.576 leads to a 99% confidence interval.

EXAMPLE 0.4.1 Suppose it is known that the heights of the male population have a normal distribution with a standard deviation of 3. Suppose in a random sample of size 20 from the population we find that the sample mean is 70.8 inches. Then a 95% confidence interval for the mean height of this population is

$$70.8 - 1.96 \frac{3}{\sqrt{20}} < \mu < 70.8 + 1.96 \frac{3}{\sqrt{20}}$$

or 69.5 inches to 72.1 inches.   ∎

A statistical test of hypothesis is a procedure for deciding between two hypotheses called the *null hypothesis* (denoted H_0) and the *alternative hypothesis* (denoted H_a). For instance, a one-sided, upper-tail test of hypothesis about the mean of a population has the form H_0: $\mu = \mu_0$, H_a: $\mu > \mu_0$. If the underlying population has a normal distribution with a known standard deviation, then the decision to reject H_0: $\mu = \mu_0$ in favor of H_a: $\mu > \mu_0$ would be made if

$$\frac{\overline{X} - \mu_0}{\sigma/\sqrt{n}} > z_{(1-\alpha)}$$

where, for an appropriately chosen α, $z_{(1-\alpha)}$ is the $100(1 - \alpha)$th percentile of a standard normal distribution. The quantity

$$Z = \frac{\overline{X} - \mu_0}{\sigma/\sqrt{n}}$$

is called the *test statistic*. This decision rule has the property that if H_0 is true, then the probability of rejecting H_0 is α. The value of α is called the *level of significance* of the test. Typically $\alpha = .05$.

EXAMPLE 0.4.2   Suppose someone asserts that the mean height of the population described in Example 0.4.1 is 70 inches. Does a sample mean of 70.8 indicate that the population mean is greater than 70, or is that a value one would expect to get in random sampling from a population with a mean of 70? The two hypotheses of interest in this case are H_0: $\mu = 70$, H_a: $\mu > 70$. We find

$$Z = \frac{70.8 - 70}{3/\sqrt{20}} = 1.19$$

Since 1.19 is less than 1.645, we conclude at level $\alpha = .05$ that there is not enough evidence to reject H_0. That is, the outcome is in line with what one would expect to get when sampling from a population with a mean of 70.   ∎

The *p-value* is the probability that the test statistic is equal to or more extreme than the one observed under the null hypothesis. The *p*-value associated with the outcome $Z = 1.19$ in Example 0.4.2 is the probability that a standard normal random variable is greater than or equal to 1.19. This probability is $p = .1170$. Values of $p \leq .05$ are usually considered *statistically significant* in the sense that one would reject the null hypothesis in favor of the alternative hypothesis.

0.5
Parametric versus Nonparametric Methods

The analysis of data often begins by considering the appropriateness of the normal distribution as a model for describing the distribution of the population. If this distribution is reasonable, or if the normal approximation is deemed adequate, then the analysis will be carried out using normal-theory methods. If the normal distribution is not appropriate, it is common to consider the possibility of a *transformation* of the data. For instance, a simple transformation of the form $Y = \log(X)$ may yield data that are normally distributed, so that normal-theory methods may be applied to the transformed data.

If neither of these approaches seems reasonable, there are two ways to proceed. It may be possible to identify the type of distribution that is appropriate—say, exponential—and then use the methods that specifically apply to that distribution. However, there may not be sufficient data to ascertain the form of the distribution, or the data may come from a distribution for which methods are not readily available. In such situations one hopes not to make untenable assumptions, and this is where *nonparametric* methods come into play.

Nonparametric methods require minimal assumptions about the form of the distribution of the population. For instance, it might be assumed that the data are from a population that has a continuous distribution, but no other assumptions are made. Or it might be assumed that the population distribution depends on location and scale parameters, but the functional form of the distribution, whether normal or whatever, is not specified. By contrast, *parametric* methods require that the form of the population distribution be completely specified except for a finite number of parameters. For instance, the familiar one-sample *t*-test for means assumes that observations are selected from a population that has a normal distribution, and the only values not known are the population mean and standard deviation. The simplicity of nonparametric methods, the widespread availability of such methods in statistical packages, and the desirable statistical properties of such methods make them attractive additions to the data analyst's tool kit.

0.6
Classes of Nonparametric Methods

Nonparametric methods may be classified according to their function, such as two-sample tests, tests for trends, and so on. This is generally how this book is organized. However, methods may also be classified according to the statistical ideas upon which they are based. Here we consider the ideas that underlie the methods discussed in this book.

Methods Based on the Binomial Distribution

In Chapter 1 we develop some simple nonparametric methods for medians and percentiles based on the binomial distribution. For instance, suppose we wish to make inferences about the median of a continuous distribution. We may base our inferences on the number of observations that fall above (or below) the median. Since the probability that an observation falls above the median is 1/2, the number of observations above the median has a binomial distribution with $p = 1/2$. This fact is exploited to perform statistical tests and to make confidence intervals for the median.

Permutation Methods

A large number of nonparametric methods can be regarded as permutation methods. Almost all of the methods in Chapters 2–7 are like this.

Consider the following study: A researcher wishes to compare the decrease in cholesterol levels in subjects given a new drug with the decrease in subjects given a placebo. Twenty people are included in the study. Ten are randomly assigned to the new drug, and the remaining ten are assigned to the placebo. The decrease in the level of cholesterol is measured for each person in the study.

If there is no difference between the new drug and the placebo, the decrease in cholesterol achieved by any subject is likely to be the same with the placebo as with the new drug. Thus, any permutation (or shuffle) of the observations between the two groups, ten to the new drug and ten to the placebo, is as likely to occur as any other permutation. On the other hand, if the new drug is effective, then greater reductions in cholesterol will tend to occur with the new drug. By comparing what we observe in the data with what we would expect to see if all permutations were equally likely to occur, we may infer whether the new drug is effective or not.

Suppose we replace the smallest observation in the data with rank 1, the next smallest with rank 2, and so on. Permutation methods based on ranks rather than on the original observations lead to classic nonparametric procedures such as the Wilcoxon rank-sum test and the Kruskal–Wallis test. Ranks are special cases of *scores*, and various scores may be used in place of the original data to construct nonparametric tests. These tests differ primarily in what is permuted—whether original observations or scores—and what statistic is computed to compare sets of observations, such as differences between means or other measures that indicate how much sets of observations differ from one another.

Bootstrap Methods

Suppose we are interested in estimating the average height of a population, and we use the sample mean of a sample of size 100 as our point estimate. We know that there is variability in the sample mean because we have randomly selected data from the population. To gauge how accurate the estimate is, it is of interest to know the variance of the sample mean.

Now statistical theory provides us this answer, but suppose this theory were not available. One way to figure out the variance would be to take repeated samples of size 100 from the population, compute the sample mean each time, and then compute the variance of these sample means. However, sampling multiple times from the population is usually not possible for reasons of cost, time, and other constraints. Fortunately, we can mimic sampling from the population by taking samples with replacement from the data themselves. This is a reasonable thing to do since the distribution of the data ought to be like the distribution of the population, and sampling with replacement assures us that the data will not be exhausted when we sample from them. Such resampling of the data is called *bootstrap sampling*. Suppose we obtain multiple bootstrap samples of size 100, compute the sample mean each time, and then compute the variance of these sample means. This bootstrap variance should be a reasonable approximation of the true variance of the sample mean.

In practice, bootstrap sampling is used in situations where statistical theory does not readily provide us with the answers. For instance, we will show how to use bootstrap sampling to do statistical tests in multiple regression analysis when errors violate the usual normal distribution assumptions. Bootstrap methods are discussed in Chapter 8 and in Section 9.1.

Smoothing and Non-Least Squares Methods

Smoothing methods involve fitting curves to data without explicitly specifying the functional forms of the curves. One can imagine, for instance, plotting a stock price versus time and then drawing a smooth curve through the data to indicate trend over time. The smoothing methods that we consider use local averaging techniques to fit curves to the data. Smoothing methods are considered in Sections 10.1 and 10.2.

Another approach to curve fitting is to specify the functional form of the curve but use non-least squares methods to estimate the coefficients of the curve. For instance, if we wish to fit a quadratic curve $y = \beta_0 + \beta_1 x + \beta_2 x^2$ to the data, then the least squares method may be adversely affected by outlying observations. Popular non-least squares methods place less weight on such outlying observations. Least squares procedures have optimal properties when the distributions of the observations are normal. The objective in using non-least squares procedures is to arrive at statistical estimates that have good properties over a range of possible distributions for the population, while not necessarily being optimal in any particular circumstance. Rank-based and other methods for this application are considered in Section 10.3.

For general reference, the following textbooks in nonparametric statistics are listed roughly according to the level of required background of the students: Conover (1999) (undergraduate), Hollander and Wolfe (1999) (undergraduate/beginning graduate), Lehmann (1975) (beginning graduate), Hettmansperger (1984) (graduate), and Randles and Wolfe (1979) (graduate). Applications of permutation tests are discussed in Good (2000). See Manly (1997) for applications of nonparametric methods in the biological sciences.

1

One-Sample Methods

A Look Ahead   In Sections 1.1 and 1.2 we present some simple nonparametric tests of hypotheses and confidence intervals based on the binomial distribution. In Section 1.3 we show why a particular nonparametric test may be preferred over a well-known test that is based on the theory of the normal distribution.

1.1
A Nonparametric Test of Hypothesis and Confidence Interval for the Median

Suppose we have a random sample from a population that has a continuous cdf $F(x)$. Let $\theta_{.5}$ denote the median of the population; that is, $\theta_{.5}$ is a value such that half the probability is less than $\theta_{.5}$ and half greater. We wish to test hypotheses of the form

$$H_0: \theta_{.5} = \theta_H, \quad H_a: \theta_{.5} > \theta_H$$

EXAMPLE 1.1.1   Suppose a certain food product is advertised to contain 75 mg of sodium per serving, but preliminary studies indicate that servings may contain more than that amount. Considering that the amount of sodium in the product varies from one serving to another, we can formulate this problem as a test of hypothesis about the median of the distribution of the amount of sodium per serving. That is, we can test

$$H_0: \theta_{.5} = 75, \quad H_a: \theta_{.5} > 75 \qquad \blacksquare$$

1.1.1 Binomial Test

Tests of hypotheses for medians are typically used in the same situations that are appropriate for tests of hypotheses for means. Indeed, if the underlying population distribution is symmetric (and if the mean of the population exists), then the mean and the median are the same. For now, we will develop a statistical test for the median. In Section 1.3 we will compare statistical properties of this test with those of a well-known test for a mean.

Let $X_1, X_2, \ldots, X_n$ denote a random sample from the population. Let B denote the number of X_i's out of n that fall above the hypothesized median θ_H. If H_0 is true, then each X_i has probability .5 of falling above θ_H, so B has a binomial distribution with probability $p = .5$. If the true median is greater than θ_H, then B has a binomial distribution with probability $p > .5$. Thus, we can decide between H_0 and H_a based on the value of B; that is, we can test

$$H_0: p = .5, \quad H_a: p > .5$$

Typically, a test for p is carried out when samples are large, in which case the normal approximation of the binomial distribution may be used. We compute the test statistic

$$Z_B = \frac{B - .5n}{\sqrt{.25n}}$$

For a test at level of significance α, we reject H_0 in favor of H_a if $Z_B > z_{(1-\alpha)}$, where $z_{(1-\alpha)}$ is the $100(1 - \alpha)$th percentile of the standard normal distribution. Of course, we may also test lower-tail and two-sided hypotheses. The test is nonparametric in the sense that no assumption need be made about the form of the population distribution, except that it is continuous.

EXAMPLE 1.1.2   Consider the problem in Example 1.1.1. Data on the sodium content in milligrams of 40 packages of a company's food product are listed in Table 1.1.1. There are 26 data values for which $X_i > 75$, so this test gives us

$$Z_B = \frac{26 - 20}{\sqrt{.25(40)}} = \frac{6}{\sqrt{10}} = 1.90$$

Since $1.90 > 1.645$, we conclude that the median of the population is greater than 75 mg at the 5% level of significance. The p-value of the test using the normal approximation is .0287.

TABLE 1.1.1

Sodium Contents (in mg) of 40 Servings of a Food Product

72.1	72.8	72.9	73.3	73.3	73.3	73.9	74.0	74.2	74.2
74.3	74.6	74.7	75.0	75.1	75.1	75.2	75.3	75.3	75.3
75.4	76.1	76.5	76.5	76.6	76.9	77.1	77.2	77.4	77.4
77.7	78.0	78.3	78.6	78.8	78.9	79.7	80.3	80.5	81.0

■

1.1.2 Confidence Interval

Now let us turn to the problem of finding a confidence interval for the median of a population. We begin by placing the observations in order. These so-called *order statistics* are denoted as $X_{(1)} < X_{(2)} < \cdots < X_{(n)}$. We wish to find an interval

$$X_{(a)} < \theta_{.5} < X_{(b)}$$

such that

$$P\left(X_{(a)} < \theta_{.5} < X_{(b)}\right) = 1 - \alpha$$

where $1 - \alpha$ is the desired probability that the interval captures the median.

In order to have $X_{(a)} < \theta_{.5}$, at least a of the observations must fall less than $\theta_{.5}$, and in order to have $\theta_{.5} < X_{(b)}$, at most $b - 1$ of the observations must fall less than or equal to $\theta_{.5}$. Since $\theta_{.5}$ is the median and since the distribution of the X's is continuous, we have

$$P\left(X < \theta_{.5}\right) = P\left(X \leq \theta_{.5}\right) = .5$$

Since the observations are independent, the probability that at least a and at most $b - 1$ of the observations fall less than $\theta_{.5}$ is given by the binomial probability with $p = .5$—namely,

$$\sum_{k=a}^{b-1} \binom{n}{k}(.5)^n$$

To construct a $100(1 - \alpha)\%$ confidence interval for $\theta_{.5}$, we choose a and b so that this sum is $1 - \alpha$. Since the binomial distribution is discrete we may not be able to find limits that give us exactly the desired confidence, so we choose an attainable level of confidence as close to $100(1 - \alpha)\%$ as possible without going under this value.

For large samples, approximate values of a and b may be found by using the normal approximation to the binomial distribution. Without using the continuity correction for this approximation, we may obtain a and b by solving for them in the equations

$$\frac{a - .5n}{\sqrt{.25n}} = -z_{(1-\alpha/2)}, \quad \frac{b - 1 - .5n}{\sqrt{.25n}} = z_{(1-\alpha/2)}$$

and rounding to the nearest integer.

EXAMPLE 1.1.3   Using the data in Example 1.1.2 and a 95% confidence level, we have

$$\frac{a - 20}{\sqrt{10}} = -1.96, \quad \frac{b - 1 - 20}{\sqrt{10}} = 1.96$$

or $a = 13.8$ and $b = 27.2$. Rounding off, we have $X_{(14)} = 75.0$ and $X_{(27)} = 77.1$ as the lower and upper confidence limits.   ∎

Another estimate of the median that applies to symmetric distributions is the median of pairwise averages of the form $(X_i + X_j)/2$. This estimate and a confidence interval based on pairwise averages are discussed in the context of paired-comparison experiments. See Exercise 14 of Chapter 4.

1.1.3 Computer Analysis

The test and confidence interval for the median are available in MINITAB. These procedures may be accessed through the "Nonparametrics" and "1-Sample Sign" menu options. An output is shown in Figure 1.1.1 for the analysis of the data in Table 1.1.1. The *p*-value for the test is computed from the binomial distribution. Three confidence intervals are given. Two of them have achieved levels of confidence that bracket the desired level of .95. The larger interval is the one we use. The other is derived from interpolation.

FIGURE 1.1.1
MINITAB Test and Confidence Interval for the Median of Data in Table 1.1.1

Sign Test for Median: C1

Sign test of median = 75.00 versus > 75.00

	N	Below	Equal	Above	P	Median
C1	40	13	1	26	0.0266	75.35

Sign CI: C1

Sign confidence interval for median

	N	Median	Achieved Confidence	Confidence interval	Position
C1	40	75.35	0.9193	(75.10, 76.90)	15
			0.9500	(75.04, 77.02)	NLI
			0.9615	(75.00, 77.10)	14

1.2
Estimating the Population cdf and Percentiles

1.2.1 Confidence Interval for the Population cdf

The cdf plays an important role in nonparametric statistics. Let us assume that we have selected a random sample $X_1, X_2, \ldots, X_n$ from a population whose cdf is the continuous function $F(x)$. The estimate of $F(x)$ is the *empirical cdf:*

$$\hat{F}(x) = \text{fraction of observations} \le x$$

The empirical cdf is a step function that takes a step at each observed data value. If the data points are distinct, the size of each step is $1/n$, and if k data points have the value x, the step size is k/n.

EXAMPLE 1.2.1   Testing of electrical and mechanical devices often involves an action such as turning a device on and off or opening and closing a device many times. The interest is in the distribution of the number of on-off or open-close cycles that occur before the device fails. The hypothetical data in Table 1.2.1 are the number of cycles (in thousands) that it takes for 20 door latches to fail. The empirical cdf for the data in Table 1.2.1 is graphed in Figure 1.2.1.

TABLE 1.2.1
Cycles Until Latch Failure (thousands)

7	11	15	16	20	22	24	25	29	33	34	37	41	42	49	57	66	71	84	90

FIGURE 1.2.1
Empirical cdf of Latch Failure Data

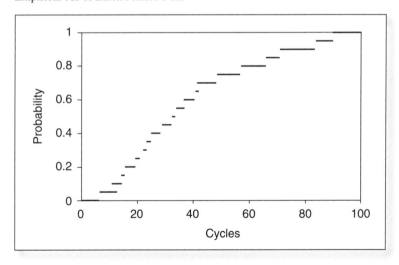

Since the number of observations for which $X_i \le x$ is a binomial random variable and since $p = P(X_i \le x) = F(x)$, the standard deviation of the empirical cdf is given by

$$\text{SD}\left(\hat{F}(x)\right) = \sqrt{\frac{F(x)\left[1 - F(x)\right]}{n}}$$

Based on the approximate normality of the binomial distribution, an approximate $100(1 - \alpha)\%$ confidence interval for $F(x)$ is given by

$$\hat{F}(x) \pm z_{(1-\alpha/2)} \sqrt{\frac{\hat{F}(x)\left[1 - \hat{F}(x)\right]}{n}}$$

EXAMPLE 1.2.2   For the data in Table 1.2.1, 14 of 20 observations are less than or equal to 45. Thus, an approximate 90% confidence interval for $F(45)$ is

$$.7 \pm 1.645 \sqrt{\frac{(.7)(.3)}{20}}$$

or $.53 < F(45) < .87$. As this result indicates, unless samples are large, the empirical cdf is not a very accurate estimate of the population cdf.   ∎

1.2.2 Inferences for Percentiles

Now consider the problem of making inferences for the $(100p)$th percentile θ_p. The special case of $p = .5$ was treated in Sections 1.1.1 and 1.1.2. The extension to an arbitrary value of p, $0 < p < 1$, is straightforward. We will consider the confidence interval for θ_p; the test of hypothesis is left as an exercise.

The confidence interval is of the form

$$X_{(a)} < \theta_p < X_{(b)}$$

where $X_{(a)}$ and $X_{(b)}$ are the ath- and bth-order statistics, respectively. The values of a and b are chosen so that

$$P\left(X_{(a)} < \theta_p < X_{(a)}\right) = 1 - \alpha$$

where $1 - \alpha$ is the desired probability that the interval captures θ_p.

In order to have $X_{(a)} < \theta_p$, at least a of the observations must fall less than θ_p, and in order to have $\theta_p < X_{(b)}$ at most $b - 1$ of the observations must fall less than or equal to θ_p. Since the observations are independent, the probability that at least a and at most $b - 1$ of the observations fall less than θ_p is given by the binomial

$$\sum_{k=a}^{b-1} \binom{n}{k} p^k (1 - p)^{n-k}$$

To construct a $100(1 - \alpha)\%$ confidence interval for θ_p, we choose a and b so that this sum is $1 - \alpha$.

EXAMPLE 1.2.3   Consider making a confidence interval for the lower quartile of the latch failure data in Table 1.2.1. For $p = .25$, the binomial probabilities from $a = 2$ to $b - 1 = 8$

add to .9347, so a 93.47% confidence interval for the lower quartile is $11 < \theta_{.25} <$ 29. Because of the discrete nature of the binomial distribution, it is not possible to select a and b so that there is an exact 95% confidence interval. ∎

For large samples, the values of a and b that yield approximately the desired level of confidence may be found by using the normal approximation to the binomial distribution. We may obtain a and b by solving for them in the equations

$$\frac{a - np}{\sqrt{np(1-p)}} = -z_{(1-\alpha/2)}, \quad \frac{b - 1 - np}{\sqrt{np(1-p)}} = z_{(1-\alpha/2)}$$

and rounding to the nearest integers.

EXAMPLE 1.2.4   Using the data in Table 1.2.1 with $p = .25$ and 90% confidence, we have

$$\frac{a - 5}{\sqrt{3.75}} = -1.645, \quad \frac{b - 1 - 5}{\sqrt{3.75}} = 1.645$$

or $a = 1.8$ and $b = 9.2$. Rounding to the nearest integers, we obtain the same interval as in Example 1.2.3. ∎

1.2.3 Computer Analysis

MINITAB has a calculator for the binomial distribution from which one can find probabilities and percentiles. It is possible to construct confidence intervals for θ_p or obtain *p*-values for testing hypotheses about θ_p using these values.

1.3
A Comparison of Statistical Tests

Suppose we have a sample from a symmetric distribution such as the normal or the Laplace. Since the mean and median are the same, any test for the mean is also a test for the median. We will compare a familiar large-sample test for the mean with the binomial test Z_B for the median given in Section 1.1.1.

Let $X_1, X_2, \ldots, X_n$ denote a random sample from a population with mean μ and standard deviation σ. We will not make any assumptions about the form of the population distribution. Instead we will rely on the central limit theorem, from which we can assert that the sample mean $\overline{X}$ has an approximate normal distribution with mean μ and standard deviation $\sigma/\sqrt{n}$ for large samples. A large-sample test statistic for testing the null hypothesis $H_0: \mu = \mu_0$ against the upper-tail alternative hypothesis $H_a: \mu > \mu_0$ when the population standard deviation σ is known is

$$Z_\mu = \frac{\bar{X} - \mu_0}{\sigma/\sqrt{n}}$$

The hypothesis H_0 is rejected in favor of H_a at level of significance α if $Z_\mu > z_{(1-\alpha)}$. We refer to the test based on Z_μ as the central limit theorem (CLT) test.

If the population standard deviation is unknown and estimated by the sample standard deviation S then we can replace σ by S in Z_μ, and again for large samples, we may use the standard normal distribution as the reference distribution for the test statistic. For small samples that have been selected from a normal population, the reference distribution is Student's t-distribution with $n - 1$ degrees of freedom. For our discussion, we will consider the large-sample statistic with σ known.

There are two statistical issues in choosing between the CLT and binomial tests. First is the Type I error, which occurs if H_0 is rejected when it is true. The probability of a Type I error should be what we claim it to be. If we claim that it is .05, then it should be .05 or at least close enough to it for practical purposes. The second issue is the power of a test. Power is the probability of rejecting H_0 if it is false. Power measures the ability of a test to detect a departure from the null hypothesis. If two tests have the correct probability of a Type I error, then the one with the greater power is the preferred test.

1.3.1 Type I Errors

Let us consider the Type I error of the CLT test. From the central limit theorem, Z_μ has an approximate standard normal distribution for large samples, so probabilities of Type I errors found by using the standard normal distribution will be approximately correct. Many studies have shown that the approximation is quite good even for moderate sample sizes and a range of population distributions. We should note that if observations come from a distribution that does not have a finite variance, such as the Cauchy, then Z_μ is inappropriate to use even for large samples.

Now consider the Type I error of the binomial test. This test may be applied to testing a hypothesis about the median of any continuous distribution, including the exponential, Laplace, uniform, or Cauchy. The distributional properties of Z_B under H_0 depend only upon the fact that B has a binomial distribution with $p = .5$ and not upon the form of the underlying continuous population distribution. The test is nonparametric in this sense. The use of the standard normal distribution as a reference distribution for Z_B depends on the normal approximation to the binomial distribution. This approximation is quite good when $p = .5$ even for moderate sample sizes. Thus, if we use the standard normal reference distribution for Z_B for large samples, the stated probability of a Type I error will be essentially correct.

1.3.2 Power

To compare the power of the two tests, we must consider the distribution of the observations under H_a. Many circumstances could lead to the null hypothesis not being true. Consider, for instance, Example 1.1.2, in which a researcher is testing the sodium contents in packages of food. The desired sodium content of 75 mg might not occur because the setting is too high on the machine that puts salt into the product. For instance, if the amount of sodium per package has a normal distribution with mean 75 mg and standard deviation 2.5 mg under the correct setting of the salt machine, it may be reasonable to assume that the distribution is normal with mean $\mu > 75$ mg and standard deviation 2.5 mg when the setting is too high.

However, the situation may be even more complicated. There may be a glitch in the machine that causes every tenth package, say, to have twice as much salt as it should. Or an incorrect setting of the machine might cause both the mean and the standard deviation to change. One test may have greater power in one circumstance and the other test may be more powerful in another. The strategy is to choose a test that has good power under the alternatives that seem plausible for the problem at hand.

EXAMPLE 1.3.1 Again refer to Example 1.1.2. Suppose the mean sodium content has increased from the desired value of 75 mg to 75.8 mg with the standard deviation remaining $\sigma = 2.5$. Table 1.3.1 gives the powers of the CLT and binomial tests under two different population distributions, the normal and the Laplace. Note that the power of the binomial test is less than that of the CLT test in the case of the normal population, but greater in the case of the Laplace. Generally, the binomial test will have higher power than the CLT test for heavier-tailed population distributions, but the opposite will be true for lighter-tailed distributions. Derivations are given at the end of this section.

TABLE 1.3.1
A Comparison of Power of the CLT and Binomial Tests $\mu_0 = 75$, $\mu = 75.8$, $\sigma = 2.5$, $\alpha = .05$

Population Distribution	Power of CLT Test	Power of Binomial Test
Normal	.65	.48
Laplace	.65	.76

∎

A lot is known about the properties of the CLT test when samples are taken from normal populations with a known variance. Statistical theory tells us that this test has the greatest power among all tests that have the same level of significance, and this advantage holds for all values of μ. The test is said to be *uniformly most powerful* in this case. See Hogg and Craig (1995). The CLT is the test to use when sampling from a normal population.

Some have mistakenly assumed that the optimal power properties of the CLT test carry over to the case of sampling from nonnormal populations. This is not the case, not even for large samples. All the central limit theorem allows us to conclude is that the power of the CLT test will be about the same for one population as another for large samples provided the population distributions have the same mean and variance. However, one cannot assume that the test will have the greatest power or nearly so for large samples taken from nonnormal populations.

1.3.3 Derivations

Derivation of Power for Samples from a Normal Population

The null hypothesis is $\mu = 75$, and we wish to find the power when a sample of size $n = 40$ is taken from a normal population with mean $\mu = 75.8$ and standard deviation $\sigma = 2.5$. We first note that the power depends on μ. The larger the value of μ, the greater the ability of the tests to detect that $\mu > 75$. For a general value of μ and for $\alpha = .05$, we have

$$\text{power of CLT test} = P\left(\frac{\overline{X} - 75}{2.5/\sqrt{40}} \geq 1.645 \;\middle|\; \mu \right)$$

$$= P\left(\frac{\overline{X} - \mu}{2.5/\sqrt{40}} \geq 1.645 - \frac{\mu - 75}{2.5/\sqrt{40}} \;\middle|\; \mu \right)$$

$$= 1 - \Phi\left(1.645 - \frac{\mu - 75}{2.5/\sqrt{40}} \right)$$

where Φ is the cdf of the standard normal distribution. If $\mu = 75.8$, we find that the power is $1 - \Phi(-.38) = .65$. For the binomial test, we need to determine p, the probability that an observation is greater than 75 for a given value of μ. Given p and using the normal approximation to the binomial distribution, we find for $\alpha = .05$

$$\text{power of binomial test} = 1 - \Phi\left(1.645 \sqrt{\frac{.25}{p(1-p)}} - \frac{p - .5}{\sqrt{p(1-p)/40}} \right)$$

If $\mu = 75.8$, we have $p = .626$, so the power is $1 - \Phi(.053) = .48$. Clearly, the CLT test is the preferred statistic for this situation.

Derivation of Power for Samples from a Laplace Population

Suppose under H_a the population has a Laplace distribution with mean $\mu = 75.8$ and standard deviation $\sigma = 2.5$. We first note that both tests will have essentially the correct probability of a Type I error: the CLT test because of the approximate normality due to the central limit theorem effect and the binomial test because of the approximate normality of the binomial distribution. Moreover, the central limit theorem suggests that the power of the CLT test is the same (essentially) as it is in the case of a normally distributed population—namely, .65 when $\mu = 75.8$.

To obtain the power of the binomial test, we must find $p = P(X > 75)$ when $\mu = 75.8$. If X has a Laplace distribution with mean μ and standard deviation σ, then $(X - \mu)/\sigma$ has the standard form of the Laplace distribution given in Table 0.2.1. For $x < \mu$, it can be shown that

$$P(X > x) = .5 + .5\left(1 - e^{-\sqrt{2}|x - \mu|/\sigma}\right)$$

Applying this formula to our example, we find $p = P(X > 75) = .682$. From the formula for the power of the binomial test in the power computation for a normal distribution, we find that the power is .76. Thus, the form of the distribution of the population can greatly affect the relative power of two tests.

Exercises

1   The data in the table are simulated exam scores. Suppose the exam was given in the semester after the course content was revised, and the previous median exam score was 70. We would like to know whether or not the median score has increased. Answer the question by applying the binomial test.

Simulated Exam Scores

79	74	88	80	80	66	65	86	84	80	78	72	71	74	86	96	77	81	76	80
76	75	78	87	87	74	85	84	76	77	76	74	85	74	76	77	76	74	81	76

2   Refer to the data in Exercise 1.

  a   Make a 90% confidence interval for the median.

  b   Make a 90% confidence interval for the 75th percentile.

  c   Make a 90% confidence interval for $F(80)$, the probability that a score is less than or equal to 80.

3   The data in the table are the yearly rainfall totals in Scranton, Pa., for the years 1951–1984.

Rainfall Totals (inches) for Scranton, Pa., 1951–1984

21.3	28.8	17.6	23.0	27.2	28.5	32.8	28.2	25.9	22.5	27.2	33.1	28.7	24.8	24.3	27.1	30.6
26.8	18.9	36.3	28.0	17.9	25.0	27.5	27.7	32.1	28.0	30.9	20.0	20.2	33.5	26.4	30.9	33.2

  a   Make a 95% confidence interval for the median.

  b   Make 90% confidence intervals for the 20th and 80th percentiles.

  c   The confidence interval procedure assumes that the observations are independent and identically distributed. Do you think this is a reasonable assumption for the rainfall data? If not, what could cause this assumption to be invalid?

4   Suppose we test the hypotheses H_0: $\theta_{.5} = 75$ versus H_a: $\theta_{.5} > 75$ and, regardless of the data, we reject H_0. What is the probability of a Type I error? What is the power of the test for values of $\theta_{.5} > 75$?

5   Suppose we assume that the population distribution under H_0 is symmetric so that $\theta_{.5} = \mu$. Without looking at the data to check the validity of this assumption, we apply the binomial test and the CLT test. Suppose it turns out that 39 data values that are equal to 75.1 and the 40th one is equal to 90.

    a   What decision is reached using the binomial test to test H_0: $\theta_{.5} = 75$ versus H_a: $\theta_{.5} > 75$?

    b   What decision is reached using the CLT test to test H_0: $\mu = 75$ versus H_a: $\mu > 75$, where the statistic is computed using the sample standard deviation S in place of the unknown population standard deviation σ?

    c   Based on the results of parts a and b, what types of distributions that satisfy the alternative hypothesis are particularly easy for the binomial test to detect in comparison to the CLT test?

    d   Replace 90 by other values such as 80, 78, and 76 that are closer to the null hypothesis. Note what happens to the value of Z_μ. Does this correspond to intuition?

6   Refer to Section 1.3.3. No computations are required to answer the following questions.

    a   What is the value of the power of the binomial test when $\mu = 75$?

    b   What happens to the power as μ gets large?

    c   How does increasing the sample size affect the power of the binomial test?

7   Suppose we test H_0: $\theta_{.5} = \theta_H$ versus $H_a > \theta_H$ using the binomial test with a sample size $n = 10$.

    a   If we reject H_0 when $B \geq 8$, use the binomial Table A1 to determine the exact probability of a Type I error.

    b   Suppose we observe a value of $B = b_{obs}$. The p-value is the probability that $B \geq b_{obs}$ given that H_0 is true. Find the p-values for $b_{obs} = 5, 6, 7, 8, 9, 10$.

Theory and Complements

8   Refer to the derivations of the power functions in Section 1.3. Evaluate and sketch the power functions of the statistics Z_μ and Z_B for values of the mean between 75 and 77 assuming that the populations have normal distributions. Using your sketch, determine the maximum difference between the power functions. Repeat this procedure for the Laplace population distribution.

2

Two-Sample Methods

A Look Ahead   The idea of permuting data among treatments as a way to do statistical inference dates back to the founding father of modern statistics, R. A. Fisher (1935). Theoreticians have developed this idea since then. However, it is the availability of high-speed computing that has enabled applied statisticians to make full use of this methodology.

In Section 2.1 we introduce a permutation test for comparing two treatments. Most of the methods in Chapters 2–7 are based on the idea behind this simple test. We use small data sets to introduce the methods so that the reader can see explicitly the steps involved in the various procedures. Applications are in no way limited to small data sets, however, given that excellent statistical software packages are available for carrying out the computations. Examples are provided to show the use of selected software.

2.1
A Two-Sample Permutation Test

We begin with a simple example. Suppose a company is trying to decide whether to augment its traditional instruction for new employees with computer-assisted instruction. Seven new employees are selected for a trial. Four are randomly assigned to the new method of instruction, and the other three are given the traditional instruction. A test is given afterward to compare the two methods. Scores for the subjects in this hypothetical experiment are listed in Table 2.1.1.

TABLE 2.1.1
Test Scores of Seven Employees for Comparison of Methods of Instruction

New Method	Smith (37), Lin (49), O'Neal (55), Zedillo (57)
Traditional Method	Johnson (23), Green (31), Zook (46)

FIGURE 2.1.1

Two-Sample *t*-Test for Data in Table 2.1.1 Using MINITAB

```
Two-sample T for C2

C1          N       Mean      StDev      SE Mean
1           4       49.50      9.00        4.5
2           3       33.3      11.7         6.7


Difference = mu (1) - mu (2)
Estimate for difference: 16.17
95% lower bound for difference: 0.53
T-Test of difference = 0 (vs >): T-Value = 2.08   P-Value = 0.046   DF = 5
Both use Pooled StDev = 10.2
```

A two-sample *t*-test of the null hypothesis of no difference between the two methods versus the one-sided alternative hypothesis that the mean of the new method is greater than the mean of the traditional method gives $t = 2.08$ and $p = .046$. Thus, we conclude that the new method produces a significantly higher mean test score than the traditional method at the 5% level of significance. Results are shown in a MINITAB printout in Figure 2.1.1.

The application of the *t*-test comes at a price. It requires three assumptions: (1) the samples are randomly selected from two infinite populations or, equivalently, the observations are independent; (2) the populations have normal distributions; and (3) the variances of the two populations are the same. In this case, there is no indication that the seven subjects have been randomly selected from any particular population, although presumably they represent typical new employees. If the data can be regarded as having been selected randomly from two populations, then there is no guarantee that the other assumptions of the *t*-test are met. If we apply the *t*-test anyway, we run the risk of misstating the *p*-value of the statistical test and therefore of declaring a result to be statistically significant when it is not.

2.1.1 The Permutation Test

Fortunately, the random assignment of the experimental subjects to the two methods provides a basis for drawing statistical inferences about the effect of the new method without the assumptions associated with the two-sample *t*-test. The argument goes as follows: If there is no difference between the two methods, then all data sets obtained by randomly assigning four of these scores to the new method and the other three to the traditional method would have an equal chance of being observed in the study. There are

$$\binom{7}{4} = \frac{7!}{4!3!} = 35$$

such two-sample data sets. Most data sets among the 35 have both large and small scores assigned to each treatment. These are the types of data sets we would expect to observe if the two treatments were equally effective in training new employees. In this case, the means of the scores of the two treatments ought to be about the same. On the other hand, if the new method has its intended effect, then the higher test scores would tend to occur with the new method. In this case, the mean of the scores of the new method would tend to be larger than the mean of the scores from the traditional method, and the difference would be larger by an amount greater than one would expect to occur by chance. We use the difference between the two means as our test statistic.

Table 2.1.2 lists the 35 possible two-sample sets along with the differences between the two means. Note that the second set, which is identified with an asterisk, is the original data in Table 2.1.1. It has a difference of means of 16.2. The *p*-value for an upper-tail test is the probability of observing a difference of means of 16.2 or greater under the assumption that the treatments do not differ—that is, under the assumption that all permutations are equally likely to occur. Since 16.2 is the second largest among the 35 differences of means, its *p*-value is 2/35 = .0571. This *p*-value is *exact* in the sense that its correctness does not depend on unverified assumptions about the distribution of the population from which these observations may have been selected. With this *p*-value a researcher would declare the results to be significant at the 10% level but not at the 5% level.

The procedure we have just described is called a two-sample *permutation test*, since it is based on permuting the observations among two groups in the same proportion as in the original sample. The distribution of the 35 differences of means is called the *permutation distribution* for the difference between two means. A histogram of this distribution is shown in Figure 2.1.2. The *permutation principle* states that the permutation distribution is an appropriate reference distribution for determining the *p*-value of a test and deciding whether or not a statistical test is statistically significant.

FIGURE 2.1.2

Permutation Distribution of Difference Between Means of Data in Table 2.1.1

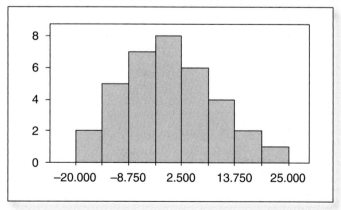

TABLE 2.1.2

All Possible Assignments of Data to New and Traditional Methods

Combined Data: 23 31 37 46 49 55 57

Permuted Samples	New Method	Traditional Method	Difference Between Means	Sum of New-Method Observations
1	46 49 55 57	23 31 37	21.4	207
2*	37 49 55 57	23 31 46	16.2	198
3	37 46 55 57	23 31 49	14.4	195
4	37 46 49 57	23 31 55	10.9	189
5	37 46 49 55	23 31 57	9.8	187
6	31 49 55 57	23 37 46	12.7	192
7	31 46 55 57	23 37 49	10.9	189
8	31 46 49 57	23 37 55	7.4	183
9	31 46 49 55	23 37 57	6.3	181
10	31 37 55 57	23 46 49	5.7	180
11	31 37 49 57	23 46 55	2.2	174
12	31 37 49 55	23 46 57	1.0	172
13	31 37 46 57	23 49 55	0.4	171
14	31 37 46 55	23 49 57	−0.8	169
15	31 37 46 49	23 55 57	−4.3	163
16	23 49 55 57	31 37 46	8.0	184
17	23 46 55 57	31 37 49	6.3	181
18	23 46 49 57	31 37 55	2.8	175
19	23 46 49 55	31 37 57	1.6	173
20	23 37 55 57	31 46 49	1.0	172
21	23 37 49 57	31 46 55	−2.5	166
22	23 37 49 55	31 46 57	−3.7	164
23	23 37 46 57	31 37 55	−4.3	163
24	23 37 46 55	31 49 57	−5.4	161
25	23 37 46 49	31 55 57	−8.9	155
26	23 31 55 57	37 46 49	−2.5	166
27	23 31 49 57	37 46 55	−6.0	160
28	23 31 49 55	37 46 57	−7.2	158
29	23 31 46 57	37 49 55	−7.8	157
30	23 31 46 55	37 49 57	−8.9	155
31	23 31 46 49	37 55 57	−12.4	149
32	23 31 37 57	46 49 55	−13.0	148
33	23 31 37 55	46 49 57	−14.2	146
34	23 31 37 49	46 55 57	−17.7	140
35	23 31 37 46	49 55 57	−19.4	137

Another statistic that may be used instead of the difference between two means is the sum of the observations for the new method (or the traditional method). This statistic is equivalent to the difference between the two means in the sense that the permutation *p*-values of the two statistics are the same. The sums are also listed in Table 2.1.2. For the original data set, the sum of the observations for the new method is 198. This is the second largest among the 35 sums, so its upper-tail *p*-value is also 2/35 = .0571. The equivalence of the difference of means and the sum of observations from one of the treatments is derived in Section 2.1.2.

2.1.2 Summary of Steps Used in a Two-Sample Permutation Test

To summarize, a two-sample permutation test is carried out as follows.

1. Randomly assign experimental units (subjects or objects) to one of two treatments with *m* units assigned to treatment 1 and *n* units assigned to treatment 2. Obtain data on the units and compute the difference between the two means, D_{obs}.

2. Permute the *m* + *n* observations between the two treatments so that there are *m* observations for treatment 1 and *n* observations for treatment 2. Obtain all possible permutations. The number of possibilities are

$$\binom{m+n}{m} = \frac{(m+n)!}{m!n!}$$

3. For each permutation of the data, compute the difference, *D*, between the mean of treatment 1 and the mean of treatment 2.

4. If the presumed effect of treatment 1 is to produce observations that are on average larger than those for treatment 2 (upper-tail test), compute the *p*-value as the proportion of *D*'s greater than or equal to D_{obs}; that is,

$$P_{\text{upper tail}} = \frac{\text{number of } D\text{'s} \geq D_{obs}}{\binom{m+n}{m}}$$

Similarly, compute the proportion of *D*'s less than or equal to D_{obs} for a lower-tail test. For a two-sided test, perform a similar procedure on the absolute values of the differences between the means; that is,

$$P_{\text{two tail}} = \frac{\text{number of } |D\text{'s}| \geq |D_{obs}|}{\binom{m+n}{m}}$$

5. If a predetermined level of significance has been set, declare the test to be statistically significant if the *p*-value is less than or equal to this level.

Now consider the test statistic that is the sum of the observations from one of the treatments. Let T_1 denote the sum of the observations from treatment 1 and T_2

the sum of the observations from treatment 2. Let T denote the sum of all the observations. Now

$$D = \frac{T_1}{m} - \frac{T_2}{n} = \frac{T_1}{m} - \frac{T - T_1}{n} = T_1\left(\frac{1}{m} + \frac{1}{n}\right) - \frac{T}{n}$$

Since T is the same for all permutations of the data, we see that D can be determined from T_1, and vice versa. Since there is a one-to-one correspondence between D and T_1, it follows that the p-value for a test based on D will be the same as the p-value for a test based on T_1. That is, D and T_1 are equivalent test statistics, and the same conclusion will be reached whichever one is used.

2.1.3 Hypotheses for the Two-Sample Permutation Test

To what group of experimental units may the inferences of a permutation test be applied? If the units are initially selected at random from larger populations of units, then inferences may be drawn about the effects of the treatments as applied to the larger populations. In this situation, let $F_1(x)$ and $F_2(x)$ be the cdf's of the two populations. The null hypothesis is

$$H_0 : F_1(x) = F_2(x)$$

That is, the two distributions are identical under the null hypothesis. It should be emphasized that the null hypothesis does not allow for the situation in which the means are the same but the variances are different for the two populations. When we say that the two treatments are the "same" under the null hypothesis, we mean that the distributions of the observations are the same.

A one-sided alternative hypothesis is

$$H_a : F_1(x) \leq F_2(x)$$

where strict inequality occurs for at least one x. Intuitively, this hypothesis denotes that the observations for treatment 1 tend to be *larger* than the observations for treatment 2. A special case of this occurs when $F_1(x) = F_2(x - \Delta)$, where $\Delta > 0$. Think of this as a situation where the distributions have the same shape but treatment 1 has a larger mean than treatment 2. Of course, we may also have the alternative hypothesis $F_1(x) \geq F_2(x)$, which would occur if the observations of treatment 1 tend to be smaller than those of treatment 2.

The two-sided alternative hypothesis is not $F_1(x) \neq F_2(x)$. This alternative would allow, for instance, the possibility that the means are the same but the variances are different. The permutation test using the difference between means as the statistic is not designed for this possibility. Rather the two-sided alternative hypothesis is

$$H_a : F_1(x) \leq F_2(x) \text{ or } F_1(x) \geq F_2(x), \text{ for all } x$$

with strict inequality occurring for at least one x. Intuitively, this statement says that the observations from treatment 1 will either tend to be larger or tend to be smaller than the observations from treatment 2, but we don't know in advance of collecting the data which direction this might go. In the case in which $F_1(x) = F_2(x - \Delta)$, the two-sided alternative hypothesis is $H_a : \Delta \neq 0$, where Δ may be thought of as the difference between the means or medians of the two populations.

It may happen that the units are not randomly selected from a larger population of units. In this case, the randomization will yield valid inferences about the effects of the treatments, but only as the treatments affect the units in the study. Drawing conclusions about a larger group of units would be a matter of scientific judgment, not statistical inference.

2.1.4 Computer Analysis

The statistical package StatXact is a simple menu-driven program that can carry out exact permutation tests as well as a variety of other tests. The output of the StatXact permutation test applied to the data in Table 2.1.1 is shown in Figure 2.1.3. Data are entered in the Case Data format. StatXact uses the sum of the observations of the first treatment as the test statistic. The Asymptotic Inference in the output is based on large-sample approximations as discussed in Section 2.10.

FIGURE 2.1.3

Two-Sample Permutation Test for Data in Table 2.1.1 Using StatXact

```
PERMUTATION TEST
[ Sum of scores from population <         1 > ]

Summary of Exact distribution of PERMUTATION statistic:

        Min        Max       Mean    Std-dev     Observed   Standardized
       137.0      207.0      170.3      16.59        198.0          1.670

Asymptotic Inference:
   One-sided p-value: Pr {    Test Statistic .GE. Observed }   =      0.0474
   Two-sided p-value: 2 * One-sided                            =      0.0949

Exact Inference:
   One-sided p-value: Pr {    Test Statistic .GE. Observed }   =      0.0571
                      Pr {    Test Statistic .EQ. Observed }   =      0.0286
   Two-sided p-value: Pr { | Test Statistic - Mean |
                       .GE. | Observed - Mean |               =      0.1143
   Two-sided p-value: 2*One-Sided                              =      0.1143
```

2.2
Permutation Tests Based on the Median and Trimmed Means

The mean is the most commonly used measure of the center of a set of numbers. It is a single value that in some sense represents the data set as a whole. Therefore, it is natural to use the difference between two means as a statistic for comparing two treatments. Unfortunately, means may be unduly influenced by unusually large or small observations.

Consider, for instance, the data in Table 2.1.1. The median of the data for the new method is the average of the two middle observations, $(49 + 55)/2 = 52$. Suppose that the 57 were changed to an 87, perhaps due to a recording error. Changing the largest observation from a 57 to an 87 has no effect at all on the median. However, the mean changes from 49.5 to 57.0, with a corresponding change in the difference between the two means.

2.2.1 A Permutation Test Based on Medians

Since medians are not affected by outliers in the data, it might be desirable to use the difference of medians instead of the difference of means to compare two treatments. The difficulty in doing this is that it is not readily apparent what reference distribution ought to be used to determine statistical significance. Fortunately, the permutation argument that we used to test for differences of means also applies to differences of medians.

Suppose that the data are as in Table 2.1.1 except the value 57 is replaced by 87. Table 2.2.1 displays the differences of the medians for those sets with the largest differences. We note that the change in the data value from 57 to 87 has no effect on

TABLE 2.2.1

Five Largest Differences Between Medians for Modified Data from Table 2.1.2

Combined Data: 23 31 37 46 49 55 87 (57 replaced by 87)

Permuted Samples	Data and (Median) of New Method	Data and (Median) of Traditional Method	Differences Between Medians
1	46 49 55 87 (52)	23 31 37 (31)	21
2*	37 49 55 87 (52)	23 31 46 (31)	21
3	37 46 55 87 (50.5)	23 31 49 (31)	19.5
4	37 46 49 87 (47.5)	23 31 55 (31)	16.5
5	37 46 49 55 (47.5)	23 31 87 (31)	9.8
Other	All other differences less than 21		

the medians of any of the permuted samples. Permuted samples 1 and 2 have a difference of medians of 21, and all other such differences are less than 21. Since the difference of medians for the original data is 21, the *p*-value for this permutation test is 2/35 = .0571.

The point that we want to emphasize here is the usefulness of the permutation approach in determining statistical significance. The researcher is free to choose a statistic that he or she feels best describes the difference between the two groups and then use the permutation approach to determine whether or not the statistic is significant. In Section 2.9 we consider in detail the issue of which statistic one ought to choose in comparing two treatments using a permutation test.

A permutation test using the difference of medians may be carried out with Resampling Stats. This is considered in Section 2.3.

2.2.2 Trimmed Means

Other measures of the center of a set of data may also be used in permutation tests. One such popular measure is the *trimmed mean*. In a trimmed mean, we delete equal numbers of the largest and smallest observations and then average the remaining observations. For instance, if we have ten observations given by 11, 23, 30, 31, 32, 36, 37, 40, 44, 54, then to get a 20% trimmed mean we delete the top and bottom observations and average the remaining eight to get a value of 34.1. As with the median, the idea is to eliminate unusually large or small observations that might otherwise unduly affect the value of the mean.

Deciding whether to use the mean, median, or trimmed mean in the permutation test requires some knowledge of the population from which the observations are selected. Generally, if the data come from an approximate normal distribution, then the difference in means should be used. If the population distribution is symmetric and the population mean is of interest, but if a few unusually large or small observations are likely, then a trimmed mean is appropriate because it trims these extreme observations. If the population distribution is asymmetric, then the median may be a more desirable indicator of the center of the data, in which case a permutation test based on the median might be appropriate.

Regardless of the statistic being used, the permutation test is exact in the sense that *p*-values obtained from the permutation distribution are correct and not dependent upon unverified assumptions about the distribution of the population.

2.3
Random Sampling the Permutations

Prior to the computer age, the practical use of permutations tests was limited by prohibitive computations. For instance, if two treatments each have 8 observations, then the number of possible two-sample data sets that can be obtained by permuting the 16 observations, 8 to a treatment, is

$$\binom{16}{8} = 12,870$$

If just two more observations are added to each treatment, then this number increases more than tenfold to

$$\binom{20}{10} = 184,756$$

Fortunately, there is a simple way to obtain an approximate p-value for the permutation test in such cases. Rather than using all the permutations, we take a random sample of the permutations and perform the steps involved in a permutation test on the randomly sampled permutations.

2.3.1 An Approximate p-Value Based on Random Sampling the Permutations

The following steps may be used to obtain an approximate p-value for a permutation test based on the difference of means. A similar procedure applies to other statistics such as the difference of medians.

1. Compute the difference of means, D_{obs}, between the two treatments for the observed data.
2. Create a vector of the $m + n$ observations.
3. Randomly "shuffle" the elements of this vector, assigning the first m to treatment 1 and the remaining n to treatment 2.
4. Compute the difference of means for the shuffled data set.
5. Repeat this procedure a predetermined number of times—say, 1000. For an upper-tail test, the fraction of the sampled mean differences that are greater than or equal to the observed difference D_{obs} is an approximate p-value for the permutation test. Similarly, we may obtain approximate lower-tail or two-tail p-values.

Accuracy of the Procedure

If among all possible permutations of the data the fraction of mean differences greater than or equal to D_{obs} is p (that is, if the true p-value is p), then the theory of the binomial distribution tells us that an approximate p-value based on R randomly selected permutations will have about a 95% chance of being within

$$\pm 2\sqrt{\frac{p(1-p)}{R}}$$

of the true *p*-value. For instance, if $p = .05$ and $R = 1000$, then the approximate *p*-value will likely be within $\pm.014$ of the true *p*-value. If this approximation is not adequate, then increasing the number of randomly selected permutations to 4000 will cut this error in half.

EXAMPLE 2.3.1 An agricultural engineer conducted an experiment to compare the effects of two tilling methods on the retention of moisture in the soil. A sprinkling system applied water to plots to simulate rainfall. The numbers of minutes it took to obtain various amounts of runoff on each plot were recorded. The hypothetical data in Table 2.3.1 represent the minutes it took to obtain a 2-liter runoff on eight plots under each of two treatments.

TABLE 2.3.1
Minutes to Observe 2-Liter Runoff

Treatment 1	59.1 60.3 58.1 61.3 65.1 55.0 63.4 67.8
Treatment 2	60.1 62.1 59.3 55.0 54.6 64.4 58.7 62.5

The observed mean difference between the two treatments is $D_{obs} = 1.675$. A particular sample of 1000 randomly selected permutations obtained by StatXact produced 208 mean differences greater than or equal to 1.675, giving an approximate one-sided *p*-value of .208. Thus, the difference is not statistically significant at the 5% level of significance. ∎

2.3.2 Computer Analysis Using Resampling Stats

Resampling Stats is a programming language that is specifically designed for permutation tests and other so-called resampling procedures. An annotated program for analyzing the data in Table 2.3.1 is shown in Figure 2.3.1. In this example, and in examples throughout, we have chosen to display complete code for the specific problem rather than snippets of code for the general case. We believe this makes it easier for the reader to follow the code and modify it for other problems. For conciseness of display, but contrary to good programming practice, we do not provide code to initialize constants involving observed statistics and sample sizes. These values are used within the body of the program and must be changed if the program is used with other data.

Three tests are computed for illustration in Figure 2.3.1: (1) the permutation test based on the difference of means, (2) the permutation test based on the sum of the observations from treatment 1, and (3) the permutation test based on the difference of medians. The "shuffle" command gives us the permuted data sets. The "score" command creates a vector whose elements are numerical values obtained from each shuffle of the data. The distribution of these numbers is an approximate permutation distribution for the statistic in question. The means, medians, and other statistics of

FIGURE **2.3.1**

Resampling Stats Code for Approximate Permutation Tests Using Difference of Means,
Sum of Observations from Treatment 1, and Difference of Medians as Test Statistics

```
'enter data for trt1 and trt2 and combine into single vector
copy (59.1 60.3 58.1 61.3 65.1 55 63.4 67.8) trt1
copy (60.1 62.1 59.3 55 54.6 64.4 58.7 62.5) trt2
concat trt1 trt2 a

'observed statistics
'diff of means is 1.675, sum of trt1 obs is 490.1, diff of medians is 1.1

'begin loop to generate 1000 permutations
repeat 1000

'permute the data vector "a" and put the results in "b"
shuffle a b

'select the first 8 shuffled observations for trt 1
take b 1,8 b1
'compute mean, sum, and median
mean b1 meanb1
sum b1 sumb1
median b1 medb1

'select the second 8 shuffled observations for trt 2
take b 9,16 b2
'compute mean and median
mean b2 meanb2
median b2 medb2

'compute the difference of the means and medians
subtract meanb1 meanb2 meandiff
subtract medb1 medb2 meddiff

'get approx permutation distributions
'keep track of differences of means, sums, and differences of medians
score meandiff meandist
score sumb1 sumdist
score meddiff meddist

'end the loop
end

'get pvalues for difference of means, sum, and difference of medians
count meandist >= 1.675 pvalmean
divide pvalmean 1000 pvalmean
count sumdist >= 490.1 pvalsum
divide pvalsum 1000 pvalsum
count meddist >= 1.1 pvalmed
divide pvalmed 1000 pvalmed

'print results
print pvalmean pvalsum pvalmed

'print output here

PVALMEAN = 0.203
PVALSUM = 0.203
PVALMED = 0.344
```

the original data were computed externally to the program in order to focus on the essential commands for doing the permutation test. Resampling Stats has commands for importing data from external files, and on the original data computations may be carried out within the program itself.

The *p*-values produced by the program are also shown in Figure 2.3.1. In this example, the *p*-values are based on 1000 randomly selected permutations. As should occur, the *p*-values for the difference of the means and the sum of the observations for treatment 1 are the same. The *p*-value for the difference of the medians is considerably larger than the *p*-value for the difference of the means in this example. None of the tests is significant at the 5% level.

2.3.3 Computer Analysis Using StatXact

StatXact offers the option of obtaining exact *p*-values or approximate *p*-values. Exact *p*-values are computed using fast algorithms for permutation tests developed by Mehta, Patel, and their co-authors. For instance, see Mehta and Patel (1983) and Mehta, Patel, and Wei (1988). If the data set is too large to make the computation of exact *p*-values feasible, an option called "Exact Using Monte Carlo" is available for obtaining approximate *p*-values. The name alludes to the random aspect of the procedure as in gambling. The default number of randomly selected permutations in StatXact is 10,000. With conventional desktop computing, this presents no problem in terms of computing time. The sampling method that we have suggested is simple random sampling. However, it is possible to improve accuracy with what is called importance sampling. This is an option in StatXact. See Mehta, Patel, and Senchaudhuri (1988) for details.

Output for the analysis of the data in Table 2.3.1 is shown in Figure 2.3.2 (on page 36). Both approximate and exact *p*-values for the permutation test are shown, where for illustration the approximate *p*-value is based on 1000 randomly selected permutations. The statistic computed by StatXact is the sum of the observations from treatment 1. The approximate *p*-value in the output is .2080 with a 95% confidence interval of (.1828, .2332). The exact *p*-value is .1967.

2.4
Wilcoxon Rank-Sum Test

Let $X_1, X_2, \ldots, X_N$ denote a set of N observations. The *rank* of X_i among the N observations, denoted $R(X_i)$, is given by

$$R(X_i) = \text{number of } X_j\text{'s} \leq X_i$$

Table 2.4.1 shows the data for a comparison of the methods of instruction given in Table 2.1.1 along with their ranks among the combined data. Here we assume that no two observations are the same, so that the ranks are distinct. In Section 2.5 we deal with the case of ties in the data.

FIGURE 2.3.2

StatXact Approximate and Exact *p*-Values of the Permutation Test for Data in Table 2.3.1

```
PERMUTATION TEST
[ Sum of scores from population <        1 > ]

Summary of Exact distribution of PERMUTATION statistic:

        Min      Max     Mean    Std-dev    Observed    Standardized
        459.9    506.9   483.4     7.538       490.1         0.8888

Asymptotic Inference:
    One-sided p-value: Pr {    Test Statistic .GE. Observed } =    0.1871
    Two-sided p-value: 2 * One-sided                          =    0.3741

Monte Carlo estimates of p-value estimates :
    One-sided: Pr { Test Statistic .GE. Observed   =      0.2080
              95.00% level of confidence        =(     0.1828,    0.2332)
    Two-sided p-value: Pr{| Test Statistic - Mean |
                    .GE. | Observed - Mean |} =      0.3960
              95.00% level of confidence        =(     0.3657,    0.4263)

Elapsed time is 0:0:0.06 (1000 tables sampled with starting seed 3425310)
```

```
PERMUTATION TEST
[ Sum of scores from population <        1 > ]

Summary of Exact distribution of PERMUTATION statistic:

        Min      Max     Mean    Std-dev    Observed    Standardized
        459.9    506.9   483.4     7.538       490.1         0.8888

Asymptotic Inference:
    One-sided p-value: Pr {    Test Statistic .GE. Observed } =    0.1871
    Two-sided p-value: 2 * One-sided                          =    0.3741

Exact Inference:
    One-sided p-value: Pr {    Test Statistic .GE. Observed } =    0.1967
                       Pr {    Test Statistic .EQ. Observed } =    0.0041
    Two-sided p-value: Pr { |  Test Statistic - Mean |
                    .GE. | Observed - Mean |       =    0.3933
    Two-sided p-value: 2*One-Sided                            =    0.3933

Elapsed time is 0:0:0.11
```

TABLE 2.4.1

Ranks of Data for Comparison of Methods of Instruction

Combined Data	23	31	37	46	49	55	57
Ranks	1	2	3	4	5	6	7
New Method Data	37	49	55	57			
Ranks	3	5	6	7			
Traditional Method Data	23	31	46				
Ranks	1	2	4				

Let W denote the sum of the ranks of the observations from one of the treatments. The *Wilcoxon rank-sum test* is a two-sample permutation test based on W.

2.4.1 Steps in Conducting the Wilcoxon Rank-Sum Test

Assume there are m observations for treatment 1 and n observations for treatment 2. Assume no two observations have the same value, so that the ranks are distinct. The procedure is as follows.

1. Combine the $m + n$ observation into one group, and rank the observations from smallest to largest. Let 1 be the rank of the smallest observation, 2 the rank of the next smallest observation, and so on. Find the observed rank sum W of treatment 1 (or treatment 2).

2. Find all possible permutations of the ranks in which m ranks are assigned to treatment 1 and n ranks are assigned to treatment 2.

3. For each permutation of the ranks, find the sum of the ranks for treatment 1 (or treatment 2).

4. Determine the upper-tail, lower-tail, or two-sided p-value as appropriate. For instance, for an upper-tail test, the p-value is

$$P_{\text{upper tail}} = \frac{\text{number of rank sums} \geq \text{observed rank sum } W}{\binom{m + n}{m}}$$

EXAMPLE 2.4.1   Table 2.4.2 contains five sets of permutations of the ranks of the data in Table 2.4.1. These correspond to the first five sets in Table 2.1.2. The sum of the ranks of the original data, denoted by an asterisk, is 21. Of the 35 possible sets of permuted ranks, two have a rank sum that is 21 or greater. Thus, the p-value of the test based on the rank sum is $2/35 = .0571$.

TABLE 2.4.2
Permuted Ranks and Rank Sums for Data in Table 2.4.1

Permuted Ranks	Ranks for New Method	Ranks for Traditional Method	Sum of Ranks for New Method
1	4 5 6 7	1 2 3	22
2*	3 5 6 7	1 2 4	21
3	3 4 6 7	1 2 5	20
4	3 4 5 7	1 2 6	19
5	3 4 5 6	1 2 7	18
Other	All other sums less than 21		

2.4.2 Comments on the Use of the Wilcoxon Rank-Sum Test

The rank sum of either treatment can be used; the choice is arbitrary. Moreover, the test could also be based on the difference of mean ranks. Let W_1 denote the sum of the ranks for treatment 1. If we have $N = m + n$ total observations in the data set, then the sum of all the ranks is

$$T = 1 + 2 + \cdots + N = \frac{N(N+1)}{2}$$

Using a similar computation to that for the difference of means in Section 2.1.2, we can show that

$$\text{difference of mean ranks} = W_1\left(\frac{1}{m} + \frac{1}{n}\right) - \frac{N(N+1)}{2n}$$

Since N is fixed for a given set of ranks, the mean difference of ranks is determined by the value of W_1. Thus, a statistical test based on the sum of ranks of one of the treatments will have the same p-value, and hence reach the same conclusion, as a test based on the difference of mean ranks.

The use of an upper-tail or lower-tail test is determined by the context of the problem. Suppose treatment 1 tends to produce observations that are larger than treatment 2. An upper-tail test is used if the statistic is the sum of the ranks of treatment 1, and a lower-tail test is used if the statistic is the sum of the ranks of treatment 2.

There are situations in which experimental units are ranked without any other numerical measurements known. For instance, an agronomist might do a visual inspection of damage to 16 plants and assign scores of 1 to 16 based on the severity of the damage. That is, the least damaged plant would receive a score of 1, the most damaged plant would receive a score of 16, and the others would be ranked accordingly in between. The Wilcoxon rank-sum test may be carried out on these ranks.

2.4.3 A Statistical Table for the Wilcoxon Rank-Sum Test

Table A3 of the Appendix contains upper-tail and lower-tail critical values for the Wilcoxon rank-sum test for samples ranging from 4 to 10 and significance levels 5%, 2.5%, and 1%. The tabled values are the ones closest to the stated level of significance without going over. To illustrate, if the sample sizes are $m = 9$ and $n = 7$, then the lower-tail and upper-tail critical values for one-sided tests with 5% level of significance are 43 and 76, respectively, where the sum is taken of the ranks of the treatment with seven observations. The exact significance level for these critical values can be shown to be $p = .045$.

EXAMPLE 2.4.2   A particular type of herbicide was tested for controlling weeds among strawberry plants. To see any potential damage that the herbicide might do to the strawberry plants, a researcher compared the dry weights of plants treated with the herbicide to the dry weights of untreated plants. Data were obtained on seven untreated and nine treated plants. It is expected that the untreated plants will have larger dry weights than the treated plants. Data and ranks are listed in Table 2.4.3. Since the ranks of the untreated plants are expected to be larger than those of the treated plants, an upper-tail test is used with the rank sum of the untreated plants. This sum is 84, which is greater than the 5% critical value of 76 for $m = 9$ and $n = 7$.

TABLE 2.4.3

Dry Weights (in kg) of Strawberry Plants

Untreated Plants	0.55 0.67 0.63 0.79 0.81 0.85 0.68
Treated Plants	0.65 0.59 0.44 0.60 0.47 0.58 0.66 0.52 0.51
Combined Data Ranked from Smallest to Largest	0.44 0.47 0.51 0.52 0.55 0.58 0.59 0.60 0.63 0.65 0.66 0.67 0.68 0.79 0.81 0.85
Ranks of Untreated Plants	5 12 9 14 15 16 13 Sum = 84
Ranks of Treated Plants	10 7 1 8 2 6 11 4 3 Sum = 52

2.4.4 Computer Analysis

StatXact has an option for the Wilcoxon rank-sum test. Using this option applied to the data in Table 2.4.3, we obtained the output shown in Figure 2.4.1. The samples were labeled so that 1 refers to the untreated plants (population 1). The exact

FIGURE 2.4.1

StatXact Output for Wilcoxon Rank-Sum Test Applied to Data in Table 2.4.3

```
WILCOXON-MANN-WHITNEY TEST
[ Sum of scores from population <      1 > ]

Summary of Exact distribution of WILCOXON-MANN-WHITNEY statistic:
      Min       Max       Mean      Std-dev      Observed       Standardized
     28.00     91.00     59.50       9.447         84.00            2.593
   Mann-Whitney Statistic =   56.00


Asymptotic Inference:
     One-sided p-value: Pr {    Test Statistic .GE. Observed }  =      0.0048
     Two-sided p-value: 2 * One-sided                           =      0.0095


Exact Inference:
     One-sided p-value: Pr {    Test Statistic .GE. Observed }  =      0.0039
                        Pr {    Test Statistic .EQ. Observed }  =      0.0013
     Two-sided p-value: Pr { |  Test Statistic - Mean |
                        .GE. | Observed - Mean |                =      0.0079
     Two-sided p-value: 2*One-Sided                             =      0.0079


Elapsed time is 0:0:0.05
```

p-value of the Wilcoxon test is .0039 for the one-sided test. The printout also refers to the Mann–Whitney statistic. It is a statistic that is equivalent to the Wilcoxon rank-sum statistic. This topic is discussed in Section 2.6.

The Wilcoxon rank-sum test can be carried out in Resampling Stats by doing the permutation test on the ranks and using the sum of the ranks of one of the treatments as the statistic. Code for analyzing the data in Table 2.4.3 is shown in Figure 2.4.2, which is just a modification of the code in Figure 2.3.1. The number of permuted samples in this example is set at 5000. Since Resampling Stats randomly samples the permutations, the *p*-value is approximate, not exact.

FIGURE 2.4.2

Resampling Stats Code for Wilcoxon Rank-Sum Test of Data in Table 2.4.3

```
'set the maximum vector size at 5000
maxsize default 5000

'put the ranks of the original combined data in a vector called "a"
copy (5 12 9 14 15 16 13 12 7 1 8 2 6 11 4 3) a

'observed rank sum for trt 1 is 84

'begin the loop to obtain 5000 randomly selected permutations
repeat 5000

'permute the rank data vector "a" and put the results in vector "b"
shuffle a b

'select the first 7 shuffled ranks for treatment 1, find the sum
take b 1,7 b1
sum b1 sumb1

'keep track of the 5000 rank sums of treatment 1
score sumb1 sumdist
end

'obtain the pvalue and print out result
count sumdist >= 84 pval
divide pval 5000 pval
print pval

'print output here

PVAL = 0.006
```

MINITAB implements the Mann–Whitney version of the Wilcoxon statistic. This is discussed in Section 2.6, where an example of the MINITAB output is shown. In S-Plus the Wilcoxon rank-sum test may be accessed from the "compare samples" menu. Exact *p*-values are available in S-Plus for small samples. For larger samples, the normal approximation discussed in Section 2.10.2 is used.

2.5
Wilcoxon Rank-Sum Test Adjusted for Ties

It is common to have ties among observations in a data set; that is, one or more observations may have the same value. In this case, the assignment of ranks to the observations is ambiguous. To resolve this ambiguity, we assign the average rank to the tied observations, and we call these *adjusted ranks*. The procedure is illustrated in Table 2.5.1 with hypothetical data.

TABLE 2.5.1
Average Ranks

Ranked Data	25	25	30	33	33	33	36	39	39	39	39
Ranks Disregarding Ties	1	2	3	4	5	6	7	8	9	10	11
Average Ranks	1.5	1.5	3	5	5	5	7	9.5	9.5	9.5	9.5

The Wilcoxon rank-sum statistic adjusted for ties, denoted W_{ties}, is the sum of the adjusted ranks for one of the treatments.

2.5.1 Steps in Conducting the Wilcoxon Rank-Sum Test Adjusted for Ties

The Wilcoxon rank-sum test adjusted for ties is carried out as follows:

1. Compute the adjusted ranks.
2. Apply the permutation test to the adjusted ranks, where the sum of the adjusted ranks for treatment 1 (or treatment 2) is the test statistic.

If the number of ties is small, then approximate critical values may be obtained from the distribution of the rank-sum statistic without ties (see Table A3 in the Appendix). For larger samples we may use the large-sample approximations in Section 2.10 to obtain approximate *p*-values.

EXAMPLE **2.5.1** Suppose an experiment has $m = 2$ observations for treatment 1 and $n = 3$ for treatment 2 with data and ranks shown in Table 2.5.2. To permute the data in such a way that all possible permutations are equally likely to occur under the null hypothesis of no treatment difference, we must distinguish between the two tied ranks. We do this by tagging these values as 3.5a and 3.5b. We then obtain all ten possible permutations of the tagged ranks. Finally, we compute the rank sums without regard to the tags. This process is shown in Table 2.5.3.

TABLE 2.5.2

Ranks of Data

Treatment 1: Data (Ranks)	11.7 (1)	12.0 (3.5)	
Treatment 2: Data (Ranks)	11.8 (2)	12.0 (3.5)	12.5 (5)

TABLE 2.5.3

Permuted Tagged Ranks for Data in Table 2.5.2

Tagged Ranks: 1 2 3.5a 3.5b 5

Permuted Sets of Ranks	Treatment 1: Ranks and (Rank Sum)		Treatment 2: Ranks and (Rank Sum)	
1	1 2	(3)	3.5a 3.5b 5	(12)
2*	1 3.5a	(4.5)	2 3.5b 5	(10.5)
3*	1 3.5b	(4.5)	2 3.5a 5	(10.5)
4	1 5	(6)	2 3.5a 3.5b	(9)
5	2 3.5a	(5.5)	1 3.5b 5	(9.5)
6	2 3.5b	(5.5)	1 3.5a 5	(9.5)
7	2 5	(7)	1 3.5a 3.5b	(8)
8	3.5a 3.5b	(7)	1 2 5	(8)
9	3.5a 5	(8.5)	1 2 3.5b	(6.5)
10	3.5b 5	(8.5)	1 2 3.5a	(6.5)

Computing the rank sum for treatment 1 and doing a lower-tail test, we find a rank sum of 4.5 for the ranks of the original data. There are three permuted sets of ranks for which the rank sum is less than or equal to 4.5. Thus, the lower-tail p-value is $3/10 = .3$.   ∎

2.5.2 Computer Analysis

StatXact does not require a separate option to adjust for ties. This is handled automatically under the Wilcoxon–Mann–Whitney option, which gives exact p-values with or without ties. Resampling Stats code for this problem is not shown, but it would be similar to the code in Figure 2.4.2 with input data being the adjusted ranks instead of ordinary ranks. Both MINITAB and S-Plus use the normal approximation discussed in Section 2.10.3 for tied data.

EXAMPLE 2.5.2   Table 2.5.4 shows the data from Example 2.3.1 along with the adjusted ranks. Only two values in the data are tied with adjusted ranks of 2.5. The Wilcoxon rank-sum test adjusted for ties has a one-sided *p*-value of .2296 obtained from StatXact. Since the number of ties is small, we may use the critical value for a 5% upper-tail test from Table A3 with $m = n = 8$. The rank sum for the treatment 1 data is 75.5. The critical value from Table A3 is 85, so the two treatments would not be declared statistically significantly different at this level of significance.

TABLE 2.5.4
Minutes to Observe 2-Liter Runoff

Combined Data and (Adjusted Ranks)	54.6 (1) 55.0 (2.5) 55.0 (2.5) 58.1 (4) 58.7 (5) 59.1 (6)
	59.3 (7) 60.1 (8) 60.3 (9) 61.3 (10) 62.1 (11) 62.5 (12)
	63.4 (13) 64.4 (14) 65.1 (15) 67.8 (16)

Treatment 1: Data	59.1	60.3	58.1	61.3	65.1	55.0	63.4	67.8
Ranks	6	9	4	10	15	2.5	13	16

Treatment 2: Data	60.1	62.1	59.3	55.0	54.6	64.4	58.7	62.5
Ranks	8	11	7	2.5	1	14	5	12

■

2.6
Mann–Whitney Test and a Confidence Interval

2.6.1 The Mann–Whitney Statistic

Let the observations from treatment 1 be denoted $X_1, X_2, \ldots, X_m$ and the observations from treatment 2 be denoted $Y_1, Y_2, \ldots, Y_n$. Let us assume that the data have no ties, so that any given observation is either strictly less than or strictly greater than any other observation. The *Mann–Whitney statistic*, denoted U, is defined as

$$U = \text{number of pairs} \left(X_i, Y_j \right) \text{ for which } X_i < Y_j$$

The null hypothesis is that the distributions of the X's and Y's are the same. A large value of U indicates that the larger observations tend to occur with treatment 2 (the Y's), and vice versa if U is small. Lower-tail and upper-tail values for the distribution of U under the null hypotheses are given in Table A4 in the Appendix. Upper-tail and lower-tail values are related by the equation $U_{\text{upper}} = mn - U_{\text{lower}}$.

EXAMPLE 2.6.1   Table 2.6.1 contains data on the numbers of hours that two brands of laptop computers function before battery recharging is necessary. There are four laptops for each brand, and $U = 12$ of the 16 pairs (X_i, Y_j) satisfy the inequality $X_i < Y_j$. Using

Table A4 for a sample of $m = 4$ and $n = 4$, we find that the upper-tail critical value for a 5% level of significance is 15, which is the value closest to the .05 level without going over. So $U = 12$ is not statistically significant at the 5% level; that is, there is no statistically significant difference between the hours before the two brands of batteries need recharging. The exact p-value for $U = 12$ from StatXact is .1714.

TABLE 2.6.1

Hours Until Recharge of Batteries

Brand 1	3.6 3.9 4.0 4.3
Brand 2	3.8 4.1 4.5 4.8
Pairs (X_i, Y_j):	(3.6, 3.8) (3.6, 4.1) (3.6, 4.5) (3.6, 4.8)
$U = 12$	(3.9, 3.8) (3.9, 4.1) (3.9, 4.5) (3.9, 4.8)
	(4.0, 3.8) (4.0, 4.1) (4.0, 4.5) (4.0, 4.8)
	(4.3, 3.8) (4.3, 4.1) (4.3, 4.5) (4.3, 4.8)

2.6.2 Equivalence of Mann–Whitney and Wilcoxon Rank-Sum Statistics

The Mann–Whitney statistic can be shown to be equivalent to the Wilcoxon rank-sum statistic in the sense that one is a linear function of the other. To see this, first let $R(Y_j)$ denote the rank of observation Y_j among the combined observations. Then

$$R(Y_j) = (\text{number of } Y's \le Y_j) + (\text{number of } X's \le Y_j)$$

For simplicity assume that the Y's have been arranged from smallest to largest; that is, $Y_1 < Y_2 < \cdots < Y_n$. Let W_2 denote the sum of the ranks of the Y's. Since the number of Y's $\le Y_j = j$, we have

$$W_2 = \sum_1^n R(Y_j) = 1 + 2 + \cdots + n + U$$

Since $1 + 2 + \cdots + n = n(n + 1)/2$, we have

$$W_2 = \frac{n(n+1)}{2} + U$$

Hence, a value of U uniquely determines a value of the Wilcoxon rank-sum statistic and vice versa. Rejection or failure to reject a hypothesis with one of the statistics leads to the same conclusion with the other.

For certain theoretical purposes, it is more convenient to use the Mann–Whitney statistic than the Wilcoxon statistic. Also, the Mann–Whitney statistic plays a useful role in the confidence interval procedure discussed in Section 2.6.3. From the point of view of statistical testing, there is no reason to prefer one over the other. In particular, any computer program that does the Wilcoxon rank-sum test may be used to do the Mann–Whitney test.

If the data contain ties, then the Mann–Whitney statistic may be computed by assigning 1/2 to each pair in which a tie exists. This procedure yields a statistic that is equivalent to the Wilcoxon statistic adjusted for ties discussed in Section 2.5.

2.6.3 A Confidence Interval for a Shift Parameter and the Hodges–Lehmann Estimate

Let $F_1(x)$ and $F_2(x)$ denote the cdf's of the X's and Y's, respectively. If there is a difference between treatments, suppose that the effect is to shift the distribution of one treatment an amount Δ to the right or left of the distribution of the other treatment. That is, assume the cdf's of treatments 1 and 2 satisfy

$$F_1(x) = F_2(x - \Delta)$$

We may think of Δ as being the difference between the means (if means exist) or the medians of the two distributions. We now construct a nonparametric confidence interval for Δ.

Theoretical Basis of the Confidence Interval

Since

$$P(X_i \le x) = F_1(x) = F_2(x - \Delta) = P(Y_j \le x - \Delta) = P(Y_j + \Delta \le x)$$

it follows that $Y_j + \Delta$ and X_i have the same probability distribution, a fact that we will use in constructing the confidence interval for Δ.

Form all *mn* pairwise differences $X_i - Y_j$, and arrange these differences from smallest to largest. Let pwd(k) be the kth smallest pairwise difference. Let k_a and k_b be integers such that $k_a < k_b$. The inequality

$$\mathrm{pwd}(k_a) < \Delta \le \mathrm{pwd}(k_b)$$

holds if and only if at least k_a and no more than $k_b - 1$ of the pairs (X_i, Y_j) satisfy the inequality

$$X_i - Y_j < \Delta$$

or equivalently

$$X_i < Y_j + \Delta$$

Since $Y_j + \Delta$ and X_i have the same probability distribution, probabilities involving such inequalities may be obtained from the U statistic. In particular,

$$P\big(\text{at least } k_a \text{ and no more than } k_b - 1 \text{ of pairs satisfy } X_i < Y_j + \Delta\big) = P\big(k_a \le U \le k_b - 1\big)$$

Thus, we have

$$P\big(\mathrm{pwd}(k_a) < \Delta \le \mathrm{pwd}(k_b)\big) = P\big(k_a \le U \le k_b - 1\big)$$

This equation is the basis for the confidence interval. We use the distribution of the U statistic to find k_a and k_b so that the right side of the equation has a desired probability—say .90 or .95. From the left side of the equation, the desired confidence interval is $\text{pwd}(k_a) < \Delta \leq \text{pwd}(k_b)$. The steps of the procedure are outlined next.

Steps in Constructing the Confidence Interval

1. Form all pairwise differences of the form $X_i - Y_j$.
2. Arrange the pairwise differences in order from smallest to largest. Let $\text{pwd}(k)$ denote the kth smallest pairwise difference.
3. To construct, say, a 90% confidence interval for Δ, find two numbers k_a and k_b such that the U statistic satisfies

$$P\left(k_a \leq U \leq k_b - 1\right) \approx .90$$

Since the distribution of U is discrete, it may not be possible to find points k_a and k_b such that the level of confidence is precisely 90%. In such cases, we would choose these points so that the level of confidence is at least 90% but as close to 90% as possible.

4. Having chosen k_a and k_b, we form the confidence interval

$$\text{pwd}(k_a) < \Delta \leq \text{pwd}(k_b)$$

Since the distributions of the X's and Y's are continuous, either of the inequalities in the confidence interval may be strict or inclusive. Make appropriate modifications for other levels of confidence.

If sample sizes allow, we may obtain k_a and k_b from Table A4 in the Appendix. For instance, if we want a 90% level of confidence, we find the lower-tail and upper-tail 5% values, $l_{.05}$ and $u_{.05}$, from the table, and choose

$$k_a = l_{.05} + 1, \quad k_b = u_{.05}$$

For sample sizes beyond the table, we may use a normal approximation of the distribution of U to find the desired lower-tail and upper-tail values. This topic is discussed in Section 2.10.

Hodges–Lehmann Estimate

The median of all the pairwise differences of the form $X_i - Y_j$ is called the *Hodges–Lehmann* estimate of Δ. The Hodges–Lehmann estimate is sometimes suggested as a nonparametric alternative to the difference of the means. It is usually computed in conjunction with the confidence interval based on pairwise differences.

EXAMPLE 2.6.2   The trace element cerium was measured in samples of granite and basalt; the amounts are listed in Table 2.6.2. The pairwise differences are shown in Table 2.6.3. From Table A4, we find $l_{.05} = 7$, so $k_a = 8$ and $u_{.05} = 29$, so $k_b = 29$. The 8th smallest difference among the 36 is 13.84 and the 29th difference is 47.96. Thus, $13.84 < \Delta \leq 47.96$ is a 90% confidence interval for Δ. The Hodges–Lehmann estimate is 30.045, which is the average of the 18th and 19th observations.

TABLE 2.6.2
Amounts of Cerium (in ppm) in Samples of Granite and Basalt

Granite (X's)	33.63	39.86	69.32	42.13	58.36	74.11
Basalt (Y's)	26.15	18.56	17.55	9.84	28.29	34.15

TABLE 2.6.3
Pairwise Differences for Data in Table 2.6.2

	33.63	39.86	69.32	42.13	58.36	74.11
26.15	7.48	13.71	43.17	15.98	32.21	47.96
18.56	15.07	21.30	50.76	23.57	39.80	55.55
17.55	16.08	22.31	51.77	24.58	40.81	56.56
9.84	23.79	30.02	59.48	32.29	48.52	64.27
28.29	5.34	11.57	41.03	13.84	30.07	45.82
34.15	−0.52	5.71	35.17	7.98	24.21	39.96

Pairwise differences in order from smallest to largest

1	−0.52	13	22.31	25	40.81
2	5.34	14	23.57	26	41.03
3	5.71	15	23.79	27	43.17
4	7.48	16	24.21	28	45.82
5	7.98	17	24.58	29	47.96
6	11.57	18	30.02	30	48.52
7	13.71	19	30.07	31	50.76
8	13.84	20	32.21	32	51.77
9	15.07	21	32.29	33	55.55
10	15.98	22	35.17	34	56.56
11	16.08	23	39.80	35	59.48
12	21.30	24	39.96	36	64.27

EXAMPLE 2.6.3   (continuation of Example 2.6.2) We may compare the confidence interval in Example 2.6.2 with the 90% confidence interval for the difference of two means given by the application of the t-distribution. This confidence interval is given by

$$\bar{X}_1 - \bar{X}_2 \pm t_{.05} S_e \sqrt{\frac{1}{m} + \frac{1}{n}}$$

where $t_{.05}$ is the upper 5% point of the t-distribution with $m + n - 2$ degrees of freedom and S_e is the "pooled" standard deviation defined by

$$S_e = \sqrt{\frac{(m-1)S_1^2 + (n-1)S_2^2}{n+m-2}}$$

In this case, $t_{.05} = 1.812$, $S_e = 13.38$, $\overline{X}_1 - \overline{X}_2 = 30.48$, and a 90% confidence interval is 30.48 ± 13.99, or 16.49 to 44.47. This confidence interval would be used if the observations come from normally distributed populations with equal standard deviations. The confidence interval based on the pairwise differences and the Hodges–Lehmann estimate would be particularly appropriate if the data have suspected outliers or the population distributions have heavy tails, such as occur with the Cauchy distribution or the Laplace distribution.   ■

2.6.4 Computer Analysis

The confidence interval procedure based on pairwise differences is implemented in StatXact. Output is shown in Figure 2.6.1. The "exact" confidence interval is exact in the sense that it gives an interval closest to the desired level of confidence without going below this level. The asymptotic interval is based on the normal approximation of the distribution of the Mann–Whitney statistic discussed in Section 2.10.5. The Hodges–Lehmann estimate is denoted as "Point Estimate of Shift."

FIGURE 2.6.1

StatXact Confidence Interval Based on Pairwise Differences and Hodges–Lehmann Estimate

```
HODGES-LEHMANN ESTIMATES OF SHIFT PARAMETER
POP_1 :          1              POP_2 :          2
Summary of WILCOXON MANN-WHITNEY statistic for POP_1


     Min       Max      Mean      Std-dev       Observed      Standardized
    21.00     57.00     39.00        6.245         56.00             2.722
     Mann-Whitney Statistic =    35.00
Point Estimate of Shift : Theta = POP_1 - POP_2 =    30.05


90.00% Confidence Interval for Theta :
          Asymptotic : (       13.84 ,       47.96)
          Exact       : (       13.84 ,       47.96)


Elapsed time is 0:0:0.05
```

MINITAB uses an algorithm in McKean and Ryan (1977) to obtain the confidence interval. The MINITAB output is shown in Figure 2.6.2. The Hodges–Lehmann estimate is denoted as "Point estimate for ETA1-ETA2." In addition to the confidence interval and Hodges–Lehmann estimate, MINITAB includes the test of the null hypothesis H_0: $\Delta = 0$ (denoted "ETA1 = ETA 2" in the output). It is possible to specify either two-sided or one-sided alternative hypotheses for this test.

The *p*-value is obtained using a normal approximation as discussed in Sections 2.10.2 and 2.10.3.

FIGURE 2.6.2

Minitab Confidence Interval Based on Pairwise Differences and Hodges–Lehmann Estimate

```
Mann-Whitney Test and CI: C1, C2
C1      N = 6       Median =        50.25
C2      N = 6       Median =        22.35
Point estimate for ETA1-ETA2 is    30.04
90.7 Percent CI for ETA1-ETA2 is (13.84,47.95)
W = 56.0
Test of ETA1 = ETA2   vs   ETA1 not = ETA2 is significant at 0.0082
```

2.7
Scoring Systems

Ranks can be thought of as scores that are used in place of the original observations in a permutation test. This leads us to consider coming up with other scores to be used in the same way. The key is to figure out a reasonable way to generate scores.

First, consider a particular way in which we might think of ranks. Suppose we take a random sample from a uniform probability distribution on the interval $[0, N + 1]$; that is, the population distribution is

$$f(w) = \frac{1}{N+1}, \quad 0 \le w \le N+1$$

Let $W_{(1)} < W_{(2)} < \cdots < W_{(N)}$ denote the order statistics of this random sample, where $W_{(1)}$ is the smallest observation, $W_{(2)}$ is the next smallest, and so on. It can be shown that the ranks are just the expected values of the $W_{(i)}$'s; that is, $E(W_{(i)}) = i$. In this vein, we consider scoring systems based on the expected order statistics of distributions other than the uniform.

2.7.1 Three Common Scoring Systems

Here we introduce three common scoring systems. The purpose in considering different scoring systems is to come up with ones that are effective in revealing differences that exist among treatments. Which one to choose in general depends on the distribution of the population from which the data are selected. Suggestions for choosing among scoring systems are considered in Section 2.9. Our purpose here is to define different scoring systems.

Normal Scores   Let $Z_1, Z_2, \ldots, Z_N$ be a random sample from a normal distribution with mean 0 and variance 1. Let $Z_{(1)} < Z_{(2)} < \cdots < Z_{(N)}$ denote the normal order statistics of the random sample. The expected value of $Z_{(i)}$,

$$S_{(i)} = E\left(Z_{(i)}\right)$$

is called the *i*th normal score.

Van der Waerden Scores   An alternative to normal scores are the van der Waerden scores defined by

$$V_{(i)} = \Phi^{-1}\left(\frac{i}{N+1}\right)$$

where Φ^{-1} denotes the inverse of the cdf of the standard normal distribution and N is the total number of observations. For instance, if $N = 10$, then $V_{(1)}$ is the value of z from the standard normal distribution that has cumulative probability 1/11. This value is $V_{(1)} = -1.335$. The van der Waerden scores are approximations to the normal scores, and generally it makes little difference which of these sets of scores are used. Table A5 in the Appendix contains van der Waerden scores for values of N from 10 to 20.

Exponential or Savage Scores   Another common scoring system is the expected values of the order statistics of the exponential distribution. For a set of N distinct observations, these scores are

$$\frac{1}{N}, \quad \frac{1}{N} + \frac{1}{N-1}, \quad \frac{1}{N} + \frac{1}{N-1} + \frac{1}{N-2}, \ldots$$

An equivalent set of scores is obtained by subtracting 1 from the expected value of the exponential order statistics. This procedure gives a set of scores, sometimes called the *Savage scores*, that sum to 0.

Whatever scoring system we use, a permutation test is applied to the scores, where the test statistic is the sum of the scores for one of the treatments. Scores for tied data are averaged, similar to the way in which we adjust ranks for ties.

EXAMPLE 2.7.1   Table 2.7.1 lists the ordered data, ranks, van der Waerden (VW) scores, and exponential scores for the cerium data in Table 2.6.2. The sums of the scores for the granite treatment are 56 for the ranks, 3.882 for the van der Waerden scores, and 9.652 for the exponential scores. A permutation test applied to the sum of the scores for each of the sets of scores gives an upper-tail *p*-value of .0022 for all of the tests. Although in this case all the tests turn out to have the same *p*-value, in general this is not the case. For instance, if we were to arbitrarily subtract 15 from all the granite observations, the one-sided *p*-values would be .0660, .0455, and .0530 for the Wilcoxon rank-sum test, the van der Waerden scores test, and the exponential scores test, respectively.

TABLE 2.7.1

Ranks, van der Waerden Scores, and Exponential Scores for Data in Table 2.6.2

Ordered Data	Treatment	Ranks	VW Scores	Exponential Scores
9.84	basalt	1	−1.426	0.083
17.55	basalt	2	−1.020	0.174
18.56	basalt	3	−0.736	0.274
26.15	basalt	4	−0.502	0.385
28.29	basalt	5	−0.293	0.510
33.63	granite	6	−0.097	0.653
34.15	basalt	7	0.097	0.820
39.86	granite	8	0.293	1.020
42.13	granite	9	0.502	1.270
58.36	granite	10	0.736	1.603
69.32	granite	11	1.020	2.103
74.11	granite	12	1.426	3.103

∎

2.7.2 Computer Analysis

StatXact implements the tests based on the van der Waerden scores, labeled as normal scores, and the Savage scores. The program automatically computes the scores from the original data and carries out the necessary computations. There is also a menu option to obtain scores explicitly. For Resampling Stats, the scores must be computed outside the program, with the exception being ranks. We may modify the Resampling Stats code in Figure 2.4.2, with ranks replaced by either van der Waerden scores or exponential scores, to obtain an approximate p-value for the corresponding nonparametric test applied to the data in Table 2.4.3.

2.8
Tests for Equality of Scale Parameters and an Omnibus Test

The tests we have discussed to this point are particularly designed to distinguish between the effects of two treatments when the observations from one of the treatments tend to be larger than the observations from the other. However, in some situations the variability of the observations for the two treatments is important. Suppose a machine for bottling a soft drink is designed to fill containers with 16 ounces of the beverage. Observations on the process may show that the data are centered around 16 as they should be, but there is excessive variability. This finding could lead an engineer to identify a problem that, if fixed, would reduce the variability while leaving the data correctly centered.

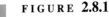

FIGURE 2.8.1

Two Distributions with Different Scale Parameters and the Same Location Parameters

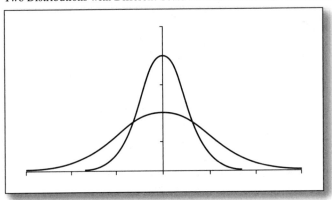

Figure 2.8.1 shows the distributions of observations of the type that we would expect to see if the treatments affect the scale parameters of the population distributions but not the location parameters. Observations from the treatment with the larger scale parameter tend to be nearer the extremes, either large or small, while observations from the treatment with the smaller scale parameter tend to be nearer the middle. Thus, a scoring procedure that places lower scores on the more extreme observations and higher scores on the middle observations should detect differences in scales. This is the basis of the Siegel–Tukey test described next.

2.8.1 Siegel–Tukey and Ansari–Bradley Test

We assume that an observation X_i from treatment 1 and an observation Y_j from treatment 2 follow the model

$$X_i = \mu + \sigma_1 \varepsilon_{ix}$$
$$Y_j = \mu + \sigma_2 \varepsilon_{jy}$$

where the ε's are independent and identically distributed random variables with a median of 0. Note the common location parameter μ for both the X's and the Y's. The null hypothesis used in the Siegel–Tukey test is

$$H_0: \sigma_1 = \sigma_2$$

It is tested against either a one-sided or a two-sided alternative hypothesis as follows.

Steps in Carrying Out the Siegel–Tukey Test

1. Arrange the observations in the combined data set from smallest to largest.
2. Assign rank 1 to the smallest observation, rank 2 to the largest observation, rank 3 to the next largest observation, rank 4 to the next smallest observation, and so on.

3. Apply the Wilcoxon rank-sum test. The smaller ranks and smaller rank sum are associated with the treatment that has the larger variability.

EXAMPLE **2.8.1**   Table 2.8.1 contains data on the amounts of liquid in randomly selected beverage containers before and after the filling process has been repaired. The rank sum of the Siegel–Tukey ranks for treatment 1 is 24. Since it is associated with the treatment that presumably has the larger variability, the rank sum should be small, so we should do a lower-tail test. Since the critical value is 19 from Table A3 in the Appendix, the test shows no significant difference in the variability of the process before and after repair. The *p*-value from StatXact is .2738. Resampling Stats may be used to analyze the data. The Siegel–Tukey ranking is used instead of the usual ranks in a program to do the Wilcoxon rank-sum test. The code is similar to that in Figure 2.4.2.

TABLE **2.8.1**

Data for Siegel–Tukey Test: Ounces of Beverage in Containers

Treatment 1 (Before Process Repair):	16.55 15.36 15.94 16.43 16.01
Treatment 2 (After Process Repair):	16.05 15.98 16.10 15.88 15.91
Combined Data: 15.36 15.88 15.91 15.94 15.98 16.01 16.05 16.10 16.43 16.55	
Siegel–Tukey Ranks: 1 4 5 8 9 10 7 6 3 2	
Sum of Ranks for Treatment 1: $2 + 1 + 8 + 3 + 10 = 24$ Lower 5% = 19	

■

Ansari–Bradley Test

One of the difficulties of the Siegel–Tukey test is that the ranking could just as well start its alternating pattern with the largest observation instead of the smallest. This in general will yield a different value of the Wilcoxon statistic, although the decision reached will generally be the same. The *Ansari–Bradley test* (Ansari and Bradley, 1960) overcomes this ambiguity by averaging the ranks obtained by doing the Siegel–Tukey ranking both ways. However, the critical values in this case can no longer be obtained from the tables for the Wilcoxon rank-sum test. See Ansari and Bradley (1960) for tables up to samples of size 10. This procedure is implemented in StatXact.

2.8.2 Tests on Deviances

Let us suppose that the location parameters of the X's and Y's are not the same, so that the models for the observations from treatment 1 and treatment 2 are

$$X_i = \mu_1 + \sigma_1 \varepsilon_{ix}$$
$$Y_j = \mu_2 + \sigma_2 \varepsilon_{jy}$$

where again the ε's are independent and identically distributed with a median of 0. The test statistic to use when the ε's have a normal distribution is the F statistic,

$F = S_1^2/S_2^2$, where S_1^2 and S_2^2 are the sample variances for the X's and Y's, respectively. However, the test is not recommended for nonnormal distributions. The lack of normality of the ε's can substantially affect the Type I error of this test, which often leads to excessive false rejections of H_0. See Conover, Johnson, and Johnson (1981).

A test that we suggest is a permutation test based on deviances. First, suppose the μ_i's are known. Form the deviances

$$\text{dev}_{ix} = X_i - \mu_1, \quad \text{dev}_{jy} = Y_j - \mu_2$$

If H_0: $\sigma_1 = \sigma_2$ is true, then the distributions of the deviances for the two treatments are the same, whereas if H_0 is not true, then the deviances for the X's tend to be either larger or smaller than the deviances of the Y's. The test statistic that we propose is the ratio of the absolute mean deviances defined by

$$\text{RMD} = \frac{\sum\limits_{i=1}^{m} |\text{dev}_{ix}|/m}{\sum\limits_{j=1}^{n} |\text{dev}_{jy}|/n}$$

Steps for a Permutation Test on Deviances (Known Location Parameters)

1. Obtain the deviances dev_{ix} and dev_{jy}, where $i = 1, \ldots, m$ and $j = 1, \ldots, n$, for the two treatments, and compute RMD from the original data, which we denote RMD_{obs}.

2. Permute the deviances among the two treatments and obtain RMD for each permutation of the deviances.

3. For small values of m and n, obtain all permutations of the data; for larger samples, obtain a random sample of R permutations.

4. For the upper-tail test H_a: $\sigma_1 > \sigma_2$, the upper-tail p-value is the fraction of the permutations in step 3 for which RMD is greater than or equal to RMD_{obs}. Similarly a lower-tail p-value may be obtained.

Remark   If a fixed level of significance α is desired, then we would reject H_0 if the p-value in step 4 is less than or equal to α. An alternative way to reach the same decision for an upper-tail test would be to obtain the upper αth critical value, RMD_α, from the permutation distribution, and if $\text{RMD}_{\text{obs}} \geq \text{RMD}_\alpha$ reject the null hypothesis. A similar procedure applies to a lower-tail test.

Modifications for a Two-Sided Test

For a two-sided test, we note that we may express the alternative hypothesis as

$$H_a: \quad \frac{\max(\sigma_1, \sigma_2)}{\min(\sigma_1, \sigma_2)} > 1$$

We form the modified RMD statistic:

$$\text{RMD}_{2\text{-sided}} = \frac{\max\left(\sum_{i=1}^{m}\left|\text{dev}_{ix}\right|/m, \sum_{j=1}^{n}\left|\text{dev}_{jy}\right|/n\right)}{\min\left(\sum_{i=1}^{m}\left|\text{dev}_{ix}\right|/m, \sum_{j=1}^{n}\left|\text{dev}_{jy}\right|/n\right)}$$

We then perform the steps above using $\text{RMD}_{2\text{-sided}}$ instead of RMD and using an upper-tail test.

Steps for a Permutation Test on Deviances (Unknown Location Parameters)

We propose an approximation to the permutation test described above when the location parameters are unknown. We replace the unknown location parameters with sample medians and proceed as before. The steps are as follows:

1. Compute the estimated deviances:

$$\hat{\text{dev}}_{ix} = X_i - \text{med}_1, \quad \hat{\text{dev}}_{jy} = Y_j - \text{med}_2$$

where med_1 and med_2 are the sample medians for the X's and Y's respectively.

2. Use the estimated RMD statistic defined by

$$\hat{\text{RMD}} = \frac{\sum_{i=1}^{m}\left|\hat{\text{dev}}_{ix}\right|/m}{\sum_{j=1}^{n}\left|\hat{\text{dev}}_{jy}\right|/n}$$

3. Carry out the steps of the permutation test above using the estimated deviances and estimated RMD.

Remark A test similar to this one is a permutation test based on the F statistic. For the F statistic, squared deviations are used instead of absolute deviations, and sample means are used instead of sample medians. Bailar (1988) investigated the properties of this test and generally found them to be desirable. However, the test does not perform well when samples are from populations that have heavy-tailed distributions like the Cauchy, a distribution for which the population mean does not exist.

EXAMPLE 2.8.2 The permutation test based on $\hat{\text{RMD}}$ was applied to the data in Table 2.8.1. The value of $\hat{\text{RMD}}$ for the data is 4.67. The 90th, 95th, and 97.5th percentiles of the permutation distribution based on 1000 randomly selected permutations were found to be 2.34, 5.18, and 6.84, respectively. Thus, the statistic is significant at the 10% level but not at the 5% level. An approximate permutation p-value is .07. Figure 2.8.2 shows Resampling Stats code for carrying out this test.

FIGURE 2.8.2

Resampling Stats Code for Permutation Test Based on $\hat{\text{RMD}}$

```
'input data and compute rmd
data (16.55 15.36 15.94 16.43 16.01) x
data (16.05 15.98 16.10 15.88 15.91) y
median x medx
median y medy
sumabsdev x medx dx
sumabsdev y medy dy
divide dx dy rmd
print rmd

'obtain permutation distribution of rmd
subtract x medx devx
subtract y medy devy
concat devx devy dev
repeat 1000
shuffle dev perm
take perm 1,5 permdevx
take perm 6,10 permdevy
sumabsdev permdevx 0 permdx
sumabsdev permdevy 0 permdy
divide permdx permdy permrmd
score permrmd permdist
end

'obtain p-value
count permdist >= rmd pvalue
divide pvalue 1000 pvalue
print pvalue

'obtain histogram and percentiles of the permutation
distribution
histogram permdist
percentile permdist (90 95 97.5) pctile
print pctile
```

■

Comments on the Type I Error of the Permutation Test Based on $\hat{\text{RMD}}$

When the location parameters are estimated from the data, the stated Type I error rate is not necessarily the actual Type I error rate. Instead, the actual error rate depends on the population distribution from which the data are selected and the sample size. In this sense, the permutation test based on $\hat{\text{RMD}}$ is not truly nonparametric. However, this effect diminishes as the sample size increases, and the test behaves more like the test based on RMD, which is a nonparametric test.

Woreck (1997) investigated the Type I error of the RMD test for equal and unequal sample sizes ranging from 5 to 20 with samples taken from normal, uniform, exponential, and Cauchy populations. Simulations were based on 1000 experiments for each combination of sample size and population distribution. The permutation tests were based on 1000 randomly selected permutations. At a presumed level of significance of .05, this test had an average Type I error of .045 (20 cases in all). For small sample sizes, the simulated error rates tended to be smaller than the presumed .05 value. For instance, for samples of $m = n = 5$, the simulated Type I errors were .011, .014, .025, and .041 for the normal, uniform, exponential, and Cauchy distributions, respectively. For larger sample sizes, the error rates were nearer the nominal .05 value. For instance, for samples of size $m = n = 20$, the simulated Type I errors were .040, .049, .065, and .048 for the normal, uniform, exponential, and Cauchy distributions, respectively.

2.8.3 Kolmogorov–Smirnov Test

Suppose it is not known how a difference between two treatments might manifest itself in the data. It might cause observations in one treatment to be larger than observations in the other, or it might affect the variability of the observations, or it might affect the shapes of the distributions in some other way. What we would like is an *omnibus test*—that is, a test designed to pick up differences among treatments regardless of the nature of the differences. The Kolmogorov–Smirnov test is appropriate for this situation.

Denote the sample cdf's for the two treatments as $\hat{F}_1$ and $\hat{F}_2$. The Kolmogorov–Smirnov statistic is the maximum absolute value of the difference between the two sample cdf's; that is,

$$\text{K-S} = \max_w \left| \hat{F}_1(w) - \hat{F}_2(w) \right|$$

EXAMPLE 2.8.3 The sample cdf's for the data in Table 2.8.1 are listed in Table 2.8.2. The value of each sample cdf is determined at each point in the combined data set, and differences in the cdf's are computed at each of these points. For instance, two of the five observations from treatment 1 and three of the five observations from treatment 2 are less than or equal to 15.98, so the difference in cdf's at $w = 15.98$ is 1/5. The value of K-S for these data is found to be 2/5, which occurs at $w = 16.10$.

TABLE 2.8.2

Computation of K-S Statistic for Data in Table 2.8.1

Combined Data	15.36	15.88	15.91	15.94	15.98	16.01	16.05	16.10	16.43	16.55
Treatment	(1)	(2)	(2)	(1)	(2)	(1)	(2)	(2)	(1)	(1)
cdf Treatment 1	1/5	1/5	1/5	2/5	2/5	3/5	3/5	3/5	4/5	1
cdf Treatment 2	0	1/5	2/5	2/5	3/5	3/5	4/5	1	1	1
Absolute Difference	1/5	0	1/5	0	1/5	0	1/5	2/5	1/5	0

■

To determine the *p*-value for the K-S statistic, we find the values of K-S for all possible permutations of the data and then determine the proportion of those values that are greater than or equal to the observed value. Large values of K-S indicate that the samples are from different populations, so the test is upper-tail.

EXAMPLE 2.8.4   For the data in Table 2.8.1, there are $\binom{10}{5} = 252$ permutations. The distinct possible values of K-S are .2, .4, .6, .8, and 1.0. The permutation distribution is shown in Table 2.8.3. The value of K-S is .4, and the *p*-value is 220/252 = .873. Tables of critical values of the K-S statistic may be found in Beyer (1968). StatXact implements the K-S test.

TABLE 2.8.3
Permutation Distribution of K-S Statistics, $m = n = 5$

K-S	.2	.4	.6	.8	1
Probability = K-S	32/252	130/252	70/252	18/252	2/252

If we know in what way the treatments are likely to affect the observations, then we should choose a test that is designed for that situation if possible. For instance, if we know that observations from one treatment will tend to be larger than observations from the other, then the Wilcoxon rank-sum test will likely have greater power at detecting differences between treatments than the Kolmogorov–Smirnov test will. The K-S test is used when such knowledge is not available.

2.8.4 Computer Analysis

The Siegel–Tukey, Ansari–Bradley, and Kolmogorov–Smirnov tests are implemented in StatXact. The asymptotic approximations for the Siegel–Tukey and Ansari–Bradley tests are based on the normal approximation for general scores discussed in Section 2.10.1. The asymptotic approximation for the Kolmogorov–Smirnov test is not considered, but a discussion may be found in Lehmann (1975). Figure 2.8.3 shows output for these three tests applied to the data in Table 2.8.1. StatXact does not allow for user-defined statistics. Thus, it is not possible to conduct the permutation test with the RMD statistic in StatXact. However, it is relatively simple to do this in Resampling Stats, as indicated by the code in Figure 2.8.2.

2.9
Selecting Among Two-Sample Tests

We now discuss the roles that Type I error and power of statistical tests play in choosing among two-sample tests. When we examine Type I error and power, it is

FIGURE 2.8.3

StatXact Output for Siegel–Tukey, Ansari–Bradley, and Kolmogorov–Smirnov Tests
Applied to Data in Table 2.8.1

```
SIEGEL-TUKEY TEST
[ Sum of scores from population <     1 > ]

        Min      Max     Mean    Std-dev     Observed      Standardized
      15.00    40.00    27.50     4.787        24.00          -0.7311

Asymptotic Inference:
  One-sided p-value: Pr {   Test Statistic .LE. Observed }   =      0.2324
  Two-sided p-value: 2 * One-sided                           =      0.4647

Exact Inference:
  One-sided p-value: Pr {   Test Statistic .LE. Observed }   =      0.2738
                     Pr {   Test Statistic .EQ. Observed }   =      0.0635
  Two-sided p-value: Pr { | Test Statistic - Mean |
                         .GE. | Observed - Mean |            =      0.5476
  Two-sided p-value: 2*One-Sided                             =      0.5476


ANSARI-BRADLEY TEST
[ Sum of scores from population <     1 > ]

        Min      Max     Mean    Std-dev     Observed      Standardized
      9.000    21.00    15.00     2.357        13.00          -0.8485

Asymptotic Inference:
  One-sided p-value: Pr { Test Statistic .LE. Observed }     =      0.1981
  Two-sided p-value: 2 * One-sided                           =      0.3961

Exact Inference:
  One-sided p-value: Pr { Test Statistic .LE. Observed }     =      0.2698
                     Pr { Test Statistic .EQ. Observed }     =      0.1190
  Two-sided p-value: Pr { | Test Statistic - Mean |
                         .GE. | Observed - Mean |            =      0.5397
  Two-sided p-value: 2*One-Sided                             =      0.5397


KOLMOGOROV-SMIRNOV TWO-SAMPLE TEST
POP_1 (F1) :        1         POP_2 (F2) :          2

      Number of Observations:
            POP_1 = 5
            POP_2 = 5
                         |F1 - F2|        F1 - F2            F2 - F1
                                     (POP_1 is larger)   (POP_2 is larger)
  Observed Statistic       0.4000        0.2000              0.4000
  Asymptotic p-value       0.8186        0.8187              0.4493
  Exact p-value            0.8730        0.8333              0.4762
  Exact Point Prob.        0.5159        0.3571              0.2976
```

common to assume that the distributions of the populations differ by a location or shift parameter Δ; that is, the cdf's for the two populations satisfy

$$F_1(x) = F_2(x - \Delta)$$

The null hypothesis to be considered is H_0: $\Delta = 0$, and the alternative hypothesis may be either one-sided or two-sided. Moreover, it is often assumed that the cumulative distribution functions are continuous so that the data have no ties. These are by no means the only conditions under which it makes sense to compare statistical tests, but they are simple conditions that can guide us about which test is appropriate under various circumstances.

Population distributions that are commonly studied are the uniform, normal, Laplace, exponential, and Cauchy. The uniform, normal, Laplace, and Cauchy distributions are symmetric and range in tail weight from light to heavy. The exponential distribution is asymmetric with heavier tails than the normal distribution.

2.9.1 The *t*-Test

We begin with the *t*-test. If the population cdf's F_1 and F_2 are normal with unknown but equal variances, then the test to use is the classical two-sample *t*-test for the difference between two means. It has the correct probability of a Type I error and has the greatest power among so-called unbiased tests (Lehmann, 1986). If the assumption of normality is violated, then one cannot necessarily assert that the *t*-test has the correct probability of a Type I error. However, numerous studies have shown that modest violations of the normality assumption have little effect on the probability of a Type I error of the *t*-test.

This robustness property for the probability of a Type I error of the *t*-test is due to the central limit theorem effect. That is, a sample mean will have an approximately normal distribution for "large" samples regardless of the form of the cdf's of the observations, provided that the population distributions have a finite variance. Although one cannot make any definitive statements about how large a sample should be before the central limit theorem effect manifests itself, in many practical cases the sample sizes may be remarkably small. For instance, if the samples are from a uniform distribution, samples of 5 to 10 are sufficient for the central limit theorem effect to be seen. If the samples are from the exponential distribution, samples of around 20 are sufficient for practical purposes to assert that the sample mean has an approximate normal distribution.

The power of the *t*-test is another issue. Some practitioners have mistakenly assumed that the central limit theorem effect will also imbue the *t*-test with optimal power properties when samples are sufficiently large. Such is not the case. The optimality of the *t*-test in terms of power can only be theoretically supported when the underlying distributions are normal. Lacking this, no such claims of optimality can be made.

2.9.2 The Wilcoxon Rank-Sum Test versus the *t*-Test

The Wilcoxon rank-sum test compares very favorably to the *t*-test. It is a nonparametric test, so the probability of a Type I error is not dependent upon the form of the distribution. It requires only that the two distributions be identical under the assumption of no treatment effect. In terms of power, there are circumstances in which the Wilcoxon test has substantial advantages over the *t*-test. This may be somewhat surprising since the Wilcoxon test replaces the original observations by ranks, perhaps making it appear that information has been disregarded. However, observations that are unusually large or small in comparison with the rest the data can have adverse effects on the power of the *t*-test. Such observations generally have far less adverse effects on tests based on ranks.

Extensive simulations were carried out by Blair and Higgins (1980) to compare the power functions of the Wilcoxon rank-sum test and the *t*-test for equal and unequal sample sizes ranging from $n + m = 12$ to $m + n = 108$ and for a number of distributions. Power values were simulated for both the Wilcoxon test and the *t*-test by letting Δ range over a sufficiently large number of values to take the power from its lowest level to approximately 1.00. To ensure that the *t*-test matched the Wilcoxon test in terms of the probability of a Type I error, critical values for the *t*-test were obtained by simulation rather than from tabled values of the *t*-distribution, which apply only when the populations have a normal distribution. Brief excerpts of their results are given in Table 2.9.1. Each entry in the table is the maximum amount by which the simulated power of the test in question exceeds the power of the other test. For instance, with the Laplace distribution and moderate sample sizes, the simulated power function of the Wilcoxon test is greater than that of the *t*-test by a maximum of .17, but the power function of the *t*-test does not exceed that of the Wilcoxon test in any of the simulated cases.

Generally speaking, the *t*-test has greater power than the Wilcoxon test for small samples and for distributions that are light-tailed—that is, distributions that

TABLE 2.9.1

Maximum Power Advantages Attained by the Wilcoxon Test and the *t*-Test for Small and Moderate Sample Sizes Drawn from Three Nonnormal Distributions

Distribution	*Statistical Test*	*Small:* $m + n = 12$	*Moderate:* $m + n = 36,$ $m + n = 108$
Uniform	Wilcoxon	.00	.01
	t	.13	.09
Laplace	Wilcoxon	.04	.17
	t	.13	.00
Exponential	Wilcoxon	.12	.44
	t	.11	.00

are not likely to produce extreme observations. The Wilcoxon test generally has greater power for moderate to large samples and for distributions that are heavier-tailed or skewed. A word of caution about applying these results: If tabled critical values of the t-distribution are used for the t-test when the sampling is from nonnormal distributions, then their validity might be in question for very small samples. Thus, on the basis of Type I error considerations, the Wilcoxon test may be preferred in such cases even though the t-test may show power advantages. For larger samples, the central limit theorem effect allows the practitioner to use the tabled values of the t-distribution, but in this case the Wilcoxon test may have power advantages over the t-test, especially when samples are selected from heavier-tailed distributions. Ranking filters out extreme observations, making differences in the central part of the data more easily detected.

2.9.3 Relative Efficiency

Suppose two statistical tests, 1 and 2, of H_0: $\Delta = 0$ versus H_a: $\Delta > 0$ are conducted at the same level of significance. Let $m_1 + n_1 = N_1$ and $m_2 + n_2 = N_2$ be the total sample sizes for tests 1 and 2, respectively, and assume $m_1/n_1 = m_2/n_2$. Let N_1 and N_2 be chosen in such a way that the two tests have the same power for a given alternative Δ. The *relative efficiency* of test 1 to test 2 is defined as

$$\text{eff}(1 \text{ vs } 2) = \frac{N_2}{N_1}$$

A ratio greater than 1 indicates that test 1 requires a smaller sample size than test 2 to achieve the same power, and hence it is more efficient.

Under rather general conditions, we may take the limit of this efficiency as $N_1 \to \infty$ and $N_2 \to \infty$ and $\Delta \to 0$. The resulting limit, which turns out to be independent of the probability of a Type I error and power, is called the *asymptotic relative efficiency* (a.r.e.) of test 1 to test 2. The a.r.e. is a measure of the large-sample efficiency of one test relative to the other. See Lehmann (1975). The a.r.e. of the Wilcoxon test to the t-test is rather remarkable. Hodges and Lehmann (1956) showed it can never be less than 0.864 but can be arbitrarily large. In other words, the Wilcoxon test can never be much less efficient than the t-test, but it has the potential of being infinitely more efficient. Table 2.9.2 gives selected a.r.e.'s of the Wilcoxon rank-sum test to the t-test. Note that for the normal distribution, where the t-test is optimal, the a.r.e. is 0.955.

The a.r.e. has also been investigated for other tests. The a.r.e. of the test based on van der Waerden scores relative to the t-test is always greater than or equal to 1, and it is equal to 1 for the normal distribution. In some cases it is more efficient than the Wilcoxon test and in some cases less. For instance, its a.r.e. is greater for the uniform, normal, and exponential distributions with a.r.e.'s ∞, 1.047, and ∞, respectively. Its a.r.e. is less for the Laplace and Cauchy distributions with a.r.e.'s 0.947 and 0.708, respectively.

TABLE 2.9.2
Asymptotic Relative Efficiency of Wilcoxon Rank-Sum Test to *t*-Test

Distribution	Efficiency
Uniform	1.0
Normal	0.955
Laplace	1.5
Exponential	3.0
Cauchy	∞

The exponential scores test is as efficient as the best parametric test for a distribution called the extreme-value distribution. This is the distribution of the random variable $Y = c \times \log(X)$, where X has the exponential distribution and $c > 0$. The a.r.e. of the exponential scores test relative to the Wilcoxon test for the extreme-value distribution is 1.33. The extreme-value distribution is a common distribution for the logarithms of survival times, and exponential scores are often used with survival data.

2.9.4 Power of Permutation Tests

We now consider permutation tests. First let us look at the issue of random sampling the permutations. Based on extensive simulations, Keller-McNulty and Higgins (1987) concluded that little is to be gained by taking more than 1600 randomly sampled permutations. These results were based on a study of several factors, including the type of test, the number of randomly selected permutations, the distributions from which samples are selected, sample sizes, and levels of significance.

Typical of their results is a comparison of the power function of the *t*-test with that of the permutation test for differences of means when samples are selected from normal distributions. Power functions of the permutation test were obtained using 100, 200, 400, 800, 1600, and 3200 randomly selected permutations. As shown in Table 2.9.3, there appears to be little difference between power functions of the permutation test and the *t*-test for samples as small as $m = n = 10$ and as few as 400 randomly sampled permutations when the probability of a Type I error is 5%. A somewhat larger number of permutations is required for tests with a Type I error of 1%. This suggests that the power function of the permutation test converges rather rapidly to its limiting value as the number of randomly selected permutations increases. The StatXact statistical package has a default of 10,000 randomly selected permutations for its Monte Carlo option, which is more than adequate for most practical purposes.

Keller-McNulty and Higgins (1987) compared the power function of the Wilcoxon rank-sum test with the power functions of various permutation tests, including those based on means, medians, and 10% trimmed means. They used 1600

TABLE 2.9.3

Maximum Difference Between Simulated Power Functions of the *t*-Test
and Permutation Tests for Differences of Means

Type I Error	Sample Size	Number of Randomly Selected Permutations					
		100	200	400	800	1600	3200
.01	m = n = 10	.087	.048	.032	.015	.014	.010
	m = n = 20	.093	.040	.021	.015	.009	.007
.05	m = n = 10	.029	.019	.017	.010	.005	.008
	m = n = 20	.012	.008	.006	.005	.008	.003

randomly selected permutations in their study. Excerpts of their results are shown in
Table 2.9.4. The entries are the maximum difference between the power function of
the permutation test and the power function of the Wilcoxon test. The letter *P* or *W*
denotes which test has the greater simulated power. We see that tail weight has an
effect on the relative power of the various tests. The permutation test based on the
mean does well in relationship to the Wilcoxon test when distributions have lighter
tails. As we move to the heavier-tailed distributions, the permutation test based on the
mean has less power than the Wilcoxon test, but the one based on the median has
greater power. None of the permutation tests is remarkably better than the Wilcoxon
test. The largest power advantage is .084 for the median version of the permutation
test when sampling is from the Cauchy distribution with $m = n = 20$.

It should be noted that the permutation test for differences of means is equivalent to a permutation version of the *t*-test. That is, one may compute either the dif-

TABLE 2.9.4

Maximum Difference Between Power Functions of Various Permutation Tests
and the Wilcoxon Rank-Sum Test; Level of Significance Is 5%
(P or W indicates which test, the permutation test or the Wilcoxon test, has greater power.)

Population Distribution	Statistic Used in Permutation Test	m = 10 n = 10	m = 20 n = 20
Normal	Mean	.045(P)	.027(P)
	Median	.031(W)	.059(W)
	10% trimmed mean	.033(P)	.009(P)
Laplace	Mean	.035(W)	.080(W)
	Median	.033(P)	.024(P)
	10% trimmed mean	.021(P)	.009(W)
Cauchy	Mean	.276(W)	.502(W)
	Median	.080(P)	.084(P)
	10% trimmed mean	.068(W)	.116(W)

ference of means or the *t*-statistic in carrying out the permutation test. Pitman (1937a) and Hoeffding (1952) showed that the ordinary two-sample *t*-test and the permutation *t*-test have the same asymptotic power under appropriate conditions. Thus, situations that favor the ordinary *t*-test will favor the permutation *t*-test. For instance, if samples are from lighter-tailed distributions, then the permutation *t*-test will tend to have greater power than the Wilcoxon text. However, for samples from heavier-tailed distributions, the Wilcoxon test will tend to have greater power than the permutation *t*-test.

2.10
Large-Sample Approximations

We present a procedure for obtaining a large-sample approximation of the permutation distribution of any statistic that can be computed as the sum of scores associated with one of the two treatments. The procedure applies to the Wilcoxon rank-sum statistic, the Wilcoxon rank-sum statistic with ties, the van der Waerden scores test statistic, and other such statistics.

2.10.1 Sampling Formulas

Let *m* and *n* denote the number of observations in treatments 1 and 2, respectively, and let $N = m + n$ be the combined sample size. Let $A_1, A_2, \ldots, A_N$ be N numbers representing ranks, normal scores, or other such scores that have been assigned to the combined observations. Let T_1 denote the sum of the A_i's associated with treatment 1.

If there is no difference between the two treatments, then any A_i is as likely to occur among the scores for treatment 1 as any other A_i. Thus, the *m* scores associated with treatment 1 occur as if they had been randomly selected without replacement from the finite population of scores $A_1, A_2, \ldots, A_N$.

From the theory of sampling from finite populations (Cochran, 1963), the expected value and variance of T_1 are given by

$$E(T_1) = m\mu$$
$$\text{var}(T_1) = \frac{mn\sigma^2}{N-1}$$

where

$$\mu = \frac{\sum_{i=1}^{N} A_i}{N}$$

and

$$\sigma^2 = \frac{\sum_{i=1}^{N}(A_i - \mu)^2}{N} = \frac{\sum_{i=1}^{N} A_i^2}{N} - \mu^2$$

If m and n are sufficiently large, then T_1 will have an approximate normal distribution. Hence, one may compute

$$Z = \frac{T_1 - E(T_1)}{\sqrt{\text{var}(T_1)}}$$

and refer the resulting statistic to the standard normal distribution to determine whether to reject the hypothesis in question.

2.10.2   Application to the Wilcoxon Rank-Sum Test

For the Wilcoxon rank-sum test without ties, the A_i's are the ranks $1, 2, \ldots, N$. Computations for μ and σ^2 involve the formulas for the sum and sum of squares of the first N integers.

$$1 + 2 + \cdots + N = \frac{N(N+1)}{2}$$

$$1^2 + 2^2 + \cdots + N^2 = \frac{N(N+1)(2N+1)}{6}$$

Thus, we have

$$\mu = \frac{1 + 2 + \cdots + N}{N} = \frac{N+1}{2}$$

and

$$\sigma^2 = \frac{1^2 + 2^2 + \cdots + N^2}{N} - \left(\frac{N+1}{2}\right)^2$$

$$= \frac{(N-1)(N+1)}{12}$$

The expected value and variance of the rank-sum statistic, W, that is computed from the treatment with m observations are given by

$$E(W) = \frac{m(N+1)}{2}$$

$$\text{var}(W) = \frac{mn(N+1)}{12}$$

EXAMPLE 2.10.1 Table 2.10.1 contains exact values and normal approximations for upper-tail probabilities, $P(W \geq w)$, for the Wilcoxon rank-sum statistic with $m = n = 6$. Here $E(W) = (6)(13)/2 = 39$, and $\text{var}(W) = (6)(6)(13)/12 = 39$. To illustrate the computation of probabilities, the normal approximation for $w = 50$ gives us

$$P(W \geq 50) \approx P\left(Z \geq \frac{50 - 39}{\sqrt{39}}\right) = P(Z \geq 1.761) = .0391$$

Another approximation can be obtained by using a continuity correction, which adjusts for the fact that we are approximating a discrete probability distribution with a continuous one. We compute the normal probability greater than 49.5 to approximate $P(W \geq 50)$; that is,

$$P(W \geq 50) \approx P\left(Z \geq \frac{49.5 - 39}{\sqrt{39}}\right) = P(Z \geq 1.681) = .0463$$

The exact upper-tail probability is .0465. Thus, even for relatively small samples, the normal approximation for the distribution of the Wilcoxon rank-sum statistic is surprisingly good.

TABLE 2.10.1
Exact Values and Normal Approximations for $P(W \geq w)$
for Wilcoxon Rank-Sum Statistic, $m = n = 6$

Selected Values of w	46	48	50	52
Exact	.1548	.0898	.0465	.0206
Approximate, Without Continuity Correction	.1312	.0748	.0391	.0187
Approximate, With Continuity Correction	.1490	.0867	.0463	.0227

EXAMPLE 2.10.2 For the data in Example 2.4.3 on the dry weights of strawberry plants, the two groups have nine and seven observations, respectively, and the rank sum is $W = 84$ for the treatment with seven observations. Using the formulas for the Wilcoxon statistic, we have $E(W) = (7)(17)/2 = 59.5$ and $\text{var}(W) = (9)(7)(17)/12 = 89.25$. The p-value for the normal approximation without continuity correction is

$$P(W \geq 84) = P\left(Z \geq \frac{84 - 59.5}{\sqrt{89.25}}\right) = .0048$$

The exact p-value can be shown to be .0039.

2.10.3   Wilcoxon Rank-Sum Test with Ties

The application of the sampling formulas to the Wilcoxon rank-sum test with ties is straightforward, as illustrated by the following example.

EXAMPLE 2.10.3   We apply the sampling formulas to the hypothetical data with ties in Table 2.10.2.

TABLE 2.10.2
Data with Ties, $m = 4$, $n = 6$

Combined Data	10	11	11	13	13	13	15	16	19	19
Ranks	1	2.5	2.5	5	5	5	7	8	9.5	9.5
Data for Treatment 1	10	11	13	13	Rank sum = 13.5					
Ranks	1	2.5	5	5						
Data for Treatment 2	11	13	15	16	19	19	Rank sum = 41.5			
Ranks	2.5	5	7	8	9.5	9.5				

The scores are the ten adjusted ranks 1, 2.5, 2.5, 5, 5, 5, 7, 8, 9.5, and 9.5. The population mean and variance of these ten numbers are

$$\mu = \frac{1 + 2.5 + \cdots + 9.5}{10} = 5.5$$

$$\sigma^2 = \frac{1^2 + 2.5^2 + \cdots + 9.5^2}{10} - 5.5^2 = 7.95$$

Therefore, the expected value and variance of the rank sum for treatment 1 are given by

$$E\left(W_{\text{ties}}\right) = (4)(5.5) = 22$$
$$\text{var}\left(W_{\text{ties}}\right) = \frac{(4)(6)(7.95)}{9} = 21.2$$

The observed value of the Wilcoxon statistic for treatment 1 is 13.5, so from the normal approximation, the lower-tail *p*-value is

$$P\left(W_{\text{ties}} \leq 13.5\right) = P\left(Z \leq \frac{13.5 - 22}{\sqrt{21.2}}\right) = .0324$$

This is the same as the asymptotic *p*-value given by StatXact. The exact *p*-value from StatXact is .0476. Output is shown in Figure 2.10.1.

MINITAB gives two approximations. Both use a continuity correction. One uses the variance formula as if the data had no ties, and the other uses the variance

FIGURE 2.10.1

StatXact Analysis of Data in Table 2.10.2

```
WILCOXON-MANN-WHITNEY TEST
[ Sum of scores from population <      1 > ]

Summary of Exact distribution of WILCOXON-MANN-WHITNEY statistic:
     Min      Max      Mean      Std-dev      Observed      Standardized
    11.00    34.00    22.00         4.604         13.50           -1.846
   Mann-Whitney Statistic =      3.500

Asymptotic Inference:
  One-sided p-value: Pr {   Test Statistic .LE. Observed } =        0.0324
  Two-sided p-value: 2 * One-sided                         =        0.0649

Exact Inference:
  One-sided p-value: Pr {   Test Statistic .LE. Observed } =        0.0476
                     Pr {   Test Statistic .EQ. Observed } =        0.0286
  Two-sided p-value: Pr { | Test Statistic - Mean |
                       .GE. | Observed - Mean |            =        0.0810
  Two-sided p-value: 2*One-Sided                           =        0.0952

Elapsed time is 0:0:0.00
```

formula of W_{ties} as computed from the sampling formulas. Without an adjustment for ties, the variance is

$$\frac{mn(N+1)}{12} = \frac{(4)(6)(11)}{12} = 22$$

Thus

$$P\left(W_{\text{ties}} \leq 13.5\right) \approx P\left(Z \leq \frac{14-22}{\sqrt{22}}\right) = .0440$$

With a continuity correction and using the variance of 21.2 as computed from the sampling formulas, we have

$$P\left(W_{\text{ties}} \leq 13.5\right) \approx P\left(Z \leq \frac{14-22}{21.2}\right) = .0412$$

MINITAB output is shown in Figure 2.10.2. Here we have chosen the lower-tail test option as indicated by "ETA1 < ETA2."

FIGURE 2.10.2

MINITAB Analysis of Data in Table 2.10.2

```
Mann-Whitney Test and CI: C1, C2
C1     N =   4     Median =        12.000
C2     N =   6     Median =        15.500
Point estimate for ETA1-ETA2 is  -3.500
95.7 Percent CI for ETA1-ETA2 is (-7.999,-0.001)
W = 13.5
Test of ETA1 = ETA2 vs ETA1 < ETA2 is significant at 0.0440
The test is significant at 0.0412 (adjusted for ties)
```

$\blacksquare$

2.10.4   Explicit Formulas for $E(W_{\text{ties}})$ and $\text{var}(W_{\text{ties}})$

Although one may apply the sampling formulas directly to the adjusted ranks for tied data, explicit formulas are available for $E(W_{\text{ties}})$ and $\text{var}(W_{\text{ties}})$. Although not necessary in practice, these formulas serve to show the effect that tied data have on the Wilcoxon rank-sum statistic. Since average ranks are used for tied observations, $E(W_{\text{ties}})$ is the same as if the data were not tied. However, the variance is adjusted downward from what it would be if the observations were not tied, as shown below.

In the combined data set, put each set of tied observations into its own group. Let k be the number of such groups, and let t_i denote the number of observations in the ith group, $i = 1, 2, \ldots, k$. For instance, the groups in Table 2.10.2 are $\{11, 11\}$, $\{13, 13, 13\}$, and $\{19, 19\}$, and $t_1 = 2$, $t_2 = 3$, and $t_3 = 2$. Compute the adjustment factor

$$\text{AF} = \frac{mn \sum_{i=1}^{k} \left(t_i^3 - t_i\right)}{12N(N-1)}$$

To obtain the variance with ties, subtract AF from the variance without ties; that is,

$$\text{var}\left(W_{\text{ties}}\right) = \frac{mn(N+1)}{12} - \text{AF}$$

A derivation follows from results in Lehmann (1975, pp. 330–333, Example 1).

EXAMPLE 2.10.4   For the data in Table 2.10.2, we have

$$\text{AF} = \frac{(4)(6)\left[\left(2^3 - 2\right) + \left(3^3 - 3\right) + \left(2^3 - 2\right)\right]}{12(10)(9)} = 0.8$$

The variance without ties is $(4)(6)(11)/12 = 22$, giving us a variance of 21.2 for the data with ties, which agrees with the computation in Example 2.10.3. Generally, unless the number of ties is significant, AF will be small. ∎

2.10.5 Large-Sample Confidence Interval Based on Mann–Whitney Test

Following the notation of Section 2.6, let the observations from treatment 1 be denoted $X_1, X_2, \ldots, X_m$ and the observations from treatment 2 be denoted $Y_1, Y_2, \ldots, Y_n$. As shown in Section 2.6, the Mann–Whitney statistic U can be expressed in terms of W_2, the sum of the ranks of the Y's, as follows:

$$W_2 = \frac{n(n+1)}{2} + U$$

Since $E(W_2) = n(N+1)/2$, it follows that

$$E(U) = \frac{n(N+1)}{2} - \frac{n(n+1)}{2} = \frac{mn}{2}$$

and

$$\mathrm{var}(U) = \mathrm{var}(W_2)$$

where var(W_2) is computed either with or without ties as appropriate. Since U is just a linear function of W_2, its distribution may be approximated by the normal distribution with $E(U)$ and var(U) as indicated.

Suppose the distribution functions of the X's and Y's satisfy

$$F_1(x) = F_2(x - \Delta)$$

as in Section 2.6.3. To find, say, a 90% confidence interval for Δ, we must first find values k_a and k_b that satisfy

$$P(k_a \leq U \leq k_b - 1) \approx .90$$

Using a normal approximation for the distribution of U, we have

$$P(k_a \leq U \leq k_b - 1) \approx P\left(\frac{k_a - E(U)}{\sqrt{\mathrm{var}(U)}} \leq Z \leq \frac{k_b - 1 - E(U)}{\sqrt{\mathrm{var}(U)}} \right) = .90$$

where Z has a standard normal distribution. Since $P(-1.645 \leq Z \leq 1.645) = .90$, it follows that

$$k_a \approx E(U) - 1.645\sqrt{\mathrm{var}(U)}$$
$$k_b \approx 1 + E(U) + 1.645\sqrt{\mathrm{var}(U)}$$

where we round to the nearest integer. After we have determined k_a and k_b, the lower and upper 90% confidence limits for Δ are the k_ath smallest and the

k_bth smallest pairwise differences of the form $X_i - Y_j$. For other levels of confidence, 1.645 is replaced by the appropriate percentiles from the standard normal distribution.

StatXact uses the normal approximation in a slightly different way. For the lower-tail value for a 90% confidence interval, use the normal approximation to find r_U such that $P(U \geq r_U) = .05$. If r is not an integer, round it up to the nearest integer r_U^* and then compute $k_a = mn - r_U^* + 1$. For the upper-tail value for a 90% confidence interval, use the normal approximation to find r_L such that $P(U \leq r_L) = .05$. If r_L is not an integer, round it down to the nearest integer r_L^* and then compute $k_b = mn - r_L^*$.

EXAMPLE 2.10.5   Refer to the cerium data in Table 2.6.2. Since $m = n = 6$ and since the data contain no ties, we have $E(U) = (6)(6)/2 = 18$, and $\text{var}(U) = (6)(6)(13)/12 = 39$. Thus,

$$k_a \approx 18 - 1.645\sqrt{39} = 7.7$$

$$k_b \approx 1 + 18 + 1.645\sqrt{39} = 29.3$$

Rounding to the nearest integer, we find $k_a = 8$ and $k_b = 29$, as obtained in Example 2.6.2. Using the StatXact method, we find

$$r_U = 18 + 1.645\sqrt{39} = 28.3$$

Rounding the result up to 29, we get $k_a = 36 - 29 + 1 = 8$. Similarly,

$$r_L = 18 - 1.645\sqrt{39} = 7.7$$

Rounding down to 7, we find $k_b = 36 - 7 = 29$. In both cases the confidence interval for Δ is the same as obtained in Example 2.6.2.   ∎

2.10.6   A Normal Approximation for the Permutation Distribution

Suppose we would like to do a permutation test based on the original data where the statistic is the sum of the observations associated with treatment 1. Let the statistic be denoted T_1. The values of μ and σ^2 required in the sampling formulas are the population mean and variance of the combined data.

EXAMPLE 2.10.6   Refer to the data in Table 2.1.1. Although the normal approximation would not usually be applied to this small a sample, it will serve to illustrate the method. The data values are 37, 49, 55, 57, 23, 31, and 46. The population mean and standard deviation of these numbers are $\mu = 42.57$ and $\sigma^2 = 137.67$. Since T_1 is the sum of the first four of these observations, $T_1 = 198$. The mean and variance of T_1 from the sampling formulas are

$$E(T_1) = (4)(42.57) = 170.28$$

$$\mathrm{var}(T_1) = \frac{(4)(3)(137.67)}{6} = 275.34$$

Using a normal approximation without continuity correction, we compute the *p*-value as

$$P(T_1 \geq 198) \approx P\left(Z \geq \frac{198 - 170.28}{\sqrt{275.34}}\right) = .0474$$

This is the asymptotic *p*-value given in the StatXact output in Figure 2.1.3. ∎

Exercises

1 A certain data set has eight distinct observations, four from each treatment, and all of the observations from treatment 1 are bigger than the observations from treatment 2. What is the one-sided *p*-value associated with the permutation test?

2 **a** Find the permutation distribution of the difference of means for the fictitious data set in the table, and find the *p*-value for the observed data.

   b Find the permutation distribution of the sum of the observations from treatment 1, and show that the *p*-value for the observed data is the same as the *p*-value in part a.

Treatment 1	10	15	50
Treatment 2	12	17	19

3 Find the permutation distribution of the difference of medians for the data in Exercise 2.

4 The carapace lengths (in mm) of crayfish were recorded for samples from two sections of a stream in Kansas.

Section 1	5	11	16	8	12	
Section 2	17	14	15	21	19	13

   a Test for differences between the two sections using a permutation test.

   b Test for differences using the Wilcoxon rank-sum test.

5 Nest heights (in meters) of two species of woodland nesting birds were measured. Test for differences between the nesting heights using the Wilcoxon rank-sum test.

Species A	5.1	9.4	7.2	8.1	8.8
Species B	2.5	4.2	6.9	5.5	5.3

6   Create a fictitious data set where the Wilcoxon rank-sum test and the two-sample *t*-test lead to different conclusions at the 5% level of significance. (*Hint:* Try a data set in which one treatment has a few very large observations in comparison with all other observations in either treatment.)

7   Students in an introductory statistics class were asked how many brothers and sisters they have and whether their hometown is urban or rural.

Number of Siblings in Rural versus Urban Areas

Rural	3 2 1 1 2 1 3 2 2 2 2 5 1 4 1 1 1 6 2 2 2 1 1
Urban	1 0 1 1 0 0 1 1 1 8 1 1 1 0 1 1 2

a   Test for a significant difference between rural and urban areas using the Wilcoxon rank-sum test.

b   Test for a significant difference using the two-sample *t*-test, and compare the results with those obtained in part a. Why are the results different?

8   Do a permutation test on the data in Exercise 7. Is the *p*-value closer to that of the Wilcoxon rank-sum test or to that of the two-sample *t*-test? What does this suggest about the relationship between the permutation test and the two-sample *t*-test?

9   Refer to the data in Exercise 2. Obtain the permutation distribution of the sum of the van der Waerden scores for treatment 1.

10   For the data in Exercise 4, test for differences between sections using van der Waerden scores.

11   Discuss how to adjust van der Waerden scores and exponential scores for ties.

12   Refer to the data in Exercise 4. Make a 90% confidence interval for Δ. Obtain the Hodges–Lehmann estimate of Δ.

13   Refer to the data in Exercise 5. Test for differences between the distributions of the nesting heights of the two species using the Kolmogorov–Smirnov test.

14   Find the permutation distribution of the K-S statistic when $m = n = 3$.

15   The simulated data in the table are from two normal distributions with the same mean and unequal variances.

Treatment 1	21.9	20.2	19.4	20.3	19.6	20.4	18.4	20.1	22.0	18.9
Treatment 2	20.2	13.8	21.8	19.2	19.6	25.5	17.0	17.6	19.5	22.2

a   Test for differences between the scale parameters using the Siegel–Tukey test.

b   Test for differences between the scale parameters using the approximate RMD permutation test.

16 **a** Obtain the exponential scores when $m = 5$ and $n = 6$. Let T denote the sum of the scores for the treatment with five observations. Find $E(T)$ and $\text{var}(T)$.

b Use the exponential scores and the large-sample approximation to analyze the data in Exercise 4.

17 Refer to the data in Exercise 7. Apply the large-sample approximation to the Wilcoxon rank-sum test to test for differences between the rural and urban groups.

18 A biologist examined the effect of a fungal infection on the eating behavior of rodents. Infected apples were offered to a group of eight rodents, and sterile apples were offered to a group of four. The amounts consumed (grams of apple/kilogram of body weight) are listed in the table. Apply several nonparametric tests to the data, including the Wilcoxon rank-sum test, the van der Waerden scores test, the exponential scores test, and the permutation test on the original data. Discuss differences in conclusions using these tests, and justify the use of one or more of these tests on these data.

Experimental Group	11, 33, 48, 34, 112, 369, 64, 44
Control Group	177, 80, 141, 332

Theory and Complements

19 Let the observations from treatment 1 be denoted $X_1, X_2, \ldots, X_m$ and the observations from treatment 2 be denoted $Y_1, Y_2, \ldots, Y_n$. Let the sample means and the sample standard deviations of the X's and Y's be denoted as $\overline{X}$, $\overline{Y}$, S_X, and S_Y, respectively. The two-sample t-statistic is defined as

$$t = \frac{\overline{X} - \overline{Y}}{S_p \sqrt{1/m + 1/n}}$$

where

$$S_p = \sqrt{\frac{(m-1)S_X^2 + (n-1)S_Y^2}{m+n-2}}$$

Express t in terms of $T_1 = \sum X_i$, $T = \sum Y_i + \sum Y_i$, and $T_* = \sum X_i^2 + \sum Y_i^2$. Since T and T^* are fixed for any permutation of the data, then t is a function of T_1. Show that t is a monotone function of T_1, and hence conclude that permutation tests based on T_1 and t are equivalent.

20 The purpose of this exercise is to show in what sense van der Waerden scores are approximations of normal scores.

a If a random variable X has a continuous, strictly increasing cdf $F(x)$, and if U has a uniform distribution on the interval $[0, 1]$, show that the distribution function of $F^{-1}(U)$ is the same as that of X.

b   Suppose that $U_{(i)}$ is the *i*th order statistic of a random sample of size N from a uniform [0, 1] distribution. Let $X_{(i)}$ denote the *i*th-order statistic of a random sample of size N from a population with distribution $F(x)$. Use part a to conclude that $E(X_{(i)}) = E(F^{-1}(U_{(i)}))$.

c   Show that $E(U_{(i)}) = i/(N + 1)$.

d   Conclude that $E(Z_{(i)}) = E\left(\Phi^{-1}(U_{(i)})\right) \approx \Phi^{-1}\left[E(U_{(i)})\right] = \Phi^{-1}\left[i/(N+1)\right]$.

e   The normal scores for $N = 10$ are –1.54, –1.00, –0.65, –0.38, –0.12, 0.12, 0.38, 0.65, 1.00, and 1.54. Compare these with the van der Waerden scores. Infer where the approximation is good and where it is not so good.

21   Consider the distribution of the Mann–Whitney statistic U under the null hypothesis that the two treatments have the same distribution. Show that $P(U \leq t) = P(U \geq mn - t)$. Conclude that the lower and upper critical values for a test at level of significance α satisfy the relationship $U_{\text{upper}} = mn - U_{\text{lower}}$.

22   The following is a sketch of the derivation of the variance of the sum of a random sample of size m without replacement from a finite population of size N. Let the numbers in the population be denoted $A_1, A_2, \ldots, A_N$, where we allow for the possibility that some of the numbers are the same. Let $X_1, X_2, \ldots, X_m$ be a random sample of size m without replacement from this population. Let $n = N - m$.

a   Note that the probability that item A_i is selected on any given draw is $1/N$. Thus, conclude that $E(X_i) = \Sigma A_i /N = \mu$ and $\text{var}(X_i) = \Sigma(A_i - \mu)^2 /N = \sigma^2$.

b   Note that the probability that item A_i is selected first and A_j is selected afterward is $1/(N)(N - 1)$. Conclude that

$$E\left[\left(X_i - \mu\right)\left(X_j - \mu\right)\right] = \frac{1}{N(N-1)} \sum_{i \neq j} (A_i - \mu)(A_j - \mu)$$

$$= \frac{1}{N(N-1)} \sum_{i=1}^{N} \sum_{j=1}^{N} (A_i - \mu)(A_j - \mu) - \frac{1}{N(N-1)} \sum_{i=1}^{N} (A_i - \mu)^2$$

$$= 0 - \frac{1}{N-1} \sigma^2$$

That is, the covariance of X_i and X_j is $-\sigma^2/(N - 1)$.

c   Let $T = \Sigma X_i$. Conclude that

$$\text{var}(T) = \sum_{i=1}^{m} \text{var}(X_i) + \sum_{i=j} \text{cov}(X_i, X_j)$$

$$= m\sigma^2 - m(m-1)\frac{\sigma^2}{N-1}$$

$$= \frac{mn}{(N-1)} \sigma^2$$

23   Suppose we have random samples from two populations with distribution functions F_1 and F_2 that are related by $F_1(x) = F_2(x - \Delta)$. Suppose a test statistic T tests $H_0: \Delta = 0$ versus

H_a: $\Delta \neq 0$. A general method for making a confidence interval for Δ follows. For a given value of Δ, subtract Δ from each of the Y's; that is, let $Y_i^* = Y_i - \Delta$. Apply the test statistic T to the data consisting of the X's and the Y^*'s. If the null hypothesis is accepted, then the value of Δ is included in the confidence interval; otherwise, it is not.

a Show that if the test is done at the 5% level of significance, this procedure gives a 95% confidence interval for Δ.

b Use the data in Exercise 4 with exponential scores to obtain a confidence interval for Δ. The values of Δ that fall in the confidence interval can be determined by trial and error using, say, StatXact to carry out the test statistic.

c Show that the confidence interval in Section 2.6 can be obtained using this procedure with the Mann–Whitney statistic.

3

K-Sample Methods

A Look Ahead The permutation methods that we developed for comparing two treatments extend naturally to the problem of comparing more than two treatments. The usual approach is first to do an overall comparison to determine whether or not differences exist among treatments. If differences are found to exist, then multiple comparison tests are done to determine which treatments differ significantly from the others. The overall tests are considered in Sections 3.1 and 3.2, and multiple comparison tests are considered in Section 3.3. We conclude Chapter 3 with a statistical procedure that may be used when there is a known ordering among treatments.

3.1
K-Sample Permutation Tests

We extend the notion of permutation tests for comparing two treatments to studies involving k treatments. It is assumed that the experimental units are assigned to the k treatments in a completely random design, or that the observations have been randomly selected from k populations. If there are differences among the treatments, it is assumed that observations from at least one treatment will tend to be larger than observations from at least one other treatment.

If the observations have been selected randomly from populations with cdf's $F_1(x), F_2(x), \ldots, F_k(x)$, then the null hypothesis to be tested is equality of distributions—that is,

$$H_o: F_1(x) = F_2(x) = \cdots = F_k(x)$$

For the alternative hypothesis, we have

$$H_a: F_i(x) \leq F_j(x) \quad \text{or} \quad F_i(x) \geq F_j(x)$$

for at least one pair (i, j), with strict inequality holding for at least one x.

A special case of H_a is the shift alternative. That is, there is a cdf $F(x)$ and location parameters $\mu_1, \mu_2, \ldots, \mu_k$ not all equal such that

$$H_a: F_i(x) = F(x - \mu_i)$$

The analysis of variance model is like this, where the μ_i's are means and F is a normal distribution with mean 0 and variance σ^2. The shift alternative may also be expressed as

$$X_{ij} = \mu_i + \varepsilon_{ij}$$

where X_{ij} is the jth observation for the ith treatment, and the ε_{ij}'s are independent and identically distributed random variables with distribution $F(\varepsilon)$. Table 3.1.1 defines the notation used in this section. The total sample size is denoted $N = n_1 + n_2 + \cdots + n_k$.

TABLE 3.1.1
One-Way Data Layout

Treatments	Observations	Sample Sizes	Means	Variances
1	$X_{11}, X_{12}, \ldots, X_{1n_1}$	n_1	$\overline{X}_1$	S_1^2
2	$X_{11}, X_{12}, \ldots, X_{1n_2}$	n_2	$\overline{X}_2$	S_2^2
...	...	...	...	...
k	$X_{11}, X_{12}, \ldots, X_{1n_k}$	n_k	$\overline{X}_k$	S_k^2

3.1.1 The *F* Statistic

Here we review the computation of the F statistic that is used in the one-way analysis of variance. It is this statistic, or its equivalent, that we use in the permutation test. The *sum of squares for treatments* is defined as

$$\text{SST} = \sum_{i=1}^{k} n_i \left(\overline{X}_i - \overline{X} \right)^2$$

where $\overline{X}$ is the mean of all the observations—namely,

$$\overline{X} = \frac{\sum_{i=1}^{k} \sum_{j=1}^{n_i} X_{ij}}{N}$$

The *mean squares for treatment* is

$$\text{MST} = \frac{\text{SST}}{k - 1}$$

The *sum of squares for error* is defined as

$$\text{SSE} = \sum_{i=1}^{k} (n_i - 1) S_i^2$$

and the *mean squares for error* is

$$\text{MSE} = \frac{\text{SSE}}{N - k}$$

The *F* statistic is given by

$$F = \frac{\text{MST}}{\text{MSE}}$$

If the observations are selected at random from normally distributed populations with equal variances, then this statistic has an *F*-distribution with $k - 1$ degrees of freedom for the numerator and $N - k$ degrees of freedom for the denominator. One may use this distribution to determine a *p*-value for the observed statistic. However, if we are unwilling to assume that the population distributions are normal, then we may carry out a *permutation F-test* instead.

3.1.2 Steps in Carrying Out the Permutation *F*-Test

1. Obtain the *F* statistic for the original data, denoted F_{obs}.
2. Obtain all possible permutations of the *N* observations among *k* treatments in which there are n_i observations in treatment i, $i = 1, 2, \ldots, k$. There are

$$\frac{N!}{n_1! n_2! \ldots n_k!}$$

such possibilities. If it is not possible to generate all permutations, then select a random sample of *R* permutations.

3. For each permutation, compute the *F* statistic defined above.
4. Obtain the *p*-value as the fraction of the *F*'s in step 3 that are greater than or equal to F_{obs}. The *p*-value is approximate if it is based on a random sample of the permutations. Note that this is always an upper-tail test.

The number of permutations is large even for modest sample sizes. For instance, for $k = 3$ treatments and $n_i = 5$ observations per treatment, there are 756,756 possibilities. We may randomly sample the permutations as follows, assuming that our programming language has the capability to randomly shuffle the elements of a vector: We place the observations in a vector of dimension *N*, randomly shuffle the observations in the vector, and then select the first n_1 for treatment 1, the next n_2 for treatment 2, and so on. A random sample of $R = 1000$ permutations is usually sufficient for most applications, but it is generally not costly in terms of computer time to obtain more than this.

EXAMPLE 3.1.1   We compare the *p*-value obtained from the standard analysis of variance with that obtained by the permutation *F*-test. Table 3.1.2 lists data for three treatments. The observations were randomly sampled from normal populations with means 15, 25, and 30, respectively, and standard deviation 9.

TABLE 3.1.2
Samples from Normal Populations, $\mu_1 = 15$, $\mu_2 = 25$, $\mu_3 = 30$, $\sigma = 9$

Treatment 1	6.08	22.29	7.51	34.36	23.68
Treatment 2	30.45	22.71	44.52	31.47	36.81
Treatment 3	32.04	28.03	32.74	23.84	29.64

A MINITAB analysis of variance output is shown in Figure 3.1.1. The observed F statistic for the original data is $F_{obs} = 3.78$, and the analysis of variance p-value is .053 as determined from the F-distribution with 2 degrees of freedom for the numerator and 12 degrees of freedom for the denominator. The exact p-value of the permutation F-test as determined from StatXact is .0513. We present the StatXact output with explanation in Section 3.1.4.

FIGURE 3.1.1
MINITAB Analysis of Variance for Data in Table 3.1.2

```
One-way ANOVA: C2 versus C1
Analysis of Variance for C2
Source      DF        SS        MS        F        P
C1          2        554.6     277.3     3.78     0.053
Error       12       880.0      73.3
Total       14      1434.6
```

Comment on the Relationship Between Permutation F-Test and One-Way Analysis of Variance F-Test

In many scientific studies, the experimental units are not drawn randomly from a larger population, but instead they are units that happen to be available at the time of the study. To avoid bias, the units are randomly assigned to the treatments. Because of the randomization, the permutation F-test may be used to test for differences among treatments, although the inferences would apply only to those units in the study. However, researchers often apply the one-way analysis of variance in this situation, as if observations were randomly selected from normally distributed populations. We consider the justification for doing this.

The close agreement between the p-values in Example 3.1.1 suggests that the permutation distribution of the F statistic may be approximated by the usual F-distribution with the appropriate degrees of freedom. To illustrate this, we obtained

10,000 values of the F statistic from randomly selected permutations of the data in Table 3.1.2. Table 3.1.3 compares the percentiles of the distribution of these F statistics with the percentiles of the F-distribution with 2 degrees of freedom for the numerator and 12 degrees of freedom for the denominator. We see close agreement except in the tails of the distribution.

This is not an isolated result. Results due to Pitman (1938) and Hoeffding (1952) show with appropriate assumptions that the permutation distribution may be approximated by the F-distribution for large samples. In our example, which has only five observations per treatment, the approximation is quite good for percentiles up to the 95th. Thus, the one-way analysis of variance F-test may be thought of as an approximation to the permutation F-test in situations in which experimental units have been randomly assigned to treatments.

TABLE 3.1.3

Comparison of Permutation Percentiles to F-Distribution Percentiles

Percentile	80	85	90	95	97.5	99
Permutation Distribution	1.8	2.2	2.8	3.8	4.8	5.9
F-distribution	1.8	2.2	2.8	3.9	5.1	6.9

3.1.3 Alternative Forms of the Permutation F Statistic

We now consider alternative forms of the permutation F statistic that we will find useful later. The total sum of squares is defined as

$$SS_{total} = \sum_{i=1}^{k} \sum_{j=1}^{n_i} \left(X_{ij} - \overline{X} \right)^2$$

From the properties of sums of squares in analysis of variance, we have

$$SS_{total} = SST + SSE$$

Now SS_{total} has the same value for all permutations of the data. Call this value C. The F statistic can be expressed as

$$F = \frac{SST/(k-1)}{(C-SST)/(N-k)}$$

Since F is an increasing function of SST, it follows that a permutation test based on SST is equivalent to one based on F. Which one to use is just a matter of computational convenience. Moreover, since

$$SST = \sum_{i=1}^{k} n_i \overline{X}_i^2 - N\overline{X}^2$$

and since $\overline{X}$ has the same value for all permutations of the data, we may base the test on

$$\text{SSX} = \sum_{i=1}^{k} n_i \overline{X}_i^2$$

3.1.4 Computer Analysis

The permutation *F*-test is implemented in StatXact. Output of the analysis of the data in Table 3.1.2 is shown in Figure 3.1.2. The option for obtaining this analysis is labeled "ANOVA with General Scores" in the StatXact menu for *k* independent samples. The observed statistic used by StatXact is $(N-1)\text{SST}/\text{SS}_{\text{total}}$. Using the values of SST and SS_{total} from the output in Figure 3.1.1, we see that the value of this statistic is $(14)(554.6)/(1434.6) = 5.412$. Since SS_{total} is unaffected by permutations of the data, a test based on this statistic is equivalent to a test based on SST or the *F* statistic. The asymptotic *p*-value comes from the fact that the distribution of $(N-1)\text{SST}/\text{SS}_{\text{total}}$ can be approximated by a chi-square distribution with $k-1$ degrees of freedom for large samples. For instance, the asymptotic *p*-value in Figure 3.1.2 is $P(\chi_{(2)}^2 \geq 5.412) = .0668$, where $\chi_{(2)}^2$ is a chi-square random variable with 2 degrees of freedom.

FIGURE 3.1.2

Permutation *F*-Test Using StatXact

```
ANOVA TEST [That the 3 populations are identically distributed]

Statistic based on the observed data :
     The Observed Statistic =        5.412

Asymptotic p-value: (based on Chi-square distribution with 2 df )
     Pr { Statistic .GE.         5.412 } =        0.0668


Exact p-value and point probability :
     Pr { Statistic .GE.         5.412 } =        0.0513
     Pr { Statistic .EQ.         5.412 } =        0.0000

Elapsed time is 0:0:8.02
```

Resampling Stats code for the data in Table 3.1.2 is shown in Figure 3.1.3. The output below the code shows an estimated *p*-value of .0518 based on 5000 randomly selected permutations. The statistic computed is SSX.

FIGURE 3.1.3

Resampling Stats Code for Permutation *F*-Test Applied to Data in Table 3.1.2

```
'set maximum vector size at 5000
maxsize default 5000

'input data from the 3 treatments and form combined data vector
copy (6.08 22.29 7.51 34.36 23.68) d1
copy (30.45 22.71 44.52 31.47 36.81) d2
copy (32.04 28.03 32.74 23.84 29.64) d3
concat d1 d2 d3 dat

'observed SSX = 11552.89

'get 5000 permutations of the data
repeat 5000
shuffle dat sdat
take sdat 1,5 sdat1
take sdat 6,10 sdat2
take sdat 11,15 sdat3

'compute treatment means and put into a vector m
mean sdat1 m1
mean sdat2 m2
mean sdat3 m3
concat m1 m2 m3 m

'compute SSX
'square of means are weighted by sample sizes
square m msqr
multiply (5 5 5) msqr wtmsqr
sum wtmsqr ssx

'keep track of values of SSX
score ssx ssxdist
end

'get p-value
count ssxdist >= 11552.89 pctssx
divide pctssx 5000 pvalssx
print pvalssx

'output displayed here

PVALSSX = 0.0518
```

3.2
The Kruskal–Wallis Test

A way to obtain a nonparametric rank test for comparing k treatments is to replace the original observations with ranks and then perform the permutation F-test on these ranks. What we are concerned about in this section is obtaining a statistic that is equivalent to the F statistic applied to ranks with a permutation distribution that may be approximated by the chi-square distribution with $k - 1$ degrees of freedom.

3.2.1 The Kruskal–Wallis Statistic

Referring to the notation in Table 3.1.1, we let R_{ij} denote the rank of observation X_{ij}. We assume the data contain no ties. The data layout for the ranks is shown in Table 3.2.1. We let $N = n_1 + n_2 + \cdots + n_k$.

TABLE **3.2.1**
Data Layout for Ranks

Treatments	Ranks	Sample Size	Means
1	$R_{11}, R_{12}, \ldots, R_{1n_1}$	n_1	$\overline{R}_1$
2	$R_{11}, R_{12}, \ldots, R_{1n_2}$	n_2	$\overline{R}_2$
...	...	...	...
k	$R_{11}, R_{12}, \ldots, R_{1n_k}$	n_k	$\overline{R}_k$

The *Kruskal–Wallis statistic* is defined as

$$ \text{KW} = \frac{12}{N(N+1)} \sum_{i=1}^{k} n_i \left(\overline{R}_i - \frac{N+1}{2} \right)^2 $$

Since the mean of all the ranks is $(N + 1)/2$, the term

$$ \sum_{i=1}^{k} n_i \left(\overline{R}_i - \frac{N+1}{2} \right)^2 $$

is the treatment sum of squares, SST, defined in Section 3.1.1 applied to ranks. The constant $C = 12/N(N + 1)$ is a scaling factor that makes it possible to use the chi-square distribution with $k - 1$ degrees of freedom to approximate the permutation distribution of KW. See Section 3.2.3 for an intuitive derivation of this approximation.

We may obtain a *p*-value for the KW statistic from the permutation distribution of this statistic. The steps for doing this are essentially the same as those for the permutation F test presented in Section 3.1.2, except that ranks are used in place of the original observations and KW is used in place of *F*. Critical values of the permutation distribution of the KW statistic are given for small sample sizes in Table A6 in the Appendix. For larger samples, an approximate *p*-value may be obtained by referring the statistic to the chi-square distribution with $k - 1$ degrees of freedom, Table A7.

We may compare critical values for the Kruskal–Wallis statistic in Table A6 with the corresponding critical values for the chi-square distribution with the appropriate degrees of freedom in Table A7. For these small samples, and for nominal levels of significance 10%, 5%, 2.5%, and 1%, most of the Kruskal–Wallis critical values are somewhat smaller than the chi-square critical values. For instance, with three treatments and samples of size four in each treatment, the 5% critical value of the KW statistic from Table A6 is 5.69, whereas the 5% chi-square critical value with 2 degrees of freedom is 5.99. In this case, rejection of the null hypothesis using the chi-square critical value of 5.99 would assure us that the exact level of significance is no greater than .05. The approximation gets better for larger samples.

EXAMPLE 3.2.1 Three preservatives and a control were compared in terms of their ability to inhibit the growth of bacteria. Samples were treated with one of the three preservatives, or left untreated for the control, and bacteria counts were made 48 hours later. The data in Table 3.2.2 are the logarithms of the counts. Actual bacteria counts would not satisfy the assumptions of analysis of variance. The variances would be unequal among the treatments. The purpose of making the logarithmic transformation is to obtain data that meet the standard analysis of variance assumptions. In terms of ranking, one may deal with either the original data or logarithms, since the ranks will be the same.

TABLE 3.2.2
Logarithms of Bacteria Counts

Control:	Data	4.302	4.017	4.049	4.176		
	Ranks	21	18	19	20		
Preservative 1:	Data	2.021	3.190	3.250	3.276	3.292	3.267
	Ranks	1	8	9	11	12	10
Preservative 2:	Data	3.397	3.552	3.630	3.578	3.612	
	Ranks	13	14	17	15	16	
Preservative 3:	Data	2.699	2.929	2.785	2.176	2.845	2.913
	Ranks	3	7	4	2	5	6

The mean ranks for the control and three preservatives are 19.5, 8.5, 15.0, and 4.5, respectively. The average of all the ranks is 11. The Kruskal–Wallis statistic is computed as

$$\mathrm{KW} = \frac{12}{21(22)}\left[4(19.5-11)^2 + 6(8.5-11)^2 + 5(15.0-11)^2 + 6(4.5-11)^2\right] = 17.14$$

The upper 1% point from the chi-square distribution with 3 degrees of freedom is 11.3, so the result is statistically significant at the 1% level.   ∎

3.2.2 Adjustment for Ties

When data are tied, we adjust the ranks using average ranks for tied data as we did in Section 2.5. We may apply a permutation *F*-test to the adjusted ranks or use any equivalent statistic such as SST applied to adjusted ranks. However, in order to maintain the chi-square approximation, we must adjust the KW statistic for ties. This is done as follows.

Let S_R^2 denote the sample variance of the combined adjusted ranks. The Kruskal–Wallis test adjusted for ties is

$$\mathrm{KW}_{\mathrm{ties}} = \frac{1}{S_R^2}\sum_{i=1}^{k} n_i\left(\bar{R}_i - \frac{N+1}{2}\right)^2$$

See Section 3.2.3 for the rationale behind this adjustment.

EXAMPLE 3.2.2   In some types of scientific studies, an expert is asked to provide a subjective judgment of the effects of treatments on experimental units, and this judgment is expressed as a score ranging, say, from 1 to 5 or 1 to 10. In such cases many ties may occur. The data in Table 3.2.3 are saltiness scores, on a scale of 1 to 5, assigned by a taste expert to samples of three food products that differ in the amounts of soy meal they contain.

The combined data include five 1's, five 2's, four 3's, three 4's, and four 5's. The five 1's are assigned an average rank of 3. The five 2's are the ordered observations 6 through 10, so they receive an average rank of 8. Likewise, the 3's, 4's, and 5's receive average ranks of 12.5, 16, and 19.5, respectively. The average ranks for the three products are 15.86, 10.31, and 6.25, respectively, and the overall average rank is 11. The sample variance of the 21 adjusted ranks is $S_R^2 = 36.9$. The Kruskal–Wallis statistic adjusted for ties is

$$\mathrm{KW}_{\mathrm{ties}} = \frac{1}{36.9}\left[7(15.86-11)^2 + 8(10.31-11)^2 + 6(6.25-11)^2\right] = 8.25$$

The upper 2.5% point from the chi-square distribution with 2 degrees of freedom is 7.38, so there is a significant difference at the 2.5% level among the products in terms of saltiness as judged by the expert.

TABLE 3.2.3

Saltiness Scores for Three Food Products

Product 1:	Data	4	5	3	4	5	5	2	
	Adjusted Ranks	16	19.5	12.5	16	19.5	19.5	8	
Product 2:	Data	3	4	5	2	3	1	1	2
	Adjusted Ranks	12.5	16	19.5	8	12.5	3	3	8
Product 3:	Data	2	1	1	2	1	3		
	Adjusted Ranks	8	3	3	8	3	12.5		

∎

An Alternative Formula for Tied Ranks

An explicit formula is available for KW_{ties}. Though not necessary, it shows how ties affect KW. Assume the tied data are arranged into g groups of like observations. We let t_i denote the number of observations in the ith group, $i = 1, 2, \ldots, g$. The Kruskal–Wallis statistic adjusted for ties is

$$KW_{ties} = \frac{KW}{1 - \dfrac{\sum_{i=1}^{g}\left(t_i^3 - t_i\right)}{N^3 - N}}$$

where KW is the Kruskal–Wallis statistic defined in Section 3.2.1 applied to adjusted ranks.

EXAMPLE 3.2.3 The data in Table 3.2.3 have $g = 5$ groups. Since the combined data set has five 1's, five 2's, four 3's, three 4's, and four 5's, we have $t_1 = 5$, $t_2 = 5$, $t_3 = 4$, $t_4 = 3$, and $t_5 = 4$. The KW statistic applied to the adjusted ranks is

$$KW = \frac{12}{21(22)}\left[7(15.86 - 11)^2 + 8(10.31 - 11)^2 + 6(6.25 - 11)^2\right] = 7.91$$

The adjustment factor for ties in the Kruskal–Wallis statistic is

$$1 - \frac{\left(5^3 - 5\right) + \left(5^3 - 5\right) + \left(4^3 - 4\right) + \left(3^3 - 3\right) + \left(4^3 - 4\right)}{21^3 - 21} = 0.96$$

Thus, $KW_{ties} = 7.91/0.96 = 8.24$, which except for rounding is the same as the value obtained in Example 3.2.2. ∎

Use of the Kruskal–Wallis Statistic

The conditions that favor the use of the Kruskal–Wallis statistic are the same as those that favor the Wilcoxon rank-sum statistic. If the data have potential outliers, if the population distributions have heavy tails, or if the population distributions are significantly skewed, then the Kruskal–Wallis statistic will generally have greater power to detect differences among treatments than the one-way analysis of variance *F*-test. On the other hand, if population distributions are normal or are light-tailed and symmetric, then the *F*-test will have greater power.

3.2.3 An Intuitive Derivation of the Chi-Square Approximation for KW

Here we give an intuitive argument to suggest the chi-square approximation for the permutation distribution of the KW statistic or the KW_{ties} statistic. Under the assumption that observations come from a normal distribution with common variance σ^2, the quantity SST/σ^2 has a chi-square distribution with $k-1$ degrees of freedom. For ranks, or adjusted ranks in the case of ties, the mean of the ranks is $(N+1)/2$, so the rank version of SST is given by

$$SST_R = \sum_{i=1}^{k} n_i \left(\overline{R}_i - \frac{N+1}{2} \right)^2$$

It is reasonable to suppose that it should be possible to find a constant C so that $C(SST_R)$ has an approximate chi-square distribution with $k-1$ degrees of freedom. Since the expected value of a chi-square random variable is equal to its degrees of freedom, we will find C so that

$$E\left[C(SST_R)\right] = k - 1$$

From the properties of finite sampling discussed in Section 2.10, we have

$$E\left(\overline{R}_i - \frac{N+1}{2} \right)^2 = \mathrm{var}\left(\overline{R}_i \right) = \frac{N - n_i}{N - 1} \frac{\sigma_R^2}{n_i}$$

where σ_R^2 is the population variance of the combined ranks or adjusted ranks. After substitution and simplification, we have

$$E(SST_R) = \sum_{i=1}^{k} \frac{N - n_i}{N - 1} \sigma_R^2 = (k-1) \frac{N\sigma_R^2}{N - 1}$$

It follows that the constant that will make $E[C(SST_R)] = k - 1$ is

$$C = \frac{N - 1}{N\sigma_R^2} = \frac{1}{S_R^2}$$

where S_R^2 is the sample variance of the ranks or adjusted ranks. In the case of no ties in the data,

$$S_R^2 = \frac{N(N+1)}{12}$$

Multiplying C by SST_R gives us the Kruskal–Wallis statistic.

3.2.4 Tests on General Scores

Suppose we have general scores $A_1, A_2, \ldots, A_N$ that are assigned to the ordered observations. These include the van der Waerden scores, exponential scores, and others. We may apply a k-sample permutation test to these scores to test for differences among treatments. We simply replace the original observations with the general scores and carry out the permutation F-test on these scores as discussed in Section 3.1. We may also use a chi-square approximation for the statistic

$$GS = \frac{1}{S_A^2} \sum_{i=1}^{k} n_i \left(\overline{A}_i - \mu_A \right)^2$$

where $\overline{A}_i$ is the average of the observed scores for the ith treatment, μ_A is the mean of the combined scores $A_1, A_2, \ldots, A_N$, and S_A^2 is the sample variance of these scores. The degrees of freedom for the chi-square approximation are $k - 1$. The rationale is the same as that for the Kruskal–Wallis test as discussed in Section 3.2.3.

3.2.5 Computer Analysis

StatXact implements the Kruskal–Wallis test and gives both exact p-values and the chi-square approximation. The chi-square approximation is available in S-Plus from its "Compare Samples" menu. The test may be carried out in Resampling Stats using randomly selected permutations to estimate the p-value. The code is essentially the same as that for the permutation F-test in Figure 3.1.3, except that original observations are replaced by ranks or adjusted ranks.

MINITAB output for the Kruskal–Wallis test is shown in Figure 3.2.1. The note regarding small samples is to warn that the chi-square approximation may not be

FIGURE 3.2.1

MINITAB Analysis of Data in Table 3.2.2 Using the Kruskal–Wallis Test

```
Kruskal-Wallis Test: C2 versus C1

Kruskal-Wallis Test on C2

C1            N        Median      Ave Rank           Z
1             4         4.113          19.5        3.05
2             6         3.258           8.5       -1.17
3             5         3.578          15.0        1.65
4             6         2.815           4.5       -3.04
Overall      21                        11.0

H = 17.14   DF = 3   P = 0.001

* NOTE * One or more small samples
```

particularly good for small samples. In fact, for 10,000 randomly selected permutations in StatXact, the *p*-value to four decimal places is .0000, and the confidence interval for the *p*-value is .0000 to .0005, so the chi-square approximation with the *p*-value .001 is adequate in this case.

3.3
Multiple Comparisons

If we have more than two treatments, a researcher will want to know which treatments differ from the others. The procedures described in Sections 3.1 and 3.2 and the usual analysis of variance *F*-test do not answer this question. All we can gather from these tests is whether or not there are differences among the treatments, but we cannot identify where the differences occur.

One possible way to compare treatments is to do pairwise tests. For instance, if we have three treatments labeled 1, 2, and 3, we would compare 1 to 2, 1 to 3, and 2 to 3 using a two-sample test such as the Wilcoxon rank-sum test. Conducting each test at the traditional 5% level of significance has an undesirable consequence, especially when the number of treatments is large. For *k* treatments, there are $k(k-1)/2$ pairwise comparisons. For instance, if $k = 7$, then there are 21 comparisons. Just by virtue of doing so many pairwise comparisons, the probability of declaring at least two treatments to be different may be considerably greater than 5% even if all the treatments are the same. If *k* is large enough, it is virtually certain that we will find at least one statistically significant difference just by chance.

The multiple comparison problem is to find a way to determine which treatments differ from others, and to do so in a way that will reduce the chance of spurious results. What we will be concerned with here is controlling the *experiment-wise error rate*. The experiment-wise error rate in making pairwise comparisons among *k* treatments is the probability of declaring at least two treatments to be different when there are no differences among the *k* treatments.

3.3.1 Three Rank-Based Procedures for Controlling Experiment-Wise Error Rate Assuming No Ties in the Data

We present three rank-based procedures for controlling the experiment-wise error rate. To motivate each rank-based procedure, we first give the corresponding procedure that is appropriate for normally distributed data. Then we show how this procedure is modified to accommodate ranking of the data. We assume that the data have no ties.

Each procedure has advocates and detractors, and we will not recommend which, if any, of these to use. Our interest is in implementing the chosen methodology. For a discussion of multiple comparison procedures from a practitioner's point of view, see Milliken and Johnson (1984). Also see Westfall and co-authors (1999). We will use the notation in the data layouts in Table 3.1.1 and Table 3.2.1.

Bonferroni Adjustment

If we wish to have an experiment-wise error rate no greater than α, then the Bonferroni adjustment is to do each of the $k(k-1)/2$ comparisons at level of significance

$$\alpha' = \frac{\alpha}{\dfrac{k(k-1)}{2}}$$

For instance, if there are four treatments and an experiment-wise error rate of .05 is desired, then each of the six pairwise comparisons would be done at the .05/6 = .0083 level of significance. If the data are normally distributed, we may use the t-test to do the pairwise comparisons. If the data are not normally distributed, we may use the Wilcoxon rank-sum test or any other nonparametric test to do the pairwise comparisons.

Fisher's Protected Least Significant Difference (LSD)

Fisher's protected LSD applies to observations that have been selected from normal distributions, but it is often applied even when the assumption of normality is violated. First, do an F-test for equality of means as in the one-way analysis of variance. If this test is significant at a desired level α, then do all pairwise t-tests at level α. This is equivalent to declaring means for treatments i and j to be statistically significantly different if

$$\left| \overline{X}_i - \overline{X}_j \right| \geq t(\alpha/2,\ \mathrm{df}) \sqrt{\mathrm{MSE}\left(\frac{1}{n_i} + \frac{1}{n_j} \right)}$$

where $t(\alpha/2, \mathrm{df})$ is the upper-tail $100\alpha/2\%$ point of a t-distribution with degrees of freedom $\mathrm{df} = N - k$. Note that MSE, defined in Section 3.1.1, is used as the estimate of the variance. The right-hand side of this inequality is termed the *least significant difference*. If the F-test is not significant, we do not do the pairwise comparisons. That is, we don't try to determine which treatments differ from the others unless the analysis of variance indicates that there are differences among the treatments. This is the "protected" part of this procedure.

A rank-based analogue of Fisher's protected LSD is to test first for equality of distributions using the Kruskal–Wallis test. If this test is significant at level α, then declare the distributions of treatments i and j to be different if

$$\left| \overline{R}_i - \overline{R}_j \right| \geq z(\alpha/2) \sqrt{\frac{N(N+1)}{12}\left(\frac{1}{n_i} + \frac{1}{n_j} \right)}$$

where $z(\alpha/2)$ is the upper-tail $100\alpha/2\%$ point of the standard normal distribution. Otherwise, do not do the pairwise tests. We note that the quantity $N(N + 1)/12$ is just the sample variance of the ranks $1, 2, \ldots, N$, which replaces MSE in the LSD formula for the difference of sample means. See Section 3.3.4 for the justification for doing this.

Tukey's HSD Procedure

Suppose populations are normally distributed and sample sizes are equal. Let $n_1 = n_2 = \cdots = n_k = n$. We consider a statistic that measures the largest difference between sample means. Let max{.} and min{.} denote the maximum and minimum, respectively, of the indicated quantity, and define the statistic Q as

$$Q = \max_{ij} \left\{ \frac{\sqrt{n}\left|\overline{X}_i - \overline{X}_j\right|}{\sqrt{\text{MSE}}} \right\}$$

$$= \frac{\sqrt{n}\left(\max_i\{\overline{X}_i\} - \min_i\{\overline{X}_i\}\right)}{\sqrt{\text{MSE}}}$$

Let $q(\alpha, k, \text{df})$ denote the upper-tail $100\alpha\%$ point of the distribution of Q when there are no differences among k treatments. Here df denotes the degrees of freedom of MSE [df $= k(n - 1)$ in this case]. If

$$\left|\overline{X}_i - \overline{X}_j\right| \geq q(\alpha, k, \text{df})\sqrt{\frac{\text{MSE}}{n}}$$

we declare treatments i and j to be different. Tukey coined the term *honest significant difference (HSD)* to denote the right-hand side of this inequality. Values of $q(\alpha, k, \text{df})$ are given in Table A8 in the Appendix for $\alpha = .05$.

An analogue based on ranks is to declare distributions i and j to be different if

$$\left|\overline{R}_i - \overline{R}_j\right| \geq q(\alpha, k, \infty)\sqrt{\frac{N(N+1)}{12n}}$$

Here we use df $= \infty$. Note again that MSE in Tukey's HSD has been replaced by the sample variance of the ranks $1, 2, \ldots, N$.

For unequal sample sizes, the *Tukey–Kramer procedure* is to declare treatments i and j to be statistically significantly different if

$$\left|\overline{X}_i - \overline{X}_j\right| \geq q(\alpha, k, \text{df})\sqrt{\frac{\text{MSE}}{2}\left(\frac{1}{n_i} + \frac{1}{n_j}\right)}$$

where df $= N - k$. The corresponding procedure for ranks is to declare treatments i and j to be different if

$$\left| \overline{R}_i - \overline{R}_j \right| \geq q(\alpha, \ k, \ \infty) \sqrt{\frac{N(N+1)}{24} \left(\frac{1}{n_i} + \frac{1}{n_j} \right)}$$

EXAMPLE 3.3.1   An important indicator of soil type is the amount of clay in the soil. Six samples of soil were selected from four locations, and the percentage of clay was determined in each sample. Data and ranks are listed in Table 3.3.1. The purpose of this example is to compare the normal-theory multiple comparison procedures with their counterparts based on ranks. From the analysis of variance, we find MST = 150.8, MSE = 27.8, and $F = 5.42$, which is significant at the 5% level of significance ($p = .0068$). The tabled t and q values for the Bonferroni adjustment, the protected LSD procedure, and the HSD procedure are $t(.025/6, 20) = 2.93$, $t(.025, 20) = 2.09$, and $q(.05, 4, 20) = 3.96$, respectively. Thus, the Bonferroni difference (BON), the LSD, and the HSD for a 5% level of significance are

$$BON = 2.93 \sqrt{27.8 \frac{2}{6}} = 8.92$$

$$LSD = 2.09 \sqrt{27.8 \frac{2}{6}} = 6.36$$

$$HSD = 3.96 \sqrt{\frac{27.8}{6}} = 8.52$$

TABLE 3.3.1
Percentages of Clay in Soil Samples from Four Locations

Location 1:	Data	26.5	15.0	18.2	19.5	23.1	17.3
	Ranks	16	2	7	9	12	5
Location 2:	Data	16.5	15.8	14.1	30.2	25.1	17.4
	Ranks	4	3	1	21	13	6
Location 3:	Data	19.2	21.4	26.0	21.6	35.0	28.9
	Ranks	8	10	14	11	23	19
Location 4:	Data	26.7	37.3	28.0	30.1	33.5	26.3
	Ranks	17	24	18	20	22	15

The means for the four locations are 19.93, 19.85, 25.35, and 30.32, respectively. Using any of the procedures, we see that the fourth location is significantly different from locations 1 and 2. All other differences are not significant.

The Kruskal–Wallis statistic is KW = 10.29, which is significant at the 5% level of significance ($p = .016$ using the chi-square approximation). We have $z(.025/6) =$

2.64, $z(.025) = 1.96$, and $q(.05, 4, \infty) = 3.63$. Thus, the rank versions of the Bonferroni adjustment, LSD, and HSD for the 5% level of significance are

$$\text{BON}_{\text{ranks}} = 2.64\sqrt{\frac{(24)(25)}{12}\frac{2}{6}} = 10.78$$

$$\text{LSD}_{\text{ranks}} = 1.96\sqrt{\frac{(24)(25)}{12}\frac{2}{6}} = 8.00$$

$$\text{HSD}_{\text{ranks}} = 3.63\sqrt{\frac{(24)(25)}{(12)(6)}} = 10.48$$

The means of the ranks for the four locations are 8.50, 8.00, 14.17, and 19.33, respectively. From any of the three procedures, location 4 is significantly different from locations 1 and 2, and there are no other significant differences.   ∎

3.3.2 Multiple Comparisons for General Scores (Including Ties)

Suppose we have general scores $A_1, A_2, \ldots, A_N$ that are assigned to the ordered observations. They include tied ranks, van der Waerden scores, exponential scores, and others. Multiple comparison formulas for general scores involve a simple modification of the normal-theory formulas for comparing means. We replace the sample means with the means of the scores, and we replace MSE with the sample variance S_A^2 of the scores and use df $= \infty$.

EXAMPLE 3.3.2   Consider the data in Table 3.2.3. The tied ranks for the three products are as follows: product 1: 16, 19.5, 12.5, 16, 19.5, 19.5, 8; product 2: 12.5, 16, 19.5, 8, 12.5, 3, 3, 8; product 3: 8, 3, 3, 8, 3, 12.5. The sample variance of the combined ranks is $S_A^2 = 36.9$. There are seven observations with product 1 and six observations with product 3, so the LSD for comparing product 1 with product 3 at the 5% level of significance is

$$1.96\sqrt{36.9\left(\frac{1}{7} + \frac{1}{6}\right)} = 6.62$$

The difference between the mean ranks for the two groups is $15.86 - 6.25 = 9.61$, so these two products are judged to be significantly different using the LSD criterion. ∎

3.3.3 Multiple Comparison Permutation Tests

Permutation versions of the Bonferroni, protected LSD, and HSD procedures are now considered. We may apply these procedures to the original data or to any set of scores we assign to the observations.

Bonferroni Adjustment for Permutation Tests

A permutation version of a multiple comparison test using Bonferroni's adjustment is carried out by performing two-sample permutation tests on each of the pairs of treatments at the adjusted level of significance $\alpha' = 2\alpha/k(k-1)$. This procedure has an experiment-wise error rate of no more than α.

Permutation LSD

For a protected LSD procedure at level of significance α, the permutation F-test is first performed at level of significance α. If the test is not significant, then no pairwise comparisons are made. If the overall test is significant, select a statistic T_{ij} for comparing the mean scores of two treatments:

1. From the $N = n_1 + n_2 + \cdots + n_k$ combined observations, obtain all possible samples without replacement of sizes n_i and n_j, or if this is not feasible, obtain an appropriately large randomly selected subset of such samples. A sample is obtained by placing all observations in a vector, randomly shuffling elements of the vector, and selecting the first n_i for treatment i and the next n_j for treatment j.

2. From these samples obtain the permutation distribution of $|T_{ij}|$. Let $t^*(\alpha)$ denote the upper-tail $100\alpha\%$ point of this distribution. Declare treatments i and j to be different if

$$|T_{ij}| \ge t^*(\alpha)$$

Note that the value of $t^*(\alpha)$ applies to all pairs of treatments that have the same sample sizes n_i and n_j.

3. A p-value for comparing treatment i with treatment j is the fraction of the permutation distribution greater than or equal to the observed value of $|T_{ij}|$.

The statistic T_{ij} may be the difference of two means, the two-sample t, the Wilcoxon rank-sum, or the like.

Permutation HSD

The permutation analogue of Tukey's honest significant difference is carried out as follows for equal sample sizes.

1. Permute the data as in the permutation F-test, and for each permutation obtain

$$Q^* = \max_{ij} |T_{ij}|$$

2. From the permutation distribution of Q^*, obtain the upper-tail $100\alpha\%$ point $q^*(\alpha)$. Declare treatments i and j to be different at the αth level of significance if

$$|T_{ij}| \ge q^*(\alpha)$$

3. A *p*-value for the multiple comparison is the fraction of the permutation distribution of Q^* greater than or equal to the observed value of $\left|T_{ij}\right|$.

4. For the special case of the two-sample *t*-statistics for T_{ij}, find the permutation distribution of

$$Q^* = \max_{ij}\left\{\frac{\left|\overline{X}_i - \overline{X}_j\right|}{\sqrt{MSE(1/n_i + 1/n_j)}}\right\}$$

In this case

$$T_{ij} = \frac{\overline{X}_i - \overline{X}_j}{\sqrt{MSE(1/n_i + 1/n_j)}}$$

3.3.4 Variance of a Difference of Means When Sampling from a Finite Population

The rationale for the formula for the rank versions of LSD is based on the variance of a difference of means when sampling is from a finite population. Suppose a finite population consists of the numbers $A_1, A_2, \ldots, A_N$. Let μ and σ denote the mean and population standard deviation of the scores, respectively. Suppose random samples of sizes n_1 and n_2 are selected from this population without replacement. Let $\overline{T}_1$ and $\overline{T}_2$ denote the sample means of the two samples. The variance of the difference of the means is given by

$$\mathrm{var}\left(\overline{T}_1 - \overline{T}_2\right) = \mathrm{var}\left(\overline{T}_1\right) + \mathrm{var}\left(\overline{T}_2\right) - 2\,\mathrm{cov}\left(\overline{T}_1, \overline{T}_2\right)$$

Now

$$\mathrm{var}\left(\overline{T}_i\right) = \left(\frac{N - n_i}{N - 1}\right)\frac{\sigma^2}{n_i}, \quad i = 1, 2$$

and

$$\mathrm{cov}\left(\overline{T}_1, \overline{T}_2\right) = -\frac{\sigma^2}{N - 1}$$

Putting these results together, we find

$$\mathrm{var}\left(\overline{T}_1 - \overline{T}_2\right) = \frac{N\sigma^2}{N - 1}\left(\frac{1}{n_1} + \frac{1}{n_2}\right)$$

The quantity $N\sigma^2/(N-1)$ is the sample variance of the scores $A_1, A_2, \ldots, A_N$.

Assuming approximate normality for the difference of sample means, we can assert that the quantity

$$Z = \frac{\overline{T}_1 - \overline{T}_2}{\sqrt{\dfrac{N\sigma^2}{N-1}\left(\dfrac{1}{n_1} + \dfrac{1}{n_2}\right)}}$$

has an approximate standard normal distribution. From this one can obtain the LSD, which is given by

$$z_{\alpha/2}\sqrt{\dfrac{N\sigma^2}{N-1}\left(\dfrac{1}{n_1} + \dfrac{1}{n_2}\right)}$$

In the case in which the scores are the ranks $1, 2, \ldots, N$, we have $\sigma^2 = (N-1)(N+1)/12$, and the LSD is given by

$$\text{LSD}_{\text{ranks}} = z_{\alpha/2}\sqrt{\dfrac{N(N+1)}{12}\left(\dfrac{1}{n_1} + \dfrac{1}{n_2}\right)}$$

A similar approach applies to the Bonferroni and Tukey procedures.

3.3.5　Computer Analysis

Resampling Stats code for carrying out multiple comparison tests on the data in Example 3.3.1 is shown in Figure 3.3.1 (see page 100). Here we find a 5% critical value of $q^*(.05) = 9.5$ for the Tukey's HSD procedure and a value of $t^*(.05) = 7.6$ for the LSD procedure. These critical values lead to the same conclusions as those in Example 3.3.1. The code also computes the multiple comparison HSD and LSD p-value for the largest difference, which is $30.32 - 19.85 = 10.47$. The p-values are .0222 for Q^* and .0054 for the LSD procedure. Here T_{ij} is the difference of means.

Multiple comparison procedures based on ranks or general scores are not implemented in either StatXact or MINITAB. The procedure PROC MULTTEST in the SAS® programming language has the capability of doing multiple comparison permutation procedures. See Westfall and co-authors (1999).

3.4
Ordered Alternatives

If treatments are not equal, then in some situations it may be possible to anticipate the direction in which the treatments differ. For instance, in a study of the effectiveness of a pain-relieving drug, a researcher may anticipate that the degree of relief will be greater for treatments that use larger doses of the drug. With prior knowledge like this we can construct statistical tests that are more powerful than the Kruskal–Wallis test and other tests that do not take advantage of such knowledge.

FIGURE 3.3.1

Resampling Stats Code for Computing Permutation HSD and LSD for Data in Example 3.3.1

```
maxsize default 5000                   'obtain upper 5% critical values
'enter data and create data vector     percentile qdist (95) pctq
data (26.5 15.0 18.2 19.5 23.1 17.3) d1   percentile lsddist (95) pctlsd
data (16.5 15.8 14.1 30.2 25.1 17.4) d2   print pctq
data (19.2 21.4 26.0 21.6 35.0 28.9) d3   print pctlsd
data (26.7 37.3 28.0 30.1 33.5 26.3) d4
concat d1 d2 d3 d4 dat                 'compute p-values for largest difference
                                       'largest difference is 10.47
'obtain permutation distribution       count qdist >= 10.47 pvalq
repeat 5000                            divide pvalq 5000 pvalq
shuffle dat sdat                       count lsddist >= 10.47 pvallsd
take sdat 1,6 sdat1                     divide pvallsd 5000 pvallsd
take sdat 7,12 sdat2                   print pvalq
take sdat 13,18 sdat3                  print pvallsd
take sdat 19,24 sdat4
mean sdat1 m1                          'print output here
mean sdat2 m2                          PCTQ      =      9.4667
mean sdat3 m3                          PCTLSD    =      7.6
mean sdat4 m4
concat m1 m2 m3 m4 mm                  PVALQ     =      0.0222
                                       PVALLSD   =      0.0054
'HSD computation
max mm maxmean
min mm minmean
subtract maxmean minmean qsim

'LSD computation
subtract m1 m2 diffm
abs diffm lsdsim

'save permuted values
score qsim qdist
score lsdsim lsddist
end

(Continue next column)
```

We are interested in alternative hypotheses in which observations from treatment 1 tend to be smaller than observations from treatment 2, and so on. If $F_i(x)$ denotes the cdf of treatment i, the alternative hypothesis may be expressed as

$$H_a: F_1(x) \geq F_2(x) \geq \cdots \geq F_k(x)$$

If we have shift alternatives—that is, if there is a cdf $F(x)$ such that $F_i(x) = F(x - \mu_i)$—then the alternative hypothesis may be stated as

$$H_a: \mu_1 \le \mu_2 \le \cdots \le \mu_k$$

Let T_{ij} be any test statistic for testing the null hypothesis $H_0: F_i(x) = F_j(x)$ against the *one-sided* alternative $H_a: F_i(x) \ge F_j(x)$, for $i < j$. A general form of a test statistic for testing the hypothesis $H_a: F_1(x) \ge F_2(x) \ge \cdots \ge F_k(x)$ is the sum of the pairwise statistics T_{ij}—that is,

$$T = \sum_{i<j} T_{ij}$$

In the discussion below, we use the same notation as in Section 3.1.

3.4.1 Jonckheere–Terpstra Test

The *Jonckheere–Terpstra statistic*, JT, is a test statistic of the form T above in which the T_{ij}'s are one-sided Mann–Whitney statistics. A p-value for JT may be obtained using the following steps:

1. Compute JT_{obs}, the observed value of JT from the original data.

2. Obtain all possible permutations of the data among the k treatments in which each treatment has n_i observations, $i = 1, 2, \ldots, k$. If this is not feasible, obtain a randomly selected subset of the permutations. Compute JT for each of these permutations.

3. To obtain the upper-tail p-value, compute the fraction of the JT's in step 2 that are greater than or equal to JT_{obs}. The upper-tail test assumes that the X's and Y's for each pairwise Mann–Whitney statistic are such that the X's are identified with the treatment anticipated to have the smaller observations and the Y's are identified with the treatment anticipated to have the larger observations.

Tables of JT values are given in Lehmann (1975) for small sample sizes.

3.4.2 Large-Sample Approximation

It is possible to use a normal approximation for the distribution of JT. When the data have no ties, the expected value of JT is

$$E(\mathrm{JT}) = \sum_{i<j} \frac{n_i n_j}{2} = \frac{N^2 - \sum_{i=1}^{k} n_i^2}{4}$$

and the variance of JT is

$$\text{var(JT)} = \frac{N^2(2N+3) - \sum_{i=1}^{k} n_i^2(2n_i+3)}{72}$$

The standardized test statistic is

$$Z = \frac{\text{JT} - E(\text{JT})}{\sqrt{\text{var(JT)}}}$$

which is referred to the standard normal distribution. Again, when we compute each of the Mann–Whitney statistics, if the X's are identified with the treatment anticipated to have the smaller observations and the Y's with the treatment anticipated to have the larger observations, Z is an upper-tail test. An adjustment for ties is available but not presented here. See Lehmann (1975) for details.

An equivalent form of JT is obtained by using the Wilcoxon rank-sum statistic instead of the Mann–Whitney statistic. The expected value in this case is

$$E(\text{JT}_{\text{Wilcoxon}}) = \sum_{i<j} \frac{n_j(n_i + n_j + 1)}{2}$$

where the rank sum of each pairwise test is taken for the treatment with observations that are expected to be larger and n_j is the sample size of this treatment. The variance is the same as in the Mann–Whitney version of JT.

EXAMPLE 3.4.1   An agronomist studied the effect of mowing height on the phosphorus content of a certain species of prairie grass. He postulated that phosphorus levels would tend to be lower in plants that have been mowed at greater heights. Hypothetical data from the experiment are listed in Table 3.4.1.

TABLE 3.4.1
Phosphorus Contents of Plants Under Four Mowing Treatments

Mowing Height	*Observations*					
Treatment 1: Unmowed	13.0	24.1	11.7	16.3	15.5	24.5
Treatment 2: 20 cm	42.0	18.0	14.0	36.0	11.6	19.0
Treatment 3: 10 cm	15.6	23.8	24.4	24.0	21.0	21.1
Treatment 4: 5 cm	35.3	22.5	16.9	25.0	23.1	26.0

The Mann–Whitney statistics are 22 (1 vs. 2), 24 (1 vs. 3), 30 (1 vs. 4), 22 (2 vs. 3), 22 (2 vs. 4), and 25 (3 vs. 4). The sum of the Mann–Whitney statistics is JT = 145. The expected value is

$$E(\text{JT}) = \frac{24^2 - 4(6)^2}{4} = 108$$

and the variance is

$$\text{var}(\text{JT}) = \frac{24^2(51) - (4)(6^2)(15)}{72} = 378$$

Thus,

$$Z = \frac{145 - 108}{\sqrt{378}} = 1.903$$

The one-sided *p*-value is .0285 with the normal approximation. The exact *p*-value from StatXact is .0299. By contrast, the Kruskal–Wallis test for these data gives a value of KW = 3.46, which is not significant (p = .3260). By incorporating the agronomist's expectation regarding the behavior of the mowing treatments, we have been able to detect a significant difference among the treatments. ∎

3.4.3 Computer Analysis

StatXact implements the JT test as one of its *k* independent-samples tests. Resampling Stats code for the data in Table 3.4.1 is shown in Figure 3.4.1. For convenience we computed the pairwise Wilcoxon rank-sum statistics instead of the pairwise Mann–Whitney statistics. In this run of 5000 randomly selected permutations, the estimated *p*-value is .0296.

FIGURE 3.4.1

Resampling Stats Code for Wilcoxon Rank-Sum Version of JT Test for Data in Table 3.4.1

```
maxsize default 5000

'enter data and create data vector
data (13.0 24.1 11.7 16.3 15.5 24.5) d1
data (42.0 18.0 14.0 36.0 11.6 19.0) d2
data (15.6 23.8 24.4 24.0 21.0 21.1) d3
data (35.3 22.5 16.9 25.0 23.1 26.0) d4
concat d1 d2 d3 d4 dat
'sum of pairwise Wilcoxon statistics is 271

'obtain permutation distribution
repeat 5000
shuffle dat sdat
take sdat 1,6 sdat1
```

(continued)

FIGURE 3.4.1

Resampling Stats Code for Wilcoxon Rank-Sum Version of JT Test for Data in Table 3.4.1
(continued)

```
take sdat 7,12 sdat2
take sdat 13,18 sdat3
take sdat 19,24 sdat4

'get ranks for pairwise Wilcoxon tests
concat sdat1 sdat2 dat_1_2
ranks dat_1_2 r_1_2
take r_1_2 7,12 w_1_2

concat sdat1 sdat3 dat_1_3
ranks dat_1_3 r_1_3
take r_1_3 7,12 w_1_3

concat sdat1 sdat4 dat_1_4
ranks dat_1_4 r_1_4
take r_1_4 7,12 w_1_4

concat sdat2 sdat3 dat_2_3
ranks dat_2_3 r_2_3
take r_2_3 7,12 w_2_3

concat sdat2 sdat4 dat_2_4
ranks dat_2_4 r_2_4
take r_2_4 7,12 w_2_4

concat sdat3 sdat4 dat_3_4
ranks dat_3_4 r_3_4
take r_3_4 7,12 w_3_4

'sum Wilcoxon ranks to get Wilcoxon version of JT
concat w_1_2 w_1_3 w_1_4 w_2_3 w_2_4 w_3_4 w_all
sum w_all w_jt

'keep track of results
score w_jt w_jtdist
end

'get pvalue and print out results
count w_jtdist >= 271 pvalue
divide pvalue 5000 pvalue
print pvalue

'print output here
PVALUE    =      0.0296
```

Exercises

1 The data are samples from three simulated distributions.

Group 1	2.9736	0.9448	1.6394	0.0389	1.2958
Group 2	0.7681	0.8027	0.2156	0.074	1.5076
Group 3	4.8249	2.2516	1.5609	2.0452	1.0959

a Apply the permutation F-test to the data.

b Compare the results in part a with the results of the usual one-way analysis of variance.

2 The following data from the National Transportation Safety Administration are the left femur loads on driver-side crash dummies for automobiles in various weight classes. Apply the permutation F-test and the ANOVA F-test to the data, and compare p-values. Does it appear that the data are normally distributed? The complete data set may be obtained from the Data and Story Library at http://lib.stat.cmu.edu/DASL/.

	Vehicle Weight Classification				
1700 lb	*2300 lb*	*2800 lb*	*3200 lb*	*3700 lb*
574	791	865	998	1154
976	1146	775	1049	541
789	394	729	736	406
805	767	1721	782	1529
361	1385	1113	730	1132
529	1021	820	742	767
	2073	1613	1219	1224
	803	1404	705	314
	1263	1201	1260	1728
	1016	205	611	
	1101	1380	1350	
	945	580	1657	
	139	1803	1143	

3 Refer to the data in Exercise 1. Apply the Kruskal–Wallis test to the data, and compare the conclusions with those obtained in Exercise 1.

4 Obtain the permutation distribution of the Kruskal–Wallis statistic when $n_1 = n_2 = n_3 = 2$.

5 An agronomist gave scores from 0 to 5 to denote insect damage to wheat plants that were treated with four insecticides. The data are given in the following table. Use the Kruskal–Wallis test with ties to test whether or not there is a difference among the treatments.

Treatment 1	0	2	1	3	1	1
Treatment 2	2	0	3	1	3	4
Treatment 3	1	3	4	2	2	1
Treatment 4	3	4	2	5	3	4

6   The data in the table are the head injury readings for driver-side crash dummies for various classes of vehicles. In alphabetical order the classes are compact (compact automobile), heavy (heavy automobile), light (light automobile), medium (medium weight automobile), MPV (multi-purpose vehicle), pickup (pickup truck), and van (passenger van). Test for differences among the groups using the Kruskal–Wallis test. Separate means using the rank versions of the LSD and HSD criteria. The data were randomly selected from a larger database from the same source as the data in Exercise 2.

			Type of Vehicle			
Compact	*Heavy*	*Light*	*Medium*	*MPV*	*Pickup*	*Van*
791	423	551	712	1345	985	805
846	541	1068	435	1269	1074	2613
1024	517	757	298	1477	742	903
1007	1328	1114	733	758	985	949
399	471	920	1200	996	1342	1183
1279	533	1809	1701	1306	1184	1051
407	863	1238	707	968	977	1387
1656	786	918	790	943	1465	1320
1036	551	1339	800	1026	892	1434
1226	1068	603	480	1564	1074	1603

7   Use the procedure outlined in Section 3.3.2 to obtain the HSD for van der Waerden scores for the data in Table 3.3.1.

8   Obtain the upper 10% and 5% critical values of the permutation version of Tukey's HSD for the data in Exercise 6.

9   Apply the Jonckheere–Terpstra statistic to the data in Exercise 2. Assume that the alternative hypothesis is that responses will tend to increase as weight increases.

10   Make up a data set for which the Jonckheere–Terpstra statistic is significant but the Kruskal–Wallis statistic is not.

Theory and Complements

11   Consider the setup illustrated in the following table for ten treatments. The population mean of treatment 1 is 50 and the population means of treatments 2 through 10 are all 5. The variance of the error is $\sigma = 1$, so that if reasonable sample sizes are taken from the ten treatments, the one way analysis of variance is virtually certain to reject the equality of means.

Normal Random Variables: $\mu_1 = 50$, $\mu_2 = \ldots = \mu_{10} = 5$, $\sigma = 1$	
Treatment 1	49.17, 48.19, 48.93
Treatment 2	4.48, 4.43, 4.79
Treatment 3	6.36, 7.03, 5.36
Treatment 4	5.34, 4.86, 6.64
Treatment 5	5.17, 6.47, 6.58
Treatment 6	3.99, 5.89, 6.61
Treatment 7	6.02, 4.96, 6.30
Treatment 8	3.98, 6.30, 4.96
Treatment 9	5.94, 6.03, 5.56
Treatment 10	6.29, 5.11, 5.93

a   Consider using the LSD to separate the means. Clearly, the LSD will identify differences between treatment 1 and the rest. However, what might happen when the LSD is applied to treatments 2 through 10? Do you think it might be likely that some differences will be found among these treatments as well? In considering this situation, comment on the shortcomings of the LSD as a multiple comparisons procedure.

b   Is it likely or unlikely that the Bonferroni method would find differences among the means of treatments 2 though 10? What about Tukey's HSD method?

12   In doing the permutation LSD in Section 3.3.3, the samples of size n_i and n_j are randomly selected without replacement from the combined $N = n_1 + n_2 + \cdots + n_k$ observations. An alternative procedure is to use only the observations from treatments i and j to compare treatments i and j as in the two-sample permutation test. Refer to the data in Exercise 6. Use both the method in Section 3.3.3 and the alternative procedure for constructing the LSD for comparing light to medium vehicles.

13   Modify the code in Figure 3.3.1 to carry out the HSD procedure based on the two-sample t-statistic. Compare PVALLSD and PVALQ to raw and adjusted p-values given by SAS® PROC MULTTEST using the permutation option.

14   In the permutation HSD procedure based on the two-sample t-statistic, replace MSE with the sample variance of the scores S_A^2. Show that this procedure is equivalent to the permutation HSD procedure based on differences of means.

4

Paired Comparisons
and Blocked Designs

A Look Ahead Pairing and blocking are experimental design techniques that enable a researcher to detect differences among treatments more easily in environments that have a lot of variability among experimental units. For instance, suppose we wish to compare two brands of a food product for taste. One way to do the experiment would be to have one group of consumers taste product A and a different group taste product B. Consumers in the experiment would be assigned randomly to either A or B, and they would give a taste score on a scale of 1 to 5, say. This is a two-sample experiment as discussed in Chapter 2. On the other hand, we could do the experiment differently. We could have each consumer taste both brands, with the order of tasting determined randomly. This is a paired-comparison experiment. Intuitively it ought to be better for determining brand preference, since each consumer has a chance to evaluate both brands. It is this type of experiment that we are concerned with in this chapter.

4.1
Paired-Comparison Permutation Test

A researcher in the area of nutrition compared two methods of obtaining dietary information. Method 1 was a 24-hour recall, in which respondents were asked to record all they ate during the last 24 hours. Method 2 was a dietary survey, in which respondents were asked questions about what they typically ate. The researcher selected a group of college women to provide dietary information using both methods. From the data provided, the researcher estimated the daily caloric intake for each method. The purpose was to see whether or not the methods differed in their estimates of caloric intake. Data for five students are given in Table 4.1.1.

If the differences have a normal distribution, which would be difficult to determine in this case, then a paired *t*-test may be used to analyze the data. The mean of the differences is 168, the standard deviation is 233, and $t = 1.61$ with a one-sided *p*-value of .0911. We consider the problem of how to analyze the data when the normality assumption may be in question. In particular, we show how to obtain the permutation distribution of the mean of the differences when the data are paired.

109

TABLE 4.1.1

Caloric Intake Measured by 24-Hour Recall and Dietary Survey

	Student				
	1	*2*	*3*	*4*	*5*
24-Hour Recall	1530	2130	2940	1960	2270
Survey	1290	2250	2430	1900	2120
Difference	240	−120	510	60	150

If there are no differences between the methods, then the two caloric values for each student have an equal chance of occurring with either method. For instance, for student 1, it is just as likely that 1290 would have occurred with method 1 and 1530 with method 2 as the other way around. Consequently, the difference could just as well have been −240 as 240. There are $2^5 = 32$ possible ways that the pairs of observations could have occurred with the two methods, and therefore there are 32 ways in which plus and minus signs could have been assigned to the differences. All of these possibilities, which are listed in Table 4.1.2, are equally likely to occur if there are no differences between the two methods.

The 32 means of differences in Table 4.1.2 form the permutation distribution for this problem. The mean of the differences for the original data in Table 4.1.1 is denoted by an asterisk in Table 4.1.2. By inspection we see that this mean is the third largest among the 32 means. Thus, the upper-tail *p*-value is $3/32 = .0938$.

4.1.1 Steps for a Paired-Comparison Permutation Test

The procedure for a paired-comparison permutation test based on *n* pairs of observations is as follows:

1. Compute the differences, D_i's, for the *n* pairs of data, and compute the mean of the differences, $\overline{D}_{obs}$.
2. For the *n* possible pairs, obtain the 2^n possible assignments of plus and minus signs to the $|D_i|$'s.
3. For each of the 2^n possibilities in step 2, compute $\overline{D}$, the mean of the differences.
4. The upper-tail *p*-value is

$$P_{\text{upper tail}} = \frac{\text{number of } \overline{D}\text{'s} \geq \overline{D}_{obs}}{2^n}$$

The expression for the lower-tail *p*-value is similar. Since the permutation distribution is symmetric, we may obtain a two-tail *p*-value by doubling the one-tail value, assuming the one-tail value is less than or equal to 1/2.

Alternative Forms of the Statistic

Let S_+ and S_- denote the sums of the positive and negative differences, respectively. For the data in Table 4.1.1, $S_+ = 240 + 510 + 60 + 150 = 960$ and $S_- = -120$. If no

TABLE 4.1.2

Paired-Comparison Permutation Distribution for Mean of Differences in Table 4.1.1

Permutation	Student 1	Student 2	Student 3	Student 4	Student 5	Mean of Differences
1	240	120	510	60	150	216
2	240	120	510	60	−150	156
3	240	120	510	−60	150	192
4	240	120	−510	60	150	12
5*	240	−120	510	60	150	168
6	−240	120	510	60	150	120
7	240	120	510	−60	−150	132
8	240	120	−510	60	−150	−48
9	240	−120	510	60	−150	108
10	−240	120	510	60	−150	60
11	240	120	−510	−60	150	−12
12	240	−120	510	−60	150	144
13	−240	120	510	−60	150	96
14	240	−120	−510	60	150	−36
15	−240	120	−510	60	150	−84
16	−240	−120	510	60	150	72
17	240	120	−510	−60	−150	−72
18	240	−120	510	−60	−150	84
19	240	−120	−510	60	−150	−96
20	240	−120	−510	−60	150	−60
21	−240	120	510	−60	−150	36
22	−240	120	−510	60	−150	−144
23	−240	120	−510	−60	150	−108
24	−240	−120	510	60	−150	12
25	−240	−120	510	−60	150	48
26	−240	−120	−510	60	150	−132
27	240	−120	−510	−60	−150	−120
28	−240	120	−510	−60	−150	−168
29	−240	−120	510	−60	−150	−12
30	−240	−120	−510	60	−150	−192
31	−240	−120	−510	−60	150	−156
32	−240	−120	−510	−60	−150	−216

differences are positive, then $S_+ = 0$, and similarly for S_-. We may base the test on either S_+ or S_- instead of $\overline{D}$. Since $S_+ - S_- = \Sigma|D_i|$, and since $\Sigma|D_i|$ is the same for all permutations, we may obtain S_- from S_+, and vice versa, so these statistics are equivalent. Since $\overline{D} = (S_+ + S_-)/n$, it may be expressed uniquely in terms of either S_+ or S_-, so it is equivalent to either.

Hypotheses

Suppose the differences can be regarded as being selected at random from a population of such differences. Let $F(x)$ denote the cdf of this population. The null hypothesis is that $F(x)$ is symmetric about 0; that is,

$$H_0: F(x) = 1 - F(-x)$$

Equivalently, the distributions of X and $W = -X$ are the same. We may have either one-sided or two-sided alternatives. The upper-tail alternative hypothesis is that the differences tend to fall more to the positive side of 0. This is expressed as

$$H_a: F(x) \leq 1 - F(-x)$$

Similarly, we have lower-tail and two-tail alternatives. A special case is the shift alternative, in which $F(x) = G(x - \Delta)$, where $G(x)$ is a distribution that is symmetric about 0. The hypotheses are $H_0: \Delta = 0$ and $H_{a:} \Delta > 0$, $H_a: \Delta < 0$, or $H_a: \Delta \neq 0$. The parameter Δ is the median of the distribution of differences.

EXAMPLE 4.1.1   Suppose twins were involved in a study of cholesterol-reducing drugs. One of the twins in each pair was given drug 1 and the other was given drug 2, where the choice was made at random. The amount by which cholesterol was reduced in each subject is shown in Table 4.1.3. There are $2^{17} = 131,072$ possible permutations of the data. The two-tail p-value from StatXact is .0250. The p-value for the paired t-test is .0215.

TABLE 4.1.3
Cholesterol Reduction for Two Drugs

Pair	Drug 1	Drug 2	Difference	Pair	Drug 1	Drug 2	Difference
1	74	63	11	10	58	38	20
2	55	58	-3	11	54	56	-2
3	61	49	12	12	53	38	15
4	41	47	-6	13	69	47	22
5	53	50	3	14	60	41	19
6	74	69	5	15	61	46	15
7	52	67	-15	16	54	47	7
8	31	40	-9	17	57	44	13
9	50	44	6				

4.1.2 Randomly Selected Permutations

If it is not feasible to obtain the means of the differences for all possible 2^n permutations, then we may randomly sample the means of the differences. This is done as follows: Let $U_1, U_2, \ldots, U_n$ be n independent random variables such that $U_i = 1$ or $U_i = -1$ with probability .5 each. Many programming languages have random number generators that enable us to obtain such numbers. Let $D_1, D_2, \ldots, D_n$ denote the differences for the n pairs of observations. A randomly selected mean of differences may be obtained as

$$\overline{D} = \frac{\sum_{i=1}^{n} U_i |D_i|}{n}$$

A suitably large selection of such values, typically 1000 to 5000, may be used to approximate the permutation distribution of the means of the differences and to obtain an approximate p-value.

For the permutation distribution of S_+, let $V_1, V_2, \ldots, V_n$ be independent random variables such that $V_i = 0$ or $V_i = 1$ with probability .5 each. Then a randomly selected value of S_+ may be obtained as

$$S_+ = \sum_{1}^{n} V_i |D_i|$$

A similar expression applies to S_-.

4.1.3 Large-Sample Approximations

The permutation distributions of the statistics $\overline{D}$, S_+, and S_- will have approximate normal distributions. The expected values and variances of these statistics under the assumption of no difference between the treatments can be derived from the expressions in Section 4.1.2.

Since $E(U_i) = 0$ and $\text{var}(U_i) = 1$, we find

$$E(\overline{D}) = 0, \quad \text{var}(\overline{D}) = \frac{1}{n^2} \sum_{1}^{n} |D_i|^2$$

The normal approximation is obtained from the statistic $Z = \overline{D} / \sqrt{\text{var}(\overline{D})}$, which is referred to the standard normal distribution.

Since $E(V_i) = 1/2$ and $\text{var}(V_i) = 1/4$, we find

$$E(S_+) = \frac{1}{2} \sum_{1}^{n} |D_i|$$

$$\text{var}(S_+) = \frac{1}{4} \sum_{1}^{n} |D_i|^2$$

A similar expression applies to S_-. For the normal approximation for S_+, we use the statistic

$$Z = \frac{S_+ - E(S_+)}{\sqrt{\text{var}(S_+)}}$$

which is again referred to the standard normal distribution.

EXAMPLE 4.1.2   For the data in Table 4.1.3, we find $S_+ = 148$, $\Sigma|D_i| = 183$, and $\Sigma|D_i|^2 = 2603$. Thus,

$$Z = \frac{148 - 183/2}{\sqrt{2603/4}} = 2.215$$

The corresponding two-tail p-value from the standard normal distribution is .0268.

∎

4.1.4 A Test for the Median of a Symmetric Population

We may apply the paired-data procedure in certain situations in which the D_i's do not arise as paired differences. Suppose the D_i's are a random sample from a symmetrically distributed population, and we wish to test the null hypothesis that the population median is 0. We may apply the paired-data procedure to the D_i's because symmetry implies that an observation is as likely to be $-D_i$ as D_i under the null hypothesis. Suppose, for instance, a random sample of students from a certain population is given a standardized test to see whether the median score of this population differs from the median score, 65, of the general population. Let X_i denote the score of the ith individual in the sample, and assume the X_i's are symmetrically distributed about 65 under the null hypothesis. Let $D_i = X_i - 65$. In this case, the D_i's are symmetrically distributed about 0, so the paired-comparison permutation test may be applied. This test is an alternative to the one-sample t-test.

4.1.5 Computer Analysis

Figure 4.1.1 shows Resampling Stats code for the paired-comparison permutation test applied to the data in Table 4.1.3, and Figure 4.1.2 shows the output. For this run of 5000 randomly selected permutations, the approximate one-tail p-value is

FIGURE 4.1.1

Resampling Stats Code for Paired-Comparison Permutation Test
Applied to Data in Table 4.1.3

```
maxsize default 5000
'absolute values of differences entered as data
copy (11 3 12 6 3 5 15 9 6 20 2 15 22 19 15 7 13) dat
'sum of positive differences for original data is 148

'generate permutation distribution
repeat 5000

'generate random 0's and 1's
sample 17 (0 1) weight

'compute s+ statistic
multiply weight dat wtdat
sum wtdat splus

'keep track of values of s+
score splus permdist

end

'compute p-value
count permdist >= 148 pvalue
divide pvalue 5000 pvalue
print pvalue

'obtain permutation distribution and histogram
histogram permdist
```

.0134, giving a two-tail value of .0268. The frequency distribution and histogram of the permutation distribution were generated by the "histogram" statement.

The paired-comparison permutation test in StatXact is accessed through the "Paired Samples" menu. Output for the analysis of the data in Table 4.1.3 is shown in Figure 4.1.3. StatXact computes the statistic S_+. The asymptotic inference uses the large-sample approximations presented in Section 4.1.3.

FIGURE 4.1.2

Output of Resampling Stats Code in Figure 4.1.1

```
PVALUE = 0.0134

Vector no. 1: PERMDIST

        Bin                             Cum
      Center      Freq      Pct         Pct
    ----------------------------------------------
          10         1      0.0         0.0
          20         7      0.1         0.2
          30        31      0.6         0.8
          40       112      2.2         3.0
          50       242      4.8         7.9
          60       365      7.3        15.2
          70       572     11.4        26.6
          80       681     13.6        40.2
          90       767     15.3        55.6
         100       674     13.5        69.0
         110       625     12.5        81.5
         120       411      8.2        89.8
         130       264      5.3        95.0
         140       157      3.1        98.2
         150        70      1.4        99.6
         160        14      0.3        99.9
         170         7      0.1       100.0

Note: Each bin covers all values within 5 of its center.
```

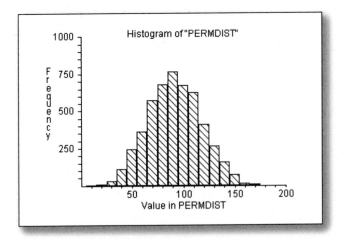

FIGURE 4.1.3
StatXact Analysis of Paired Data in Table 4.1.3

```
PERMUTATION TEST FOR TWO RELATED SAMPLES

Summary of Exact distribution of PERMUTATION TEST statistic:
    Min          Max          Mean       Std-dev       Observed    Standardized
   0.0000       183.0        91.50        25.51          148.0          2.215

Asymptotic Inference:
    One-sided p-value: Pr {   Test Statistic .GE. Observed }  =     0.0134
    Two-sided p-value: 2 * One-sided                          =     0.0268

Exact Inference:
    One-sided p-value: Pr {   Test Statistic .GE. Observed }  =     0.0125
                       Pr {   Test Statistic .EQ. Observed }  =     0.0014
    Two-sided p-value: Pr { | Test Statistic - Mean |
                        .GE.  | Observed - Mean |             =     0.0250
    Two-sided p-value: 2*One-Sided                            =     0.0250
```

4.2
Signed-Rank Test

We now develop a nonparametric test for paired-comparison experiments based on ranks. The ranking of data for a paired-comparison experiment is done differently than the ranking for a two-sample experiment. Assume that the differences of the pairs have no ties and that none of the differences is 0. We rank the absolute values of the differences, and then we attach the signs of the differences to the ranks. These are called *signed ranks*. Table 4.2.1 shows signed ranks for the data in Table 4.1.1.

TABLE 4.2.1
Signed Ranks for Data in Table 4.1.1

	Student				
	1	*2*	*3*	*4*	*5*
Differences	240	−120	510	60	150
Ranks of Absolute Values	4	2	5	1	3
Signed Ranks	4	−2	5	1	3

If the two treatments have the same effect, then any difference is as likely to be positive as it is to be negative, so any possible assignment of plus and minus signs to the ranks is as likely to occur as any other. On the other hand, suppose treatment

1 tends to produce larger observations than treatment 2, so that the differences tend to be positive. Then there should be more positive signed ranks than negative ones, and the larger ranks should be on the positive side. Thus, if we take the sum of the positive signed ranks, it should be larger than what we would expect if the treatments are the same.

The signed-rank statistic is the sum of the positive signed ranks, denoted SR_+. The *Wilcoxon signed-rank test* is essentially the paired-comparison permutation test of Section 4.1 applied to signed ranks.

4.2.1 The Wilcoxon Signed-Rank Test without Ties in the Data

The steps in carrying out the Wilcoxon signed-rank test are as follows. We assume the differences have no ties and no differences are 0.

1. Obtain the signed ranks, and compute SR_{+obs}, the observed value of the sum of the positive signed ranks.

2. Compute SR_+ for all 2^n possible assignments of plus and minus signs to the ranks of the absolute values of the differences.

3. The upper-tail p-value is the fraction of the SR_+'s that are greater than or equal to SR_{+obs}. We compute the lower-tail p-value in a similar way. The two-tail p-value is twice the one-tail p-value, assuming the one-tail value is less than or equal to 1/2. Table A9 in the Appendix contains selected values of $P(SR_+ \geq c)$.

If it is not possible to obtain all possible permutations, we may modify the method described in Section 4.1.2 for S_+ to obtain a random sample of the SR_+'s. In the modification, the absolute values of the differences are replaced with the ranks.

EXAMPLE 4.2.1   The 32 possible sets of signed ranks for the observations in Table 4.2.1 are listed in Table 4.2.2 along with SR_+ for each set. For the original data—denoted by an asterisk—we have $SR_+ = 13$, which is the third largest sum among the values of SR_+. Thus, the p-value is 3/32 = .0938.   ∎

4.2.2 Large-Sample Approximation

Following the procedure presented in Section 4.1.3, we let $V_1, V_2, \ldots, V_n$ be independent random variables, each taking on values 1 and 0 with equal probability. We may represent SR_+ under the null hypothesis of no difference between the treatments as

$$SR_+ = \sum_{i=1}^{n} i V_i$$

TABLE 4.2.2
Signed Rank-Permutation Distribution for Data in Table 4.2.1

Permutation	Student 1	Student 2	Student 3	Student 4	Student 5	SR_+
1	4	2	5	1	3	15
2	4	2	5	1	−3	12
3	4	2	5	−1	3	14
4	4	2	−5	1	3	10
5*	4	−2	5	1	3	13
6	−4	2	5	1	3	11
7	4	2	5	−1	−3	10
8	4	2	−5	1	−3	7
9	4	−2	5	1	−3	10
10	−4	2	5	1	−3	8
11	4	2	−5	−1	3	9
12	4	−2	5	−1	3	12
13	−4	2	5	−1	3	10
14	4	−2	−5	1	3	8
15	−4	2	−5	1	3	6
16	−4	−2	5	1	3	9
17	4	2	−5	−1	−3	6
18	4	−2	5	−1	−3	9
19	4	−2	−5	1	−3	5
20	4	−2	−5	−1	3	7
21	−4	2	5	−1	−3	7
22	−4	2	−5	1	−3	3
23	−4	2	−5	−1	3	5
24	−4	−2	5	1	−3	6
25	−4	−2	5	−1	3	8
26	−4	−2	−5	1	3	4
27	4	−2	−5	−1	−3	4
28	−4	2	−5	−1	−3	2
29	−4	−2	5	−1	−3	5
30	−4	−2	−5	1	−3	1
31	−4	−2	−5	−1	3	3
32	−4	−2	−5	−1	−3	0

It follows that

$$E(SR_+) = \frac{1}{2} \sum_{i=1}^{n} i = \frac{n(n+1)}{4}$$

Similarly,

$$\text{var}(SR_+) = \sum_{i=1}^{n} \frac{i^2}{4} = \frac{n(n+1)(2n+1)}{24}$$

We may then compute

$$Z = \frac{SR_+ - E(SR_+)}{\sqrt{\text{var}(SR_+)}}$$

and refer the resulting value to the standard normal distribution.

EXAMPLE 4.2.2   For the data in Table 4.2.1, we find

$$E(SR_+) = \frac{(5)(6)}{4} = 7.5$$

$$\text{var}(SR_+) = \frac{(5)(6)(11)}{24} = 13.75$$

Although we would not usually use a normal approximation on a sample as small as five, in fact the normal approximation is reasonably good even in this case. For an observed value of $SR_+ = 13$, we have

$$Z = \frac{13 - 7.5}{\sqrt{13.75}} = 1.48$$

$$P(Z \geq 1.48) = .069$$

Using a continuity correction, we find an approximate p-value of

$$P(SR_+ \geq 13) \approx P\left(Z \geq \frac{12.5 - 7.5}{\sqrt{13.75}}\right) = .088$$

which compares favorably with the actual p-value of .0938.   ∎

4.2.3 Adjustment for Ties

Two types of ties can occur with paired data. One is when the absolute values of two differences are the same. The other is when the observations from two treatments are the same, giving a difference of 0. We consider two methods of ranking data in computing adjusted signed ranks for tied data.

The first method of ranking is called *ranking with zeros*. The absolute values of the differences, including the zeros, are ranked from the smallest to the largest. The average rank is assigned to differences that are tied. Plus and minus signs are attached to ranks that correspond to positive and negative differences, respectively, while 0 is attached to the ranks of differences that are 0. Table 4.2.3 contains ten hypothetical differences involving ties along with their signed ranks.

TABLE 4.2.3
Differences and Signed Ranks Involving Ties Using Ranking with Zeros

Differences	−5	−3	−3	0	0	2	4	4	4	5
Ranks of Absolute Values of Differences	9.5	4.5	4.5	1.5	1.5	3	7	7	7	9.5
Signed Ranks	−9.5	−4.5	−4.5	0	0	3	7	7	7	9.5

An alternative method of handling zeros is simply to ignore them. We call this *ranking without zeros*. For the data in Table 4.2.3, only the nonzero differences are ranked, with average ranks given to the nonzero tied values. The results are listed in Table 4.2.4. The effective sample size for the data in Table 4.2.4 is 8, not 10. This is the method implemented in StatXact, MINITAB, and S-Plus.

TABLE 4.2.4
Differences and Signed Ranks Involving Ties Using Ranking without Zeros

Differences	−5	−3	−3	0	0	2	4	4	4	5
Ranks of Absolute Values of Differences	7.5	2.5	2.5	#	#	1	5	5	5	7.5
Signed Ranks	−7.5	−2.5	−2.5	#	#	1	5	5	5	7.5

Pratt (1959) noted a potential problem with ranking without zeros. He produced two data sets of the same size in which the one with smaller differences showed significance but the one with the larger differences showed nonsignificance when he used ranking without zeros. Generally, if there are not too many zeros, both methods will give about the same result.

Carrying Out the Signed Rank Test with Ties

Suppose m of the n differences are nonzero. Let $R_1, R_2, \ldots, R_m$ denote the ranks of the absolute values of these differences, where the ranks may be determined from either ranking with zeros or ranking without zeros. The steps for carrying out the Wilcoxon signed-rank test with ties are the same as those for the paired-comparison permutation test in Section 4.1.1, except that the R_i's are used instead of the $|D_i|$'s. The statistic is SR_+, the sum of the positive signed ranks.

If it is not possible to compute all possible values of SR_+, we may take a random sample of these values. A randomly selected value of SR_+ may be expressed as

$$SR_+ = \sum_1^m V_i R_i$$

where the V_i's are independent random variables that take on values 0 and 1 with equal probability. We may approximate the permutation distribution of SR_+ by taking a suitably large sample of these values. An approximate upper-tail p-value for the test statistic is the fraction of these SR_+'s that are greater than or equal to SR_{+obs}, the observed sum of the positive signed ranks. A similar procedure applies to the lower-tail p-value.

Since $E(V_i) = 1/2$ and $var(V_i) = 1/4$ under the null hypothesis of no difference between treatments, we find

$$E(SR_+) = \frac{1}{2}\sum_1^m R_i$$

$$var(SR_+) = \frac{1}{4}\sum_1^m R_i^2$$

The permutation distribution of the statistic SR_+ may be approximated by a normal distribution with the above mean and variance.

EXAMPLE 4.2.3   For the data in Table 4.2.3, where the ranking is done with zeros, the expected value and variance are

$$E(SR_+) = \frac{1}{2}(9.5 + 4.5 + \cdots + 7 + 9.5) = 26$$

$$var(SR_+) = \frac{1}{4}(9.5^2 + 4.5^2 + \cdots + 7^2 + 9.5^2) = 94.25$$

For the data in Table 4.2.4, where the ranking is done without zeros, the expected value and variance are

$$E(SR_+) = \frac{1}{2}(7.5 + 2.5 + \cdots + 5 + 7.5) = 18$$

$$var(SR_+) = \frac{1}{4}(7.5^2 + 2.5^2 + \cdots + 5^2 + 7.5^2) = 50.25$$ ∎

EXAMPLE 4.2.4   The cholesterol data in Table 4.1.3 have no ties that are zero but three nonzero ties. Table 4.2.5 shows the differences and the signed ranks. In this example, the ranking is the same with zeros and without zeros. We find $E(SR_+) = 76.5$ and $var(SR_+) = 445.5$. The observed value of $SR_+ = 123$. Using the normal approximation, we find

$$Z = \frac{123 - 76.5}{\sqrt{445.5}} = 2.203$$

The one-tail *p*-value is .0138. The exact *p*-value is .0128.

TABLE 4.2.5
Differences and Signed Ranks for Data in Table 4.1.3

Drug 1	Drug 2	Difference	Signed Rank
74	63	11	9
55	58	−3	−2.5
61	49	12	10
41	47	−6	−5.5
53	50	3	2.5
74	69	5	4
52	67	−15	−13
31	40	−9	−8
50	44	6	5.5
58	38	20	16
54	56	−2	−1
53	38	15	13
69	47	22	17
60	41	19	15
61	46	15	13
54	47	7	7
57	44	13	11

∎

Alternative Expressions for $E(\text{SR}_+)$ and $\text{var}(\text{SR}_+)$ with Ties

Here we present an alternative expression for the expected value and variance of SR_+ when the data have ties and ranking is done with zeros. Let t_0 denote the number of differences that are 0. Assume the nonzero ties of the absolute values of the differences form *g* groups, and let t_i denote the number of ties in group *i*. For *n* differences, we have

$$E(\text{SR}_+) = \frac{n(n+1) - t_0(t_0 + 1)}{4}$$

$$\text{var}(\text{SR}_+) = \frac{n(n+1)(2n+1) - t_0(t_0 + 1)(2t_0 + 1)}{24} - \frac{\sum\limits_{i=1}^{g}(t_i^3 - t_i)}{48}$$

If the ranking is done without zeros, we replace *n* with *m*, the number of non-zero differences, and we drop terms involving t_0. See Lehmann (1975, Example 2 of the Appendix) for a derivation.

EXAMPLE 4.2.5   For the data in Table 4.2.3, we have $t_0 = 2$. Three groups have nonzero ties, with two 3's, three 4's, and two 5's. Thus, $t_1 = 2$, $t_2 = 3$, and $t_3 = 2$. The expected value and variance of SR_+ for these data, where we use ranking with zeros, are

$$E(SR_+) = \frac{(10)(11) - (2)(3)}{4} = 26$$

$$\text{var}(SR_+) = \frac{(10)(11)(21) - (2)(3)(5)}{24} - \frac{(2^3 - 2) + (3^3 - 3) + (2^3 - 2)}{48} = 94.25$$

If we use ranking without zeros, then

$$E(SR_+) = \frac{(8)(9)}{4} = 18$$

$$\text{var}(SR_+) = \frac{(8)(9)(17)}{24} - \frac{(2^3 - 2) + (3^3 - 3) + (2^3 - 2)}{48} = 50.25 \qquad \blacksquare$$

4.2.4 Computer Analysis

The analysis of the data in Table 4.2.5 using the StatXact implementation of the Wilcoxon signed-rank test is shown in Figure 4.2.1. When dealing with ties, StatXact uses ranking without zeros.

FIGURE 4.2.1
Wilcoxon Signed-Rank Test Applied to Data in Table 4.2.5 Using StatXact

```
WILCOXON SIGNED RANK TEST

Summary of Exact distribution of WILCOXON SIGNED RANK statistic:
       Min          Max          Mean      Std-dev     Observed Standardized
     0.0000        153.0         76.50       21.11        123.0        2.203


Asymptotic Inference:
    One-sided p-value: Pr {   Test Statistic .GE. Observed }  =      0.0138
    Two-sided p-value: 2 * One-sided                          =      0.0276


Exact Inference:
    One-sided p-value: Pr {   Test Statistic .GE. Observed }  =      0.0128
                       Pr {   Test Statistic .EQ. Observed }  =      0.0009
    Two-sided p-value: Pr { | Test Statistic - Mean |
                       .GE. | Observed - Mean |               =      0.0256
    Two-sided p-value: 2*One-Sided                            =      0.0256
```

Resampling Stats code follows the pattern in Figure 4.1.1 except that the absolute values of the signed ranks are used in place of the absolute values of the actual differences. MINITAB and S-Plus have options for doing the Wilcoxon signed-rank test. Both use normal approximations with a continuity correction to determine p-values, and they use ranking without zeros in the case of ties. MINITAB labels the Wilcoxon signed-rank test as a one-sample test and assumes that the differences D_i are entered for analysis.

4.3
Other Paired-Comparison Tests

4.3.1 Sign Test

The sign test for paired comparisons is based on the number of positive differences, which we denote SN_+. Assume for a moment that the differences contain no ties. If the two treatments in the paired comparison have the same effect, then a difference has probability .5 of being positive, and SN_+ has a binomial distribution with $p = .5$. If there is a difference between treatments, then SN_+ tends to be larger or smaller than one would expect of a binomial random variable with $p = .5$. The upper-tail p-value for an observed value $SN_+ = k$ is

$$P(SN_+ \geq k) = \sum_{i=k}^{n} \binom{n}{i} (.5)^n$$

Table A10 in the Appendix has upper-tail probabilities for the sign test up to samples of size 20. Lower-tail values may be computed as

$$P(SN_+ \leq k) = 1 - P(SN_+ \geq k+1)$$

We may also use a normal approximation to the binomial distribution; that is, we compute

$$Z = \frac{SN_+ - .5n}{\sqrt{.25n}}$$

and refer the resulting value to the standard normal distribution. For smaller samples, we may use a continuity correction.

EXAMPLE 4.3.1   Among the 17 differences for the cholesterol data in Table 4.1.3, 12 are positive. The upper-tail p-value from Table A10 is

$$P(SN_+ \geq 12) = \sum_{i=12}^{17} \binom{17}{i} (.5)^{17} = .072$$

Using the normal approximation, we find $Z = 1.70$ and $P(Z \geq 1.70) = .045$. With a continuity correction, we have

$$P\left(SN_+ \geq 12\right) \approx P\left(Z \geq \frac{11.5 - 8.5}{\sqrt{4.25}}\right) = .072$$ ∎

If any differences are 0, then the procedure is to ignore these data and analyze the remaining differences as above. For instance, if there are originally 20 differences and 2 are equal to 0, then the sign test is applied to the 18 nonzero differences. Conclusions then apply to those pairs for which a difference exists. Note that if there a lot of zeros, then the most important conclusion of the study may be that differences do not exist among the majority of the pairs.

It is not necessary to have numerical data on the treatments to apply the sign test. For instance, consumers may be asked to express their preference for one of two brands of cola—say, A and B. A plus sign may be recorded if A is favored and a minus sign recorded if B is favored. If no preference is indicated, then the difference is recorded as a 0. Once the plus and minus values have been determined, the sign test is carried out as above. Again, conclusions would apply only to those individuals who express a preference for one cola over the other.

4.3.2 General Scoring Systems

Suppose the differences are nonzero, and assume the differences contain no ties. We may assign scores other than ranks to the absolute values of the differences. Let the scores be denoted $A_{(1)} < A_{(2)} < \cdots < A_{(n)}$, which are assumed to be positive and increasing as the absolute values of the differences increase.

After the scores are assigned to the absolute values of the differences, the signs of the differences are attached to the scores. The statistic is the sum of the positive scores, SA_+. The reference distribution is the permutation distribution determined by assigning all possible 2^n combinations of plus and minus signs to the scores and obtaining the distribution of the sum of the positive scores. The procedure is the same as that for the signed-rank test, except that general scores take the place of ranks.

The expected value and variance, assuming no ties, are given by

$$E\left(SA_+\right) = \frac{1}{2} \sum_{i=1}^{n} A_i, \quad var\left(SA_+\right) = \frac{1}{4} \sum_{i=1}^{n} A_i^2$$

As an approximation for large sample sizes, one may compute

$$Z = \frac{SA_+ - E\left(SA_+\right)}{\sqrt{var\left(SA_+\right)}}$$

which is referred to the standard normal distribution.

The scoring procedure for ties in the data is essentially the same as the procedure for ranks. We assign average scores to the tied data, and we may do the scoring either including or excluding differences that are 0. Plus and minus signs are assigned to the scores of the nonzero differences, and 0 is assigned to the scores of the differences that are 0. The expected value and variance formulas above apply, where the scores in question are those that are attached to nonzero differences, and n is the number of nonzero differences.

4.3.3 Selecting Among Paired-Comparison Tests

One may obtain the asymptotic relative efficiencies of paired-comparison tests under various assumptions about the probability distribution of the differences. The alternatives that are typically considered are those that have the effect of shifting the distribution of the differences an amount Δ to the right or left of the origin. In this situation, as in the two-sample case, one can expect the signed-rank test to perform better than the t-test for paired data when the distribution of the differences has heavy tails. However, if the distribution has lighter tails, the t-test will do better. For instance, the asymptotic relative efficiency of the signed-rank test to the t-test is 0.955 for the normal distribution, but it is 1.5 for the heavier-tailed double exponential distribution.

The sign test is generally not as efficient as the signed-rank test. For instance, in the case of the normal distribution the asymptotic relative efficiency of the sign test to the signed-rank test is 2/3. The sign test, however, may do better in situations where there is a tendency for occasional large positive or negative differences relative to the other differences. Such is the case with the Cauchy distribution, where the efficiency of the sign test relative to the signed-rank test is 1.3. See Blair and Higgins (1985) for a simulation study that compares the power of the signed-rank test with that of the paired t-test.

4.4
A Permutation Test for a Randomized Complete Block Design

Blocking is a technique that is used when the experimental units to which treatments are to be applied are not homogeneous, or when the conditions under which the experiment is to be conducted cannot be held constant throughout. For instance, suppose the treatments are three methods of applying nitrogen to wheat (liquid, granular A, and granular B), and suppose the soil in the experimental area is not homogeneous but differs in fertility and moisture. An area is subdivided in such a way that the soil within each subdivision is homogeneous or nearly so. Each of these subdivisions is called a *block*. Each block is then divided into three plots, and the nitrogen treatments are assigned randomly to the plots. Table 4.4.1 shows a typical randomization for four blocks and three treatments. This is a special case of a *randomized complete block design*.

TABLE 4.4.1

Typical Randomization for a Randomized Complete Block Design
with Three Treatments and Four Blocks

Block 1	Block 2	Block 3	Block 4
Treatment 2	Treatment 1	Treatment 2	Treatment 3
Treatment 1	Treatment 3	Treatment 3	Treatment 2
Treatment 3	Treatment 2	Treatment 1	Treatment 1

Randomized complete block designs have three defining features: (1) Experimental units are divided into blocks in such a way that units or experimental conditions within blocks are homogeneous. (2) Blocks have the same number of experimental units as there are treatments. (3) The treatments are randomly assigned to experimental units within blocks. Such designs are used in a wide variety of settings. Nutrition experiments on animals may involve blocking according to the weights of the animals. Medical studies on humans may involve blocking on a medical profile that assesses the general health of the subject before going into the study. Time may be a blocking factor if uncontrollable temporal variables such as humidity or temperature can affect the outcome. If the experiment must be carried out at more than one site, location could be a blocking factor. If different personnel or laboratories are involved in an experiment, these may be the blocking factors.

4.4.1 *F* Statistic for Randomized Complete Block Designs

Assume there are k treatments and b blocks. An observation for treatment i and block j is denoted X_{ij}. The data along with the treatment means, block means, and the overall mean are schematically displayed in Table 4.4.2.

TABLE 4.4.2

Observations and Means for a Randomized Complete Block Design

Treatments	Blocks 1	2	...	b	Means
1	X_{11}	X_{12}	...	X_{1b}	$\overline{X}_{1.}$
2	X_{21}	X_{22}	...	X_{2b}	$\overline{X}_{2.}$
...	...	...	...	...	...
k	X_{k1}	X_{k2}	...	X_{kb}	$\overline{X}_{k.}$
Means	$\overline{X}_{.1}$	$\overline{X}_{.2}$	...	$\overline{X}_{.b}$	$\overline{X}$

Suppose the observations follow the model

$$X_{ij} = \mu + t_i + b_j + \varepsilon_{ij}$$

where μ is an overall effect, the t_i's are the treatment effects, the b_j's are the block effects, and the ε_{ij}'s are independent and identically distributed random variables with median 0. In the case in which the ε_{ij}'s are normally distributed with mean 0 and common variance σ^2, we use the F statistic

$$F = \frac{b \sum_{i=1}^{k} \left(\overline{X}_{i.} - \overline{X} \right)^2 / (k-1)}{\sum_{i=1}^{k} \sum_{j=1}^{b} \left(X_{ij} - \overline{X}_{i.} - \overline{X}_{j.} + \overline{X} \right)^2 / \left[(k-1)(b-1) \right]}$$

to test the hypotheses

$$H_0: \ t_1 = t_2 = \cdots = t_k$$
$$H_a: \ \text{Not all } t_i \text{ are the same}$$

This statistic has an F-distribution with $k - 1$ degrees of freedom for the numerator and $(k - 1)(b - 1)$ degrees of freedom for the denominator.

4.4.2 Permutation F-Test for Randomized Complete Block Designs

If we are unwilling to assume that the ε_{ij}'s have a normal distribution, then we may carry out a permutation F-test for the indicated hypotheses. The procedure is as follows.

1. Compute the F statistic, F_{obs}, for the original data.
2. Permute the observations *within* each of the blocks, doing this for all the blocks. (Do not permute across blocks.) There are $(k!)^b$ possibilities.
3. For each of the possibilities in step 2, compute the F statistic for the randomized complete block design.
4. Since large values of F indicate a difference among treatments, obtain the upper-tail p-value as

$$P_{\text{upper tail}} = \frac{\text{number of } F\text{'s} \ge F_{obs}}{(k!)^b}$$

Alternatively, we may use the statistic

$$\text{SST}^* = \sum_{i=1}^{k} \left(\overline{X}_{i.} - \overline{X} \right)^2$$

or the statistic

$$\text{SSX}^* = \sum_{i=1}^{k} \left(\overline{X}_{i.} \right)^2$$

since either is equivalent to F in a permutation test. The verification of this is similar to that used in Section 3.1.3 to derive alternative statistics for the k-sample permutation test.

If it is not possible to obtain all the permutations, then we may randomly sample them. Each permuted data set is generated by randomly shuffling the observations within blocks. For instance, if there are four blocks and three treatments per block, then we can generate one randomly permuted data set by randomly shuffling the four sets of three numbers. We can find an approximate p-value by repeating this procedure a sufficient number of times and computing the fraction of the F's that are greater than or equal to F_{obs}.

EXAMPLE 4.4.1   Different types of farm machinery have different effects on the compaction of soil and thus may affect yields differently. Table 4.4.3 shows yield data from a randomized complete block design in which four different types of tractors were used in tilling the soil. The blocking factor is location of the fields.

TABLE 4.4.3
Yield Data for a Randomized Complete Block Design
with Four Treatments and Six Blocks

Treatments	Block 1	Block 2	Block 3	Block 4	Block 5	Block 6
1	120	208	199	194	177	195
2	207	188	181	164	155	175
3	122	137	177	177	160	138
4	128	128	160	142	157	179

The observed SSX* is 110,241.63, which has a p-value of .051 based on 5000 randomly selected permutations using Resampling Stats. The corresponding p-value using an F-test with degrees of freedom 3 for the numerator and 15 for the denominator is .058.   ∎

4.4.3 Multiple Comparisons

We show a permutation HSD procedure based on differences of means. We may use other statistics such as the paired-t or Wilcoxon signed-rank instead. Permute the blocked data and obtain

$$Q^* = \max_{ij} \left| \overline{X}_{i.} - \overline{X}_{j.} \right|$$

for each of the permutations. From the permutation distribution of Q^*, obtain the upper-tail $100\alpha\%$ point $q^*(\alpha)$. Declare treatments i and j to be different if the observed treatment means satisfy

$$\left| \overline{X}_{i.} - \overline{X}_{j.} \right| \ge q^{*}(\alpha)$$

A permutation multiple comparison p-value for comparing the ith mean to the jth mean is the fraction of the permutation distribution of Q^{*} greater than or equal to the observed value of $\left| \overline{X}_{i.} - \overline{X}_{j.} \right|$.

EXAMPLE 4.4.2 For the data in Table 4.4.3, we obtained approximate 10%, 5%, 2.5%, and 1% upper-tail critical values for Q^{*} using Resampling Stats to generate 5000 randomly selected permutations. These values are 35.7, 39.3, 41.8, and 44.8, respectively. The means of the four treatments are 182.2, 178.3, 151.8, and 149.0, respectively. Since the largest difference observed in the original data is 182.2 – 149.0 = 33.2, none of the differences is declared significant by Tukey's procedure at these levels of significance. If the assumption of a normal distribution for the ε_{ij}'s is valid, then upper-tail 10%, 5%, 2.5%, and 1% critical values for Tukey's HSD are 34.7, 40.0, 45.0, and 51.2, respectively. Again, no differences are declared significant at these levels of significance. ∎

4.4.4 Computer Analysis

The permutation test is not available in StatXact for randomized complete block designs. Resampling Stats code for a randomized complete block design is shown in Figure 4.4.1. Here, we computed the statistics SSX* and Q^{*}. A p-value is given for SSX*, and percentiles are given for Q^{*}.

FIGURE 4.4.1
Resampling Stats Code for Analysis of Data in Table 4.4.3 Using a Permutation Test for a Randomized Complete Block Design

```
maxsize default 5000                        'permute within blocks
                                            shuffle b1 sb1
'enter data by blocks                       shuffle b2 sb2
copy (120 207 122 128) b1                   shuffle b3 sb3
copy (208 188 137 128) b2                   shuffle b4 sb4
copy (199 181 177 160) b3                   shuffle b5 sb5
copy (194 164 177 142) b4                   shuffle b6 sb6
copy (177 155 160 157) b5
copy (195 175 138 179) b6                   'get treatment means
                                            add sb1 sb2 sb3 sb4 sb5 sb6 trtsum
'observed ssx* = 110241.63                  divide trtsum 6 m

'get permutation distribution               'compute ssx*
repeat 5000                                 square m msqr
                                            sum msqr ssx

(continued next column)                     (continued on next page)
```

FIGURE 4.4.1

Resampling Stats Code for Analysis of Data in Table 4.4.3 Using a Permutation Test
for a Randomized Complete Block Design *(continued)*

```
'compute Q* = max diff mean          'get percentiles for Q*
max m maxmean                        percentile qdist (90 95 97.5 99)
min m minmean                        qpct
subtract maxmean minmean q           print qpct

'keep track of ssx* and q*           'print results here
score ssx ssxdist
score q qdist                        PVAL     =      0.051

end                                  QPCT     =      35.667      39.333
                                     41.833   44.75
'get p-value for ssx*
count ssxdist >= 110241.63 pval
divide pval 5000 pval
print pval

(Continued next column)
```

4.5
Friedman's Test for a Randomized Complete Block Design

Suppose data have been taken according to a randomized complete block design as in Section 4.4. *Friedman's test* for such data essentially involves ranking the observations within blocks and then applying the permutation F-test for a randomized complete block design to the ranks. A computational form of the statistic whose distribution may be approximated by a chi-square distribution with $k - 1$ degrees of freedom is what is generally referred to as Friedman's statistic. We now develop this statistic.

The notation used for the ranks of the observations and the means of the ranks is shown in Table 4.5.1. The ranks within block j are denoted $R_{1j}, R_{2j}, \ldots, R_{kj}$. If there are no ties among the observations within block j, then these ranks are $1, 2, \ldots, k$. If the blocks do contain ties, then we adjust the ranks by using average ranks for the tied observations. In either case, the average of the ranks within blocks is $(k + 1)/2$. Observations that are tied across blocks do not affect the ranking, since ranking is done within blocks. Table 4.5.2 illustrates the ranking that is done when there are no ties within blocks, and Table 4.5.3 illustrates ranking when there are ties within blocks.

TABLE **4.5.1**

Ranks and Means of Observations in a Randomized Complete Block Design

Treatments	Blocks				Means
	1	*2*	. . .	*b*	*Means*
1	R_{11}	R_{12}	. . .	R_{1b}	$\overline{R}_1$
2	R_{21}	R_{22}	. . .	R_{2b}	$\overline{R}_2$
. . .		. . .		. . .	. . .
k	R_{k1}	R_{k2}	. . .	R_{kb}	$\overline{R}_k$

4.5.1 Friedman's Test without Ties

Assume observations within blocks are distinct, so that the ranks are 1, 2, . . . , *k*. The Friedman statistic is

$$\text{FM} = \frac{12b}{k(k+1)} \sum_{i=1}^{k} \left(\overline{R}_i - \frac{k+1}{2} \right)^2$$

The term

$$\text{SSR} = \sum_{i=1}^{k} \left(\overline{R}_i - \frac{k+1}{2} \right)^2$$

is, except for a constant, the numerator of the *F* statistic for the randomized complete block design but computed using ranks. The constant $C = 12b/k(k+1)$ allows us to approximate the permutation distribution of the statistic with a chi-square distribution with $k - 1$ degrees of freedom. See Section 4.5.4 for details about the choice of the constant *C*.

The steps for carrying out Friedman's test are essentially the same as those for the permutation *F*-test for the randomized complete block design presented in Section 4.4.2, except that ranks within blocks are used in place of the original observations, and FM or an equivalent statistic is used instead of *F*. If the sample sizes are large enough, we may approximate the permutation distribution of FM with a chi-square distribution with $k - 1$ degrees of freedom. Tables of $P(\text{FM} \geq c)$ for selected values of *k*, *b*, and *c* are given in Table A11 of the Appendix.

EXAMPLE **4.5.1**   The ranks of the data in Table 4.4.3 are shown in Table 4.5.2. The average ranks for the four treatments are 3.50, 2.50, 2.17, and 1.83, respectively. Here $k = 4$ and $b = 6$. The overall mean of the ranks is 2.5. Thus, the Friedman statistic is

$$\text{FM} = \frac{(12)(6)}{(4)(5)} \left[(3.50 - 2.50)^2 + (2.50 - 2.50)^2 + (2.17 - 2.50)^2 + (1.83 - 2.50)^2 \right]$$

$$= 5.61$$

The p-value from the chi-square distribution with 3 degrees of freedom is .13. The exact p-value from StatXact is .1268.

TABLE 4.5.2

Data and Ranks for Yields in Table 4.4.3

Treatments	Block 1	Block 2	Block 3	Block 4	Block 5	Block 6
1	120 (1)	208 (4)	199 (4)	194 (4)	177 (4)	195 (4)
2	207 (4)	188 (3)	181 (3)	164 (2)	155 (1)	175 (2)
3	122 (2)	137 (2)	177 (2)	177 (3)	160 (3)	138 (1)
4	128 (3)	128 (1)	160 (1)	142 (1)	157 (2)	179 (3)

■

4.5.2 Adjustment for Ties

Since ranking is done within blocks, no adjustment for ties is needed when ties occur across blocks. Within each block we assign the average of the ranks to tied observation. We may then carry out the permutation F-test on these ranks using the method presented in Section 4.4.2. To use a chi-square approximation, we need an adjustment to the Friedman statistic.

Let S_{Bj}^2 denote the sample variance of the adjusted ranks within block $j, j = 1, 2, \ldots, b$. The Friedman statistic adjusted for ties is

$$\text{FM}_{\text{ties}} = \frac{b^2}{\sum\limits_{j=1}^{b} S_{Bj}^2} \sum_{i=1}^{k} \left(\overline{R}_i - \frac{k+1}{2} \right)^2$$

See Section 4.5.4 for an intuitive justification of this formula.

EXAMPLE 4.5.2   The data in Table 4.5.3 illustrate the assignment of tied ranks in a randomized complete block design with four treatments and three blocks. The sample variances of the ranks in blocks 1, 2, and 3 are 1.33, 1.00, and 1.67, respectively, with a sum of 4. The means of ranks of the treatments are 1.5, 1.83, 2.83, and 3.83, respectively. Thus,

$$\text{FM}_{\text{ties}} = \frac{9}{4} \left[(1.5 - 2.5)^2 + (1.83 - 2.5)^2 + (2.83 - 2.5)^2 + (3.83 - 2.5)^2 \right] = 7.5$$

The p-value from the chi-square distribution with 3 degrees of freedom is .0576. The exact p-value from StatXact is .0417.

TABLE **4.5.3**

Randomized Complete Block with Ties and Ranks

Treatment	Block 1	Block 2	Block 3
1	100 (1.5)	80 (2)	50 (1)
2	100 (1.5)	80 (2)	60 (2)
3	150 (3.5)	80 (2)	80 (3)
4	150 (3.5)	90 (4)	90 (4)

■

An Alternative Formula

An alternative expression for FM_{ties} is common. Within each block, group the observations that are tied. Let t_{ij} denote the number of tied observations in the ith group within the jth block. Let g_j denote the number of groups of tied observations within the jth block. The Friedman's statistic adjusted for ties is

$$FM_{ties} = \frac{FM}{1 - \sum_{j=1}^{b} \sum_{i=1}^{g_j} \frac{t_{ij}^3 - t_{ij}}{bk(k^2 - 1)}}$$

where FM is Friedman's statistic without ties.

EXAMPLE 4.5.3 Consider the data in Table 4.5.3. For block 1, we have two groups of ties with $t_{11} = 2$, and $t_{21} = 2$. For block 2, we have one group of ties with $t_{12} = 3$. Block 3 has no tied observations, although two of the observations match observations in block 2. The denominator of the Friedman statistic adjusted for ties is

$$1 - \frac{2(2^3 - 2) + (3^3 - 3)}{(3)(4)(4^2 - 1)} = 0.8$$

The Friedman statistic without ties is

$$FM = \frac{(12)(3)}{(4)(5)}\left[(1.5 - 2.5)^2 + (1.83 - 2.5)^2 + (2.83 - 2.5)^2 + (3.83 - 2.5)^2\right] = 6.00$$

Thus, the statistic adjusted for ties is

$$FM_{ties} = \frac{6.00}{0.8} = 7.5$$

which is the same result as in Example 4.5.2.

■

Use of Friedman's Test

Friedman's test is generally not as powerful as either the F-test or the permutation F-test for randomized complete block designs. If there are only two treatments, then Friedman's test is equivalent to a two-sided sign test, which generally has low power except when extreme observations tend to occur. When the observations satisfy the usual analysis of variance assumptions as discussed in Section 4.4.1, then the asymptotic relative efficiency of the test to the F-test is as low as 0.64 for the case of two treatments but reaches a limiting value of 0.955 as the number of treatments increases. Thus, Friedman's test is generally more effective for a larger number of treatments.

4.5.3 Cochran's Q and Kendall's W

Suppose we have an experiment in which the response to the treatment is either success (1) or failure (0). For instance, suppose three vaccines, A, B, and C, are being considered for preventing infection in dogs. The vaccine is judged to be successful if the dog develops the desired antibodies. If the dogs are put in blocks of three according to weight, then the experiment can be regarded as a randomized complete block design with 0–1 responses. The data may be analyzed using Friedman's test with ties. This special case is known as *Cochran's Q test.*

Another use of Friedman's test has to do with agreement of judges in such things as sporting events. Suppose five participants in a sport are rated by three judges, and we would like to test whether or not the judges agree. Friedman's test may be applied in this situation where the judges are the blocks and the rankings 1–5 of the participants are the responses. The null hypothesis of Friedman's test is that the judges do not agree, in the sense that the assignment of ranks to the participants could just as well have been determined randomly. A rejection of the null hypothesis indicates a level of agreement among the judges in the sense that the judges tend to rate the same individuals either low or high. A statistic equivalent to Friedman's test in this context is *Kendall's W*, which is a measure of *concordance* or agreement among the judges. See Conover (1999, p. 382).

4.5.4 Chi-Square Approximation for FM or FM_{ties}

Following along the lines set in Section 3.2.3, we find a constant C such that $E[C(\text{SSR})] = k - 1$. The statistic $C(\text{SSR})$ will have a permutation distribution that may be approximated by the chi-square distribution with $k - 1$ degrees of freedom.

Let σ_{Bj}^2 denote the population variance of the ranks in block j, $j = 1, 2, \ldots, k$. It can be shown that

$$E(\text{SSR}) = \sum_{i=1}^{k} E\left(\overline{R}_i - \frac{k+1}{2} \right)^2 = \frac{k}{b^2} \sum_{j=1}^{b} \sigma_{Bj}^2$$

Thus, the constant C such that $E[C(\text{SSR})] = k - 1$ is

$$C = \frac{b^2(k-1)}{k\sum\limits_{j=1}^{b} \sigma_{Bj}^2}$$

Since $\sigma_{Bj}^2 = k\sigma_{Bj}^2/(k-1) = S_{Bj}^2$, the statistic $C(\text{SSR})$ is given by FM_{ties}. In the special case in which the observations within blocks have no ties, we have $\sigma_{Bj}^2 = (k-1)(k+1)/12$. This gives us $C = 12b/k(k+1)$ and leads to the statistic FM.

4.5.5 Computer Analysis

StatXact, MINITAB, and S-Plus have menu options for carrying out Friedman's test. MINITAB and S-Plus use the chi-square approximation, whereas StatXact provides both exact p-values and p-values from the chi-square approximation. Figure 4.5.1 shows output of the analysis of the data in Table 4.5.3 using all three of these software packages.

The Resampling Stats code in Figure 4.4.1 may be modified for Friedman's test. Simply replace the original observations by the ranks. Note that $\text{SSX}^* = \Sigma_{i=1}^{k} \overline{R}_i^2$ if ranks are used in place of original observations. This statistic is equivalent to Friedman's statistic.

FIGURE 4.5.1

StatXact, MINITAB, and S-Plus Output for Analysis of Data in Table 4.5.3

```
StatXact

FRIEDMAN TEST
[ That 4 treatments have identical effects in 3 informative blocks]

Statistic based on the observed 4 by 3 Two-way layout (x) :
     FR(x)  : Friedman Statistic    =          7.500

Asymptotic p-value: (based on Chi-square distribution with 3 df )
     Pr { FR(X) .GE.      7.500 } =       0.0576

Exact p-value and point probability :
     Pr { FR(X) .GE.      7.500 } =       0.0417
     Pr { FR(X) .EQ.      7.500 } =       0.0417
```

(continued on next page)

FIGURE 4.5.1

StatXact, MINITAB, and S-Plus Output for Analysis of Data in Table 4.5.3 *(continued)*

```
MINITAB

Friedman test for obs by trt blocked by block

S = 6.00   DF = 3   P = 0.112
S = 7.50   DF = 3   P = 0.058 (adjusted for ties)

                       Est     Sum of
trt           N      Median    Ranks
1             3       68.75      4.5
2             3       73.75      5.5
3             3       86.25      8.5
4             3       96.25     11.5

Grand median    =      81.25

S-Plus

        Friedman rank sum test

data:  OBS and TRT and BLOCK from data set SDF6
Friedman chi-square = 7.5, df = 3, p-value = 0.0576
alternative hypothesis: two.sided
```

4.6
Ordered Alternatives for a Randomized Complete Block Design

Just as the Jonckheere–Terpstra test presented in Section 3.4.1 allows us to take advantage of ordered alternatives in a completely random design, it is possible to take advantage of ordered alternatives in a randomized complete block design.

We assume that the model

$$X_{ij} = \mu + t_i + b_j + \varepsilon_{ij}$$

holds as defined in Section 4.4.1. The ordering among the treatments may be expressed as either of these alternative hypotheses:

$$H_a: t_1 \leq t_2 \leq \cdots \leq t_k$$
$$H_a: t_1 \geq t_2 \geq \cdots \geq t_k$$

with strict inequality holding for at least two treatments. Intuitively, observations from treatment 1 tend to be smaller than those from treatment 2, and so on, or the other way around.

4.6.1 Page's Test

Let us assume that observations have been ranked as in Friedman's test. Let R_i denote the sum of the ranks for the ith treatment, $i = 1, \ldots, k$. *Page's statistic* is defined as

$$PG = \sum_{i=1}^{k} iR_i$$

We may obtain the permutation distribution of PG using the same permutation procedure as with Friedman's test. The only difference is that PG is computed for each permutation instead of FM. PG is a measure of the association between the presumed order of the treatments and the rank sum of the treatments. A large value of PG indicates that the treatment responses tend to increase as the index of the treatments increases, whereas a small value indicates that the treatment responses tend to decrease as the index of the treatments increases.

For a large-sample approximation, we need the mean and variance of the permutation distribution of PG under the assumption that $t_1 = t_2 = \cdots = t_k$. These are given by

$$E(PG) = \frac{bk(k+1)^2}{4}$$

$$\text{var}(PG) = \frac{(k-1)k(k+1)}{12} \sum_{j=1}^{b} S_{Bj}^2$$

where S_{Bj}^2 is the sample variance of the ranks or adjusted ranks in the jth block, $j = 1, \ldots, b$. If there are no ties within blocks, we have $S_{Bj}^2 = k(k+1)/12$, and the variance for no ties is

$$\text{var}(PG) = \frac{b(k-1)k^2(k+1)^2}{144}$$

The statistic

$$Z = \frac{PG - E(PG)}{\sqrt{\text{var}(PG)}}$$

may be referred to the standard normal distribution to determine statistical significance. See Hollander and Wolfe (1999) for a derivation of the mean and variance of PG with no ties. The derivation with ties is similar to this.

EXAMPLE 4.6.1   Suppose the data are yields under the four treatments in Table 4.5.2. Assume we anticipate that yields will decrease as we go from treatment 1 to 4. The rank sums for the four treatments are 21, 15, 13, and 11, respectively. Page's statistic for these data is

$$PG = 21 + 2(15) + 3(13) + 4(11) = 134$$

The exact *p*-value for a lower-tail test is .0128 from StatXact. We find $E(PG)$ = 6(4)(25)/4 = 150, and since there are no ties within blocks, var(PG) = 6(3)(16)(25)/144 = 50. Thus, $Z = -2.263$ with a *p*-value of .0118. Either the exact test or the approximate test shows statistical significance at the traditional 5% level. By contrast, the *p*-value using Friedman's test is .1268. In general, if the presumption of an ordering among treatments is correct, Page's test will have greater power to detect differences than Friedman's test. ∎

4.6.2 Computer Analysis

Page's test is implemented in StatXact. Figure 4.6.1 shows output for the analysis of the data in Table 4.5.2. Resampling Stats code for Page's test is a simple modification of the code for randomized complete block designs. Details are shown in Figure 4.6.2.

FIGURE 4.6.1

Page's Test for Data in Table 4.5.2 Using StatXact

```
PAGE TEST
[ That 4 treatments have identical effects in each of 6 blocks]

Statistic based on the observed 4 by 6 two-way layout(x) :

          Mean       Std-dev  Observed(PA(x))  Standardized(PA*(x))
          150.0        7.071       134.0               -2.263

Asymptotic p-value:
    One-sided: Pr { PA*(X) .LE.      -2.263 }  =         0.0118
    Two-sided: 2 * One-sided                   =         0.0237

Exact p-value:
    One-sided: Pr {  PA*(X)   .LE.    -2.263 } =         0.0128
               Pr {  PA*(X)   .EQ.    -2.263 } =         0.0044
    Two-sided: Pr { |PA*(X)|  .GE.     2.263 } =         0.0256
```

FIGURE 4.6.2

Resampling Stats Code for the Analysis of Data in Table 4.5.2 Using Page's Test

```
maxsize default 5000
'enter ranks by blocks
copy (1 4 2 3) b1
copy (4 3 2 1) b2
copy (4 3 2 1) b3
```

(continued on next page)

FIGURE 4.6.2

Resampling Stats Code for the Analysis of Data in Table 4.5.2 Using Page's Test *(continued)*

```
copy (4 2 3 1) b4
copy (4 1 3 2) b5
copy (4 2 1 3) b6

'observed pg = 134

'get permutation distribution
repeat 5000

'permute ranks within blocks
shuffle b1 sb1
shuffle b2 sb2
shuffle b3 sb3
shuffle b4 sb4
shuffle b5 sb5
shuffle b6 sb6

'get rank sums for treatments
add sb1 sb2 sb3 sb4 sb5 sb6 ranksum
'compute page's statistic
multiply (1 2 3 4) ranksum wtsum
sum wtsum pg

'keep track of pg
score pg pgdist
end

'get lower-tail p-value for pg
count pgdist <= 134 pval
divide pval 5000 pval
print pval

'print results here

PVAL    =      0.0126
```

Exercises

1 For the data in the table, obtain the permutation distribution of the mean of the differences.

Pair	1	2	3	4
Treatment 1	100	250	50	80
Treatment 2	112	240	58	82

2   Students in an introductory statistics course were asked at the beginning of the course and at the end of the course the extent to which they disagreed or agreed with the statement: "Statistics is important to my major area of study." They responded on a scale of 1 to 5 with (1) strongly disagree, (2) moderately disagree, (3) neutral, (4) moderately agree, and (5) strongly agree. Data are shown in the table. Use a paired-comparison permutation test to determine whether responses to this question changed significantly from the beginning to the end of the semester.

Student	1	2	3	4	5	6	7	8	9	10	11	12	13	14	15	16	17	18	19
Before	2	3	4	4	3	1	3	4	4	5	3	4	2	2	4	3	4	2	2
After	2	4	4	4	4	4	3	5	4	4	4	5	4	2	5	5	4	1	2

3   Obtain the permutation distribution of the signed-rank statistic SR_+ for the data in Exercise 1.

4   Measurements of a blood enzyme LDH were taken on seven subjects before fasting and after fasting. Is there a significant difference between the LDH readings before and after fasting? Test using Wilcoxon's signed-rank statistic. Compare the p-value of the exact test with that of the normal approximation.

Subject	1	2	3	4	5	6	7
Before	89	90	87	98	120	85	97
After	76	101	84	86	105	84	93

5   Apply the Wilcoxon signed-rank test adjusted for ties to the data in Exercise 2.

6   Apply the sign test to the data in Exercises 2 and 4.

7   The data in the table are dry matter contents (in kilograms) of hay obtained from experimental plots. The experiment was designed as a randomized complete block, and the treatments are three cutting dates.

Date	Blocks					
	1	*2*	*3*	*4*	*5*	*6*
Sept. 1	1.5	2.1	1.9	2.8	1.4	1.8
Sept. 15	1.8	2.0	2.0	2.7	1.6	2.3
Sept. 30	1.9	2.5	2.5	2.6	2.1	2.4

a   Test for differences among cutting dates using a permutation test for a randomized complete block design.

b   Analyze using ANOVA for a randomized complete block and compare with part a.

8   Obtain the 90th and 95th percentiles of the permutation distribution of $\max_{ij}\left|\overline{X}_{i.} - \overline{X}_{j.}\right|$ (Section 4.4), and determine whether or not there are significant differences between the possible pairs of sampling dates for the data in Exercise 7.

9   Apply Friedman's test to the data in Exercise 7.

10   Four judges ranked each of six contestants in a diving contest. Use Friedman's test to test for agreement among the judges' rankings.

		Ranks of Contestants				
Contestant	A	B	C	D	E	F
Judge 1	4	3	1	2	6	5
Judge 2	5	2	3	1	6	4
Judge 3	2	5	1	3	4	6
Judge 4	3	4	2	1	5	6

11   The pH levels of beef carcasses should decrease over time. The data in the table are the pH levels of six beef carcasses. Use Page's test to see whether pH levels decline over time.

Carcass	Time 1	Time 2	Time 3	Time 4	Time 5	Time 6
1	6.81	6.16	5.92	5.86	5.80	5.39
2	6.68	6.30	6.12	5.71	6.09	5.28
3	6.34	6.22	5.90	5.38	5.20	5.46
4	6.68	5.24	5.83	5.49	5.37	5.43
5	6.79	6.28	6.23	5.85	5.56	5.38
6	6.85	6.51	5.95	6.06	6.31	5.39

12   Three varieties of soybeans were tested for yields in a randomized complete block design. The varieties are known to have varying levels of susceptibility to iron deficiency. The data are average chlorosis scores based on a scale of 1 to 9, with 1 being the best plant condition and 9 the worst. Use an appropriate procedure to test for differences among the varieties.

Variety	Block 1	Block 2	Block 3	Block 4
Susceptible	4.5	6.3	5.7	5.3
Moderately resistant	3.5	4.3	3.3	5.0
Resistant	1.0	1.3	1.0	1.7

Theory and Complements

13   This exercise is to be done in conjunction with Exercise 14. Assume there are no ties among the differences in a signed-rank test. Put the absolute values of the differences in increasing order, and let $R_{(i)}$ denote the ith signed rank, where $|R_{(i)}| = i$. Consider all sums of the form $R_{(i)} + R_{(j)}$, where $i \le j$. Show that SR_+ equals the number of such sums that are positive. For example, suppose the signed ranks are 1, –2, 3, –4, and 5. The number of sums such that $R_{(i)} + R_{(5)}$ is positive for $i \leq 5$ is 5. The number of sums such that $R_{(i)} + R_{(4)}$ is positive for $i \leq 4$ is 0, and so on, so the number of sums such that $R_{(i)} + R_{(j)}$ is positive for $i \leq j$ is $5 + 3 + 1 = SR_+$. Obtain a similar expression for SR_-.

14   Let $d_{(i)}$ denote the ith difference in a paired-comparison experiment. Assume $0 < |d_{(1)}| < |d_{(2)}| < \cdots < |d_{(n)}|$. Assume the distribution function of the $d_{(i)}$'s is $G(x) = F(x - \Delta)$, where Δ is the median and $F(x)$ is symmetric about 0. A Hodges–Lehmann estimate of Δ is the median of pairs of the form

$$\frac{d_i + d_j}{2}, \quad i \leq j$$

Our interest is in making a confidence interval for Δ. The procedure is as follows:

Compute all pairs $(d_{(i)} + d_{(j)})/2$, $i \leq j$, and let $A_{(1)} < A_{(2)} < \cdots < A_{(M)}$ denote these values placed in order from smallest to largest, where $M = n(n + 1)/2$. The confidence interval will be $A_{(L)} \leq \Delta < A_{(U)}$ for appropriate choices of L and U. For the upper limit, we note that $\Delta < A_{(U)}$ if and only if at most $U - 1$ terms $(d_{(i)} + d_{(j)})/2$ are less than or equal to Δ, or equivalently at least $M - U + 1$ terms are greater than Δ. Now $(d_{(i)} + d_{(j)})/2 > \Delta$ if and only if $(d_{(i)} - \Delta) + (d_{(j)} - \Delta) > 0$. The probability that at least $M - U + 1$ terms of the form $(d_{(i)} - \Delta) + (d_{(j)} - \Delta)$ are greater than 0 is the same as the probability that $SR_+ \geq M - U + 1$, from Exercise 13. (Note that the distribution of $d_{(i)} - \Delta$ is the same as the null distribution of no treatment effect in the paired comparison experiment.) Thus, for a 95% confidence interval, choose U such that $.975 = P(\Delta < A_{(U)}) = P(SR_+ \geq M - U + 1)$. Similarly, choose L so that $.025 = P(\Delta < A_{(L)}) = P(SR_+ \geq M - L + 1)$. This procedure is implemented in StatXact. The procedure may also be used to make a confidence interval for the median of a symmetric distribution where the d_i's are replaced by the sample values x_i's selected from the population.

5

Tests for Trends and Association

A Look Ahead Studying the relationship between two variables is fundamental in statistics. For instance, we may be interested in measuring the strength of the relationship between quantitative variables, such as high school grade point average and college grade point average. Nonparametric correlation and regression methods may be used to analyze such data. We may also be interested in the relationship between variables that are qualitative, such as eye color and hair color. This type of data may be analyzed using methods for contingency tables. In Sections 5.1, 5.2, and 5.3, we consider nonparametric approaches to testing for correlation between two variables. In Sections 5.4–5.8, we consider the problem of testing for association in contingency tables.

5.1
A Permutation Test for Correlation and Slope

Suppose we have pairs of observations (X_i, Y_i), $i = 1, \ldots, n$, where the data may arise in one of two ways. (1) In *bivariate sampling*, the pairs are selected at random from a bivariate population of such pairs. For instance, we might measure the height X and weight Y of each of n individuals selected at random from a population of individuals. (2) In *fixed-X sampling*, the pairs are obtained in an experiment where the values of X are fixed by the experimenter and one or more values of Y are obtained for each X. An example is placing a steel beam under a predetermined load X and measuring its deflection Y. We use both the correlation coefficient and the slope of the least squares line as measures of the relationship between X and Y.

5.1.1 The Correlation Coefficient

When pairs (X, Y) are selected at random from a population of such pairs, it is common to consider the correlation coefficient as a measure of the strength of the linear relationship between X and Y. The population correlation coefficient is defined as

$$\rho = \frac{E\left[(X - \mu_x)(Y - \mu_y)\right]}{\sigma_x \sigma_y}$$

where μ_x and μ_y are the expected values of the X's and Y's, respectively, and σ_x and σ_y are their respective standard deviations. The value of ρ ranges between -1 and 1. If ρ is at the extremes, then the linear relationship between X and Y is perfect in the sense that there are constants a and b such that $Y = a + bX$. If $\rho = 0$, then there is no linear relationship between X and Y, and we say that X and Y are *uncorrelated.*

The sample correlation coefficient or the *Pearson product-moment correlation*, which is an estimate of ρ, is

$$r = \frac{\sum_{i=1}^{n}\left(X_i - \overline{X}\right)\left(Y_i - \overline{Y}\right)}{\sqrt{\sum_{i=1}^{n}\left(X_i - \overline{X}\right)^2 \sum_{i=1}^{n}\left(Y_i - \overline{Y}\right)^2}}$$

where $\overline{X}$ and $\overline{Y}$ are the sample means of the X's and Y's, respectively. We use r to test the null hypothesis

$$H_0: \rho = 0$$

against one-sided or two-sided alternative hypotheses. If the (X, Y) pairs can be assumed to be random samples from a bivariate normal population, then the statistic

$$t_{\text{corr}} = \sqrt{\frac{n-2}{1-r^2}}\, r$$

has a t-distribution with $n - 2$ degrees of freedom under H_0, and this distribution is used as the reference distribution for rejecting or not rejecting the null hypothesis.

5.1.2 Slope of Least Squares Line

The *linear regression model* is defined as

$$Y_i = \beta_0 + \beta_1 X_i + \varepsilon_i$$

where the ε's are independent and identically distributed random variables with mean 0 and finite variance. In fixed-X sampling, the model describes how we believe Y_i behaves for the fixed value of X_i. In bivariate sampling, it is a statement about the conditional distribution of Y_i given X_i. That is, the model describes how we believe Y_i would behave if we could fix X_i.

The least squares estimates of the intercept $\hat{\beta}_0$ and the slope $\hat{\beta}_1$ are the values that minimize the sum of squares

$$\text{SSE} = \sum_{i=1}^{n}\left(Y_i - \hat{\beta}_0 - \hat{\beta}_1 X_i\right)^2$$

These values are

$$\hat{\beta}_1 = \frac{\sum_{i=1}^{n}\left(X_i - \overline{X}\right)\left(Y_i - \overline{Y}\right)}{\sum_{i=1}^{n}\left(X_i - \overline{X}\right)^2}$$

$$\hat{\beta}_0 = \overline{Y} - \hat{\beta}_1\overline{X}$$

To test whether or not there is a significant relationship between X and Y, we test the null hypothesis

$$H_0: \beta_1 = 0$$

against one-sided or two-sided alternative hypotheses. If the ε's are normally distributed, the test statistic is

$$t_{\text{slope}} = \sqrt{\frac{\sum_{i=1}^{n}\left(X_i - \overline{X}\right)^2}{\text{MSE}}}\,\hat{\beta}_1$$

where MSE = SSE/$(n-2)$. This statistic has a t-distribution with $n-2$ degrees of freedom when $\beta_1 = 0$.

For bivariate sampling in which the conditional distribution of Y given X satisfies the linear regression model, it can be shown that the slope and correlation are related by the equation

$$\beta_1 = \rho\frac{\sigma_Y}{\sigma_X}$$

The slope of the least squares line and the Pearson correlation are related by the equation

$$\hat{\beta}_1 = r\frac{S_Y}{S_X}$$

where S_Y and S_X are the sample standard deviations of the Y's and X's, respectively. Moreover, it can be shown that $t_{\text{corr}} = t_{\text{slope}}$; that is, the test statistic for correlation is equivalent to the test statistic for slope.

5.1.3 The Permutation Test

Suppose the linear regression model holds. If $\beta_1 = 0$ or if $\rho = 0$, then X does not affect the value of Y. Thus, an observed value of Y_i is as likely to appear with X_j as it is to appear with X_i for any value of $j \neq i$. To illustrate, suppose we have four pairs of observations (1, 5), (2, 7), (3, 9), and (4, 8). There are 4! ways in which the Y's may appear with the X's, and all are equally likely to occur when $\beta_1 = 0$ or if $\rho = 0$. Table 5.1.1 lists all 24 ways of assigning the four Y's to the four X's along with the slopes and correlations. The slope and correlation of the original data are 1.1 and .63, respectively. We see that three slopes are greater than or equal to 1.1, and three correlations are greater than or equal to .83, so $p = 3/24 = .125$.

TABLE 5.1.1

Permutation Distribution of Slopes and Correlation

$X_1 = 1$ Y_1	$X_2 = 2$ Y_2	$X_3 = 3$ Y_3	$X_4 = 4$ Y_4	Slope	Correlation
5	7	8	9	1.3	.98
5	7	9	8	1.1	.83
5	8	7	9	1.1	.83
5	8	9	7	0.7	.53
5	9	7	8	0.7	.53
5	9	8	7	0.5	.38
7	5	8	9	0.9	.68
7	5	9	8	0.7	.53
7	8	5	9	0.3	.23
7	8	9	5	−0.5	−.38
7	9	5	8	−0.1	−.08
7	9	8	5	−0.7	−.53
8	5	7	9	0.5	.38
8	5	9	7	0.1	.08
8	7	5	9	0.1	.08
8	7	9	5	−0.7	−.53
8	9	5	7	−0.7	−.53
8	9	7	5	−1.1	−.83
9	5	7	8	−0.1	−.08
9	5	8	7	−0.3	−.23
9	7	5	8	−0.5	−.38
9	7	8	5	−1.1	−.83
9	8	5	7	−0.9	−.68
9	8	7	5	−1.3	−.98

Steps for a Permutation Test for Slope or Correlation

The steps involved in carrying out the permutation test for the slope when given n pairs of data (X_i, Y_i), $i = 1, \ldots, n$, are given next. The steps for the test of correlation are the same except that r is computed instead of the slope.

1. Compute the slope of the least squares line $\hat{\beta}_{1,\text{obs}}$ from the original data.
2. Permute the Y's among the X's in the $n!$ possible ways.
3. For each permutation, compute the slope $\hat{\beta}_1$ of the least squares line.
4. For an upper-tail test, the p-value is

$$P_{\text{upper tail}} = \frac{\text{number of } \hat{\beta}_1\text{'s} \geq \hat{\beta}_{1,\text{obs}}}{n!}$$

Similarly, we may obtain lower-tail and two-tail p-values.

If it is not possible to list all $n!$ assignments of the Y's to the X's, then we may randomly sample the permutations. This is done by randomly shuffling the Y's and matching the shuffled values with $X_1, X_2, \ldots, X_n$. For each permutation, we obtain the slope or correlation and repeat this procedure a desired number of times. The upper-tail p-value is the fraction of the randomly sampled slopes that are greater than or equal to the observed slope, and similarly for the lower-tail and two-tail p-values.

EXAMPLE 5.1.1   A researcher took blood samples from 18 healthy rabbits and made counts of the heterophils (X) and lymphocytes (Y). A plot of the data in Figure 5.1.1 shows a positive correlation between X and Y. The data are listed in Table 5.1.2.

The values of the slope and the correlation coefficient for the original data are $\hat{\beta}_1 = 1.27$ and $r = .86$, respectively. From the StatXact printout in Figure 5.1.2, no values of $\hat{\beta}_1$ and r in a random sample of 10,000 permutations exceeded these values. Thus, the p-value based on this sample of permutations is less than .0001. The

FIGURE 5.1.1
Plot of Heterophils versus Lymphocytes

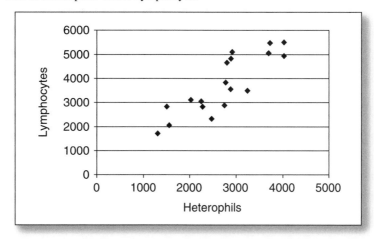

TABLE 5.1.2

Heterophils and Lymphocytes of 18 Healthy Rabbits

Rabbit	Heterophils	Lymphocytes
1	2276	2830
2	3724	5488
3	2723	2904
4	4020	5528
5	4011	4966
6	2035	3135
7	1540	2079
8	1300	1755
9	2240	3080
10	2467	2363
11	3700	5087
12	1501	2821
13	2907	5130
14	2898	4830
15	2783	4690
16	2870	3570
17	3263	3480
18	2780	3823

FIGURE 5.1.2

StatXact Output for Permutation Test of Correlation Applied to Data in Table 5.1.2

```
PEARSON'S CORRELATION TEST
Correlation Coefficient estimates based on 18 observations.

   Coefficient        Estimate         ASE1     95.00% Confidence Interval
   -----------        --------         ----     --------------------------
   Pearson's R         0.8572         0.05351   (     0.7523,      0.9621)

Asymptotic p-values (for testing no association t-distribution with 16 df):
        One-sided: Pr { Statistic .GE. Observed   } =       0.0000
        Two-sided: 2 * One-sided                   =       0.0000

Monte Carlo estimate of p-value with 99.00% Confidence Interval:
     Pr { Statistic  .GE.  Observed  } =      0.0000 (      0.0000,      0.0005)
     Pr {|Statistic| .GE. |Observed| } =      0.0000 (      0.0000,      0.0005)

Elapsed time is 0:0:0.50 (10000 tables sampled with starting seed 31190)
```

t statistic for $\hat{\beta}_1$ and r is 6.66, which has a p-value of less than .0001 using the t-distribution with 16 degrees of freedom as reference. From the work of Pitman (1937b) and Hoeffding (1952), we can expect under appropriate conditions and for sufficiently large samples to obtain the same conclusion concerning the significance of the correlation using either the permutation test or the t-test. ■

Alternative Statistics for the Correlation and Slopes

We note that the denominator of the expression for $\hat{\beta}_1$ does not vary with permutations of the Y's. Moreover, the numerator may be expressed as

$$\sum_{i=1}^{n} X_i Y_i - n\overline{X}\,\overline{Y}$$

and only the term

$$S_{XY} = \sum_{i=1}^{n} X_i Y_i$$

varies with the permutations of the Y's among the X's. Thus, a test statistic for correlation may be based on S_{XY}. Page's statistic presented in Section 4.6.1 is a special case of S_{XY} where the X_i's are the integers 1, 2, . . . , n and the Y_i's are the sums of ranks for the ith treatment.

5.1.4 Large-Sample Approximation for the Permutation Distribution of r

To use a normal approximation for the permutation distribution of r, we need the mean and variance of the permutation distribution. Under the assumption of no relationship between X and Y, the mean of the permutation distribution of r is 0 and the variance is given by

$$\mathrm{var}(r) = \frac{1}{n-1}$$

assuming that the variances of the X's and the variance of the Y's are not 0. It is interesting that this does not depend on the particular X's or Y's in the data. The distribution of $Z = r\sqrt{n-1}$ is approximately standard normal.

EXAMPLE **5.1.1** Using the Resampling Stats code in Figure 5.1.3, we obtained selected percentiles of the permutation distribution of r for the data in Table 5.1.2. We converted the percentiles to approximate standard normal percentiles using the relationship $Z = r\sqrt{n-1}$, where $n = 18$. The correspondence to standard normal percentiles is quite good, as shown in Table 5.1.3.

TABLE 5.1.3

Comparison of Percentiles of the Permutation Distribution of $Z = r\sqrt{n-1}$
with Standard Normal Percentiles for Data in Table 5.1.2

Percentile	90	95	97.5	99
r	.31	.39	.47	.54
$Z = r\sqrt{17}$	1.28	1.61	1.94	2.23
Standard Normal	1.28	1.65	1.96	2.33

Derivation of Variance of r

To derive the variance of r, we write the correlation as

$$r = \frac{\sum\limits_{i=1}^{n}(X_i - \bar{X})Y_i}{n\sigma_X\sigma_Y}$$

where σ_X and σ_Y are the population variances of $X_1, \ldots, X_n$ and $Y_1, \ldots, Y_n$, respectively. The variances are constant when the Y's are permuted among the X's. We may regard the X's as being fixed and the Y's as being randomly selected without replacement from $Y_1, \ldots, Y_n$ and assigned to the X's. Thus,

$$\text{var}\left(\sum_{i=1}^{n}(X_i - \bar{X})Y_i\right) = \sum_{i=1}^{n}(X_i - \bar{X})^2\sigma_Y^2 + \sum_{i\neq j}(X_i - \bar{X})(X_j - \bar{X})\,\text{cov}(Y_i, Y_j)$$

Since $\text{cov}(Y_i, Y_j) = -\sigma_Y^2/(n-1)$, and since

$$\sum_{i\neq j}(X_i - \bar{X})(X_j - \bar{X}) = -n\sigma_X^2$$

it follows that

$$\text{var}\left(\sum_{i=1}^{n}(X_i - \bar{X})Y_i\right) = n\sigma_X^2\sigma_Y^2 + \frac{n\sigma_X^2\sigma_Y^2}{n-1} = \frac{n^2\sigma_X^2\sigma_Y^2}{n-1}$$

Therefore, $\text{var}(r) = 1/(n-1)$.

5.1.5 Computer Analysis

StatXact implements the permutation test for the Pearson correlation under "Measures of Association (Ordinal)." Resampling Stats code for obtaining the p-value and selected percentiles of the permutation distribution of r is shown in Figure 5.1.3. Here we have included code to show how to read data from an external text file.

FIGURE 5.1.3

Resampling Stats Code for Permutation Test of Correlation Applied to Data in Table 5.1.2

```
maxsize default 5000

'read data from .txt file, first column = x, second column = y
read file "rabbit.txt" x y

'correlation of original data is .86

'permute data

repeat 5000
shuffle y sy
corr x sy r

'keep track of correlations
score r rdist
end

'get p-value and selected percentiles and print results
count rdist >= .86 pval
divide pval 5000 pval
percentile rdist (90 95 97.5 99) pctile
print pval
print pctile

'print results here

Line 4: 18 records (0 missing values) read from rabbit.txt

PVAL    =         0

PCTILE  =     0.31011   0.3941   0.46833   0.54445
```

5.2
Spearman Rank Correlation

Consider the (X, Y) pairs $(0, 0)$, $(2, 16)$, $(3, 81)$, $(4, 256)$, and $(6, 1296)$. The data are not linear, so the Pearson correlation between X and Y is not 1. In fact, it is .84. However, in some sense there is a perfect association between the X's and Y's because Y is functionally related to X by the equation $Y = X^4$. We may measure the extent to which Y increases, or decreases, with X by ranking the X's and Y's and then computing the correlation between the ranks. In this case, the ranks of the Y's are 1, 2, . . . , 5, aligning perfectly with the ranks of the X's, so the correlation of the ranks is 1.

Suppose we have pairs of observations (X_i, Y_i), $i = 1, \ldots, n$. Let $R(X_i)$ denote the rank of X_i among the X's and $R(Y_i)$ denote the rank of Y_i among the Y's. Average ranks are assigned to tied observations among the X's, and similarly for the Y's. The *Spearman rank correlation*, denoted r_s, is obtained by applying the formula for the Pearson correlation coefficient to the pairs $(R(X_i), R(Y_i))$, $i = 1, \ldots, n$. That is, the Spearman rank correlation is the correlation between the ranks of the X's and the ranks of the Y's.

5.2.1 Statistical Test for Spearman Rank Correlation

The statistical significance of the Spearman rank correlation is determined by applying the permutation test for correlation presented in Section 5.1.3 to the ranked pairs. The null hypothesis is that there is *no association* between the X's and the Y's in the sense that all possible assignments of the $R(Y_i)$'s to the $R(X_i)$'s are equally likely to occur. For a *positive association*, larger values of the $R(Y_i)$'s tend to occur with larger values of the $R(X_i)$'s, and for a *negative association*, larger values of the $R(Y_i)$'s tend to occur with smaller values of the $R(X_i)$'s.

Table A12 in the Appendix contains values of $P(r_s \geq c)$ for $c > 0$, and $n = 4, \ldots,$ 10. We may use this table to determine p-values for the upper-tail test of positive association between the X's and Y's. For a lower-tail test, p-values may be determined from the relationship $P(r_s \leq -c) = P(r_s \geq c)$. For two-sided p-values, double the one-sided values.

5.2.2 Large-Sample Approximation

Under the null hypothesis of no association between the X's and Y's, the expected value of r_s is 0, and from the results of Section 5.1.4, the variance is

$$\text{var}(r_s) = \frac{1}{n-1}$$

For large samples, the statistic

$$Z = \frac{r_s}{\sqrt{\text{var}(r_s)}} = r_s\sqrt{n-1}$$

has an approximate normal distribution, and this may be used as the reference distribution in testing hypotheses about r_s.

EXAMPLE 5.2.1   Table 5.2.1 lists the ranks of the heterophil and lymphocyte data in Table 5.1.2. We find a Spearman rank correlation of $r_s = .897$. The value of Z for these data is

$.897\sqrt{17} = 3.70$. The two-sided p-value for $Z = 3.70$ is .0002, showing a statistically significant association between the two variables. In 10,000 randomly selected permutations using StatXact, none gave a value of $r_s \geq .897$, so the estimated p-value is less than .0001.

TABLE 5.2.1

Ranks of Heterophils and Lymphocytes for Data in Table 5.1.2

Rabbit	Ranks of Heterophils	Ranks of Lymphocytes
1	6	5
2	16	17
3	8	6
4	18	18
5	17	14
6	4	8
7	3	2
8	1	1
9	5	7
10	7	3
11	15	15
12	2	4
13	13	16
14	12	13
15	10	12
16	11	10
17	14	9
18	9	11

Adjustment for Ties

Average ranks are assigned to ties among the X's, and average ranks are assigned to ties among the Y's. We consider two equivalent versions of the Spearman rank correlation adjusted for ties. The first is simply to apply the Pearson correlation to the ranks adjusted for ties. An exact test is obtained by doing the permutation test in Section 5.1.3 on these ranks. For a normal approximation, we use a mean of 0 and a variance of $1/(n - 1)$ and compute the statistic $Z = r_s \sqrt{n-1}$, which is referred to the standard normal distribution. This is the method we recommend for ties.

An alternative formula for the Spearman rank correlation with ties is a variant of an alternative formula without ties. It is complicated, but it appears in the literature. In the case of no ties, we may compute r_s using the alternative formula

$$r_s = 1 - \frac{6D}{n(n^2 - 1)}$$

$$D = \sum_{i=1}^{n} \left[R(X_i) - R(Y_i) \right]^2$$

See Exercise 13 at the end of the chapter.

Now let D denote the same formula applied to ranks adjusted for ties. Let s_i denote the number of observations in the ith group of tied observations among the X's, and let t_i denote the number of observations in the ith group of tied observations among the Y's. The Spearman correlation is given by

$$r_s = \frac{1 - 6D/n(n^2 - 1) - C_1}{C_2}$$

where

$$C_1 = \frac{\sum \left(s_i^3 - s_i \right) + \sum \left(t_i^3 - t_i \right)}{2n(n^2 - 1)}$$

and

$$C_2 = \sqrt{\left[1 - \frac{\sum \left(s_i^3 - s_i \right)}{n(n^2 - 1)} \right]\left[1 - \frac{\sum \left(t_i^3 - t_i \right)}{n(n^2 - 1)} \right]}$$

After r_s is computed this way, we may use the statistic $Z = r_s \sqrt{n-1}$ and refer it to the standard normal distribution to determine statistical significance.

EXAMPLE 5.2.2   Consider a vector of ranks of X's given by (1, 2, 3, 4) and a vector of ranks of Y's given by (1, 2.5, 2.5, 4), which has a tie between the second and third Y. Now the Pearson correlation between these two sets of ranks is .9487. To use the alternative formula, we have $D = 0^2 + 0.5^2 + 0.5^2 + 0^2 = 0.5$ and $t_1 = 2$. Thus, $C_1 = 6/120 = 0.05$, $C_2 = \sqrt{0.9}$, and

$$r_s = \frac{1 - 6(0.5)/4(4^2 - 1) - 0.05}{\sqrt{0.9}} = \sqrt{0.9} = .9487 \qquad \blacksquare$$

EXAMPLE 5.2.3   Two judges each rated the projects of ten contestants at a science fair. The data are given in Table 5.2.2. The Pearson correlation between the ranks of the X's and the ranks of the Y's is $r_s = .375$. Using the large sample approximation, we find $Z = .375\sqrt{9} = 1.125$. The two sided p-value is .26. The exact p-value from StatXact is .2832. Thus, the association between the scores of the judges is not statistically significant.

TABLE **5.2.2**

Scores of Ten Projects at a Science Fair

	Project									
	1	*2*	*3*	*4*	*5*	*6*	*7*	*8*	*9*	*10*
Judge A (*X*)	8	8	7	8	5	6	6	9	8	7
Ranks of *X*'s	7.5	7.5	4.5	7.5	1	2.5	2.5	10	7.5	4.5
Judge B (*Y*)	7	8	8	5	6	4	5	8	6	9
Ranks of *Y*'s	6	8	8	2.5	4.5	1	2.5	8	4.5	10

Now consider the alternative formula for computing r_s. For *X*'s, there are three groups of tied observations: two 6's, two 7's, and four 8's. Thus, $\Sigma(s_i^3 - s_i) = 72$. The *Y*'s also have three groups of tied observations: two 5's, two 6's, and three 8's. Thus, $\Sigma(t_i^3 - t_i) = 36$. It follows that $C_1 = 0.0545$ and $C_2 = 0.945$. We find $D = (7.5 - 6)^2 + \cdots + (4.5 - 10)^2 = 97.5$. Thus,

$$r_s = \frac{1 - 6(97.5)/990 - 0.0545}{0.945} = 0.375$$

There is no apparent advantage to using the alternative formula for the Spearman correlation, but it does serve to remind us that if r_s is expressed in terms of *D*, then this expression must be adjusted for ties. ∎

Caution in Using the Spearman Correlation

There are instances in which one is interested in testing whether or not a trend over time is statistically significant. For instance, we may wonder whether or not ocean temperatures are increasing over time at a certain location. If we can postulate a model of the form

$$Y_i = \beta_0 + \beta_1 X_i + \varepsilon_i$$

where the ε's are independent and identically distributed random variables with mean 0, then it would be acceptable to test for the significance of the Spearman correlation as we did above. However, if we are dealing with time-dependent data, the assumption of independence of the ε's may not be tenable, and this could invalidate the use of the Spearman statistic and the permutation statistic for correlation.

For instance, an unusually high deviation in ocean temperature above or below the median in one year may carry over to the next year irrespective of any long-term trends as indicated by a nonzero β_1. In particular, if the ε for one year has a strong positive correlation with ε's of the previous year, the Type I error rate of the Spearman correlation will be significantly inflated, leading one to conclude falsely that $\beta_1 \neq 0$. This case is discussed by Keller-McNulty and McNulty (1987).

5.2.3 Computer Analysis

StatXact provides an exact test for the Spearman correlation. The asymptotic approximation in StatXact is based on t_{corr} as defined in Section 5.1.1. MINITAB requires that the observations be ranked so that Pearson's correlation may be applied to the ranks. It also uses t_{corr} in the computation of *p*-values. Resampling Stats code for testing the Spearman correlation is a modification of the code for Pearson's correlation as illustrated in Figure 5.1.3. The only difference is that ranks replace the original observations.

5.3
Kendall's Tau

Suppose we have the heights and weights of two individuals—say, 68 in. and 150 lb. for individual 1, and 72 in. and 165 lb. for individual 2. Now we note that as the weight increases in going from individual 1 to individual 2, the height also increases. We say that the pairs (68, 150) and (72, 165) are concordant, since the *X*'s and *Y*'s change in the same direction. In contrast, if the weights and heights were (68, 160) and (72, 155), then these pairs would be discordant, since the heights and weights change in opposite directions. Because generally weights tend to increase as heights increase, we would expect to see more concordant pairs than discordant pairs among a sample of individuals. This idea leads us to define a measure of association between two variables based on counts of concordant and discordant pairs.

Let (X_i, Y_i), $i = 1, \ldots, n$, denote *n* pairs of observations that have been selected randomly from a continuous bivariate population. Since the population is continuous, the probability that there are ties among the *X*'s or ties among the *Y*'s is 0. The pairs (X_i, Y_i) and (X_j, Y_j) are said to be *concordant* if $X_i < X_j$ implies $Y_i < Y_j$. The pairs are said to be *discordant* if $X_i < X_j$ implies $Y_i > Y_j$. Equivalently, the pairs (X_i, Y_i) and (X_j, Y_j) are concordant if $(X_i - X_j)(Y_i - Y_j) > 0$ and discordant if $(X_i - X_j)(Y_i - Y_j) < 0$. Thus, for a concordant pair, the changes in the *X*'s and *Y*'s are in the same direction (either both increasing or both decreasing), and for a discordant pair, the changes are in opposite directions. We say there is a *positive association* between the *X*'s and *Y*'s if pairs are more likely to be concordant than discordant, and there is a *negative association* if pairs are more likely to be discordant than concordant. If pairs are just as likely to be concordant as discordant, then there is *no association* between the *X*'s and *Y*'s.

We define *Kendall's tau* as

$$\tau = 2P\left[\left(X_i - X_j\right)\left(Y_i - Y_j\right) > 0\right] - 1$$

The term $P[(X_i - X_j)(Y_i - Y_j) > 0]$ is the probability that (X_i, Y_i) and (X_j, Y_j) are concordant, and τ is a rescaling of this measure so that it ranges between -1 and 1. If there is no association, then

$$P\left[\left(X_i - X_j\right)\left(Y_i - Y_j\right) > 0\right] = \frac{1}{2}$$

and $\tau = 0$. If the pairs are certain to be concordant, then $\tau = 1$; if the pairs are certain to be discordant, then $\tau = -1$.

5.3.1 Estimating Kendall's Tau with No Ties in the Data

Suppose the data have no ties. Define U_{ij} as

$$U_{ij} = \begin{cases} 1, & \left(X_i - X_j\right)\left(Y_i - Y_j\right) > 0 \\ 0, & \left(X_i - X_j\right)\left(Y_i - Y_j\right) < 0 \end{cases}$$

Let

$$V_i = \sum_{j=i+1}^{n} U_{ij}$$

Now V_i is the number of pairs (X_j, Y_j) that are concordant with (X_i, Y_i) for $j \geq i + 1$, and

$$\frac{\sum\limits_{i=1}^{n-1} V_i}{\binom{n}{2}}$$

is the fraction of concordant pairs in the data set. The estimate of τ is

$$r_\tau = 2 \frac{\sum\limits_{i=1}^{n-1} V_i}{\binom{n}{2}} - 1$$

If $R(X_i)$ denotes the rank of X_i among the X's and $R(Y_i)$ denotes the rank of Y_i among the Y's, then the pairs (X_i, Y_i) and (X_j, Y_j) are concordant or discordant according as $(R(X_i), R(Y_i))$ and $(R(X_j), R(Y_j))$ are concordant or discordant. That is, r_τ may be computed from knowledge of the ranks of the X's and Y's.

It is convenient to arrange the X's in increasing order. Then determining concordance is just a matter of determining how many of the pairs of Y's are in increasing order. To illustrate, Table 5.3.1 lists the heights and weights of four college men. The values of X have been arranged from smallest to largest. The values of V_i are given in the table along with the data. Of the six pairs of Y's, four are in increasing order, so $r_\tau = 2(4/6) - 1 = .33$.

TABLE 5.3.1
Heights and Weights of Four College Men

	Student			
	1	*2*	*3*	*4*
Height (*X*)	68	70	71	72
Weight (*Y*)	153	155	140	180
V_i	2	1	1	Not computed

To test H_0: $\tau = 0$ against one-sided or two-sided alternative hypotheses, we may carry out a permutation test. The permutation distribution of r_τ is determined by permuting the *Y*'s among the *X*'s as in determining the permutation distribution of r in Section 5.1.3. Equivalently, the permutation distribution may be determined by permuting the ranks of the *Y*'s among the ranks of the *X*'s. To illustrate, Table 5.3.2 shows the 4! = 24 permutations of the ranks of the height and weight data and the corresponding values of r_τ. Nine values of r_τ are equal to or greater than the value of .33 for the original data denoted by the asterisk. Thus, the *p*-value is 9/24 = .375.

Selected critical values for the upper tail of the distribution of r_τ are given in Table A13 of the Appendix for samples of sizes 4 to 10. If the sample size is large, we may obtain a random sample of the permutations and construct an approximate permutation distribution for r_τ.

5.3.2 Large-Sample Approximation

If the *n* pairs of observations have no ties, we have

$$E(r_\tau) = 0, \quad \text{var}(r_\tau) = \frac{4n + 10}{9(n^2 - n)}$$

We may then use the statistic

$$Z = \frac{r_\tau}{\sqrt{\text{var}(r_\tau)}}$$

and refer the value to the standard normal distribution. For a derivation, see Hollander and Wolfe (1999).

EXAMPLE 5.3.1   For the rabbit data in Table 5.2.1, we have $r_\tau = .73$ and

$$\text{var}(r_\tau) = \frac{4(18) + 10}{9(18^2 - 18)} = 0.03$$

Thus, $Z = .73/\sqrt{0.03} = 4.21$, and the upper-tail *p*-value is 0 to four decimals.   ∎

TABLE **5.3.2**
Permutation Distribution of Kendall's τ for $n = 4$

$R(X_1)$ 1	$R(X_2)$ 2	$R(X_3)$ 3	$R(X_4)$ 4	
$R(Y_1)$	$R(Y_2)$	$R(Y_3)$	$R(Y_4)$	r_τ
1	2	3	4	1.00
1	2	4	3	.67
1	3	2	4	.67
1	3	4	2	.33
1	4	2	3	.33
1	4	3	2	.00
2	1	3	4	.67
2	1	4	3	.33
2	3	1	4	.33*
2	3	4	1	.00
2	4	1	3	.00
2	4	3	1	−.33
3	1	2	4	.33
3	1	4	2	.00
3	2	1	4	.00
3	2	4	1	−.33
3	4	1	2	−.33
3	4	2	1	−.67
4	1	2	3	.00
4	1	3	2	−.33
4	2	1	3	−.33
4	2	3	1	−.67
4	3	1	2	−.67
4	3	2	1	−1.00

5.3.3 Adjustment for Ties in the Data

For ties in either the X's or the Y's in the pairs (X_i, Y_i) and (X_j, Y_j), we have $(X_i - X_j)$ $(Y_i - Y_j) = 0$. In this case, we set $U_{ij} = 1/2$, but otherwise the formula for r_τ is the same as before. We may obtain the permutation distribution for this statistic, or we may use a normal approximation. For a relatively small number of ties, we may use the variance in the untied case. If the number of ties is substantial, the variance must be adjusted downward.

Let s_i denote the number of ties in the ith group of ties among the X's and t_i denote the number of ties in the ith group of ties among the Y's. Define terms A, B, and C as follows:

$$A = \frac{\sum s_i(s_i - 1)(2s_i + 5) + \sum t_i(t_i - 1)(2t_i + 5)}{18}$$

$$B = \frac{\left[\sum s_i(s_i - 1)(2s_i - 2)\right]\left[\sum t_i(t_i - 1)(t_i - 2)\right]}{9n(n-1)(n-2)}$$

$$C = \frac{\left[\sum s_i(s_i - 1)\right]\left[\sum t_i(t_i - 1)\right]}{2n(n-1)}$$

The variance with ties is

$$\text{var}(r_{\tau\,\text{ties}}) = \text{var}(r_\tau) - \frac{4}{n^2(n-1)^2}(A - B - C)$$

where $\text{var}(r_\tau)$ is the variance of r_τ without ties.

EXAMPLE 5.3.2   Table 5.3.3 is a display of the data in Table 5.2.2, but the data have been rearranged so that the scores of judge A are in increasing order. In addition, V_i is shown for each of the pairs. We find

$$r_{\tau\,\text{ties}} = \frac{2(27.5)}{45} - 1 = .22$$

TABLE 5.3.3

Scores of Ten Projects at a Science Fair

	Project									
	5	*6*	*7*	*3*	*10*	*4*	*9*	*1*	*2*	*8*
Judge A (X)	5	6	6	7	7	8	8	8	8	9
Judge B (Y)	6	4	5	8	9	5	6	7	8	8
V_i	5.5	7.5	6.5	1.5	0.0	2.5	2.0	1.5	0.5	

The values of s_i are 2, 2, and 4, while the values of t_i are 2, 2, and 3. Thus,

$$A = \frac{(18 + 18 + 156) + (18 + 18 + 66)}{18} = 16.33$$

$$B = \frac{(0 + 0 + 24)(0 + 0 + 6)}{6480} = 0.02$$

$$C = \frac{(2 + 2 + 12)(2 + 2 + 6)}{180} = 0.89$$

$$\text{var}(r_{\tau\,\text{ties}}) = \frac{50}{810} - \frac{4}{8100}(16.33 - 0.02 - 0.89) = 0.054$$

We find $Z = .22/\sqrt{0.054} = 0.95$ with $p = .17$. The exact p-value from StatXact is .1877.   ∎

5.3.4 Computer Analysis

StatXact provides an exact test for Kendall's tau. The Resampling Stats code for Kendall's tau is similar to that for the Spearman correlation. There is no predefined function for the computation of Kendall's tau, but it may be computed from other commands available in the programming language. MINITAB and S-Plus do not include Kendall's tau in their menu of predefined statistical procedures.

5.4
Permutation Tests for Contingency Tables

Suppose individuals in a study have been placed into categories according to two characteristics. A sociologist might categorize individuals according to marital status and religious affiliation to determine whether there is an association between the two. A medical researcher might categorize individuals according to the amount of aspirin consumption and the presence or absence of coronary heart disease in order to determine whether aspirin can prevent heart attacks. A *two-way contingency table* displays the counts of individuals who fall into each of the categories that are determined by the two characteristics being studied.

The data layout for an $R \times C$ contingency table is shown in Table 5.4.1. Here n_{ij} denotes the number of individuals who have been observed to have row characteristic i and column characteristic j, and n denotes the total number of individuals.

TABLE 5.4.1
Observations in an $R \times C$ Contingency Table

	Column 1	Column 2	. . .	Column c	Row Totals
Row 1	n_{11}	n_{12}	. . .	n_{1c}	$n_{1.}$
Row 2	n_{21}	n_{22}	. . .	n_{2c}	$n_{2.}$
$\vdots$	$\vdots$	$\vdots$		$\vdots$	$\vdots$
Row r	n_{r1}	n_{r2}	. . .	n_{rc}	$n_{r.}$
Column Totals	$n_{.1}$	$n_{.2}$	. . .	$n_{.c}$	n

5.4.1 Hypotheses to Be Tested and the Chi-Square Statistic

We now consider two cases. Case 1: All n individuals are selected at random from a population and cross-classified according to row and column characteristics. Think of randomly selecting individuals and classifying them according to marital status and religious affiliation. Case 2: A fixed number $n_{i.}$ is selected according to row characteristic i, $i = 1, \ldots, r$, and classified according to the column characteristic. For instance, the rows could represent different doses of aspirin given to selected subjects in a planned experiment, and the columns could represent the presence or absence of a heart attack.

Define the expected cell proportions as

$$p_{ij} = \frac{E(n_{ij})}{n}$$

The row and column proportions are

$$p_{i.} = \sum_{j=1}^{c} p_{ij}, \; p_{.j} = \sum_{i=1}^{r} p_{ij}$$

In Case 1, the null hypothesis of interest is independence. Specifically,

$$H_0: p_{ij} = p_{i.}p_{.j}$$

for all i and j. The alternative hypothesis H_a is not H_0, in the sense that the equality fails to hold for at least one row and column. A rejection of the null hypothesis indicates a statistical dependence between the row and column factors.

In Case 2, we define the conditional probability of column j given row i as

$$p_{j|i} = \frac{p_{ij}}{p_{i.}}$$

The null hypothesis of interest is

$$H_0: p_{j|i} = p_{j|i'}$$

for all i, i', and j. That is, for any given column, the conditional probabilities from row to row are all the same. This hypothesis is sometimes referred to as homogeneity of row distributions. The alternative is that for at least one column, the conditional probabilities from row to row are not all the same.

The null hypothesis of independence in Case 1 and the null hypothesis of homogeneity in Case 2 are equivalent, although the sampling situations are different. In either case, we term the hypothesis test a *test of association* between the row and column factors. The null hypothesis is no association.

The *chi-square statistic*, which is used for both cases, is defined as

$$\chi^2 = \sum_{i=1}^{r} \sum_{j=1}^{c} \frac{(n_{ij} - e_{ij})^2}{e_{ij}}$$

where

$$e_{ij} = \frac{n_{i.}n_{.j}}{n}$$

If the expected cell frequencies e_{ij} are 5 or greater, then χ^2 has an approximate chi-square distribution with $(r-1)(c-1)$ degrees of freedom. However, when the expected cell frequencies are smaller than this, the chi-square approximation may no longer be sufficiently accurate, and this is especially so in tables with a large number of cells for which there are no responses. In such cases, a permutation chi-square test may be used.

5.4.2 Permutation Chi-Square Test

To illustrate the permutation chi-square test, consider the following example. Suppose seven patients are included in a study to compare two methods of relieving postoperative pain. Three are allowed to control the amount of pain-relief medicine themselves according to their level of pain. The other four are given a physician-prescribed level of medicine according to standard practice. Afterward the patients are asked to evaluate the effectiveness of their pain-relief treatment, with responses being not satisfied, somewhat satisfied, and very satisfied. The data are shown in Table 5.4.2.

TABLE 5.4.2
Satisfaction with Pain-Relief Treatment

	Not Satisfied	Somewhat Satisfied	Very Satisfied
Physician-Prescribed	2	2	0
Self-Administered	0	1	2

Let N, S, and V stand for not satisfied, somewhat satisfied, and very satisfied, respectively. These labels appear with the patients as follows, where the subscript denotes the patient identification.

Physician-prescribed: N_1, N_2, S_3, S_4

Self-administered: S_5, V_6, V_7

If there is no difference between the treatments, then all possible assignments of the seven labels, four to one treatment and three to the other, are equally likely to occur. There are 35 such possibilities. The assignment is done in exactly the same way as the assignment of observations to the two treatments in the two-sample permutation test in Section 2.1.

For each of the 35 possible assignments, we compute the chi-square statistic. As it turns out, the 35 possibilities yield eight distinct two-way tables and five distinct chi-square values. These results are shown in Table 5.4.3 along with the frequency of occurrence of each possibility.

The contingency table for the original data is denoted by an asterisk in Table 5.4.3. It has a chi-square value of 4.278. There are 11 out of 35 possible assignments of the labels to the treatments that give contingency tables with chi-square

TABLE 5.4.3
Contingency Tables and Chi-Square Values for Data in Table 5.4.2

Table	Treatment	N	S	V	χ^2	Frequency
1	Physician	1	2	1	0.194	12
	Self	1	1	1		
2	Physician	1	1	2	2.236	6
	Self	1	2	0		
3	Physician	2	1	1	2.236	6
	Self	0	2	1		
4	Physician	0	2	2	4.278	3
	Self	2	1	0		
5*	Physician	2	2	0	4.278	3
	Self	0	1	2		
6	Physician	0	3	1	4.958	2
	Self	2	0	1		
7	Physician	1	3	0	4.958	2
	Self	1	0	2		
8	Physician	2	0	2	7.000	1
	Self	0	3	0		

values at least this big. Thus, the *p*-value is $11/35 = .31$. The probability that a chi-square random variable with 2 degrees of freedom is at least as big as 4.278 is .12, indicating that the chi-square approximation would not be appropriate in this case.

An Alternative Way to Permute the Data

For the purpose of constructing the permutation distribution, we can interchange the roles of the rows and columns. For instance, if we let P denote physician-prescribed and S denote self-administered, we could view the data in Table 5.4.2 as follows.

Not satisfied: P_1, P_2

Somewhat satisfied: P_3, P_4, S_5

Very satisfied: S_6, S_7

To do the permutation test with this alternative labeling, we would obtain all possible assignments of the seven labels P and S to the three columns; two to the not-satisfied column, three to the somewhat-satisfied column, and two to the very-satisfied column. There are

$$\frac{7!}{2!3!2!} = 210$$

possible assignments of the seven objects this way. It can be shown that the possible chi-square statistics are the same as those obtained by assigning observations to rows, and these possibilities occur in the same frequencies as those obtained by assignment to rows. Thus, the *p*-values are the same regardless of which way the random assignment is done.

For example, if we do the assignment to columns, consider the proportion of contingency tables of the 210 possibilities that have the configuration in Table 5.4.2. Of the four letter P's, there are

$$\binom{4}{2} = 6$$

ways to assign two to the not-satisfied column and two to the somewhat satisfied column. There are

$$\binom{3}{1} = 3$$

ways to assign one letter S to the somewhat-satisfied column and two letter S's to the very-satisfied column. Thus, the proportion of contingency tables with the configuration in Table 5.4.2 is

$$\frac{(6)(3)}{210} = \frac{3}{35}$$

which is the same as the proportion obtained by assignment of observations to rows as shown in Table 5.4.3. See Exercise 15 at the end of the chapter for a general formula.

5.4.3 Multiple Comparisons in Contingency Tables

If there is indeed statistical dependence in a two-way table or if homogeneity of row distributions is rejected, then it may be of interest to determine where differences among proportions exist. For instance, one may test whether conditional probabilities in one row are different from conditional probabilities in another row. In the earlier example, the physician may wish to know whether the proportion of patients who are very satisfied under one of the treatments is statistically significantly different from the proportion who are very satisfied under the other treatment, and similarly for the other column categories. We consider the problem of comparing these conditional probabilities.

For each pair of conditional row probabilities $p_{j|i}$ and $p_{j|i'}$ that we are interested in comparing, we compute the Z statistic for two proportions

$$Z = \frac{\hat{p}_{j|i} - \hat{p}_{j|i'}}{\sqrt{\bar{p}(1-\bar{p})\left(\dfrac{1}{n_i} + \dfrac{1}{n_{i'}}\right)}}$$

where

$$\hat{p}_{j|i} = \frac{n_{ij}}{n_i}, \quad \hat{p}_{j|i'} = \frac{n_{i'j}}{n_{i'}}, \quad \bar{p} = \frac{n_i\hat{p}_{j|i} + n_{i'}\hat{p}_{j|i'}}{n_i + n_{i'}}$$

Suppose k such comparisons are of interest to us giving us the k Z-statistics for two proportions $Z_1, Z_2, \ldots, Z_k$. The statistic upon which we base our multiple comparison procedure is

$$Q^* = \max_i |Z_i|$$

Steps in the Multiple Comparison Permutation Test for Proportions

1. Identify the comparisons of conditional row probabilities that we wish to make and compute the corresponding Z statistics for two proportions, which we denote $Z_{1obs}, Z_{2obs}, \ldots, Z_{kobs}$.

2. Permute the observations as we did with the permutation chi-square statistic. For each permutation, find the Z statistics, $Z_1, Z_2, \ldots, Z_k$, that correspond to the row proportions we wish to compare. Compute $Q^* = \max_i |Z_i|$.

3. Obtain the permutation distribution of Q^*. Let $q^*(\alpha)$ denote the upper $100\alpha\%$ point of the distribution of Q^*. Declare the two proportions involved in the computation of Z_{iobs} to be statistically significantly different at level of significance α if

$$|Z_{iobs}| \geq q^*(\alpha)$$

4. The p-value for the ith comparison is the fraction of the permutation distribution of Q greater than or equal to $|Z_{iobs}|$.

The procedure is analogous to the permutation version of Tukey's HSD.

EXAMPLE 5.4.1   Suppose we wish to compare row percentages for the data in Table 5.4.2. The proportions and Z statistics for two proportions are shown in Table 5.4.4. For instance, for the somewhat-satisfied category, we have

$$Z = \frac{.5 - .33}{\sqrt{\left(\dfrac{3}{7}\right)\left(\dfrac{4}{7}\right)\left(\dfrac{1}{4} + \dfrac{1}{3}\right)}} = 0.44$$

TABLE 5.4.4
Row Proportions for Data in Table 5.4.2

	Not Satisfied	Somewhat Satisfied	Very Satisfied
Physician-prescribed	.5	.5	0
Self-administered	0	.33	.67
Z statistics for two proportions	1.45	0.44	−1.93

The Z statistics and the value of Q^* for each of the eight distinct contingency tables in Table 5.4.3 are shown in Table 5.4.5. The permutation distribution of Q^* takes on only five distinct values in this case. There are no values that could be used as a critical value for a test at exactly the 5% level of significance. The closest value is $Q^* = 2.65$, which is the critical value for a test at the $1/35 = .03$ level of significance. If we test for a difference between proportions at the 3% level, then none of the Z statistics for the original data equals or exceeds 2.65, so no significant differences would be declared between the proportions in the three categories. Since the observed Z statistic for the not-satisfied column is 1.45, and since 23 out of 35 cases have values of $Q^* \geq 1.45$, the multiple comparison p-value is $23/35 = .66$. The multiple comparison p-values for the somewhat-satisfied and the very-satisfied columns are 1.00 and .23, respectively.

TABLE 5.4.5
Z Statistics and Permutation Distribution of Q for Data in Table 5.4.3

Table	Not Satisfied Z_1	Somewhat Satisfied Z_2	Very Satisfied Z_3	Q	Frequency
1	−0.24	0.44	−0.24	0.44	12
2	−0.24	−1.10	1.45	1.45	6
3	1.45	−1.10	−0.24	1.45	6
4	−1.93	0.44	1.45	1.93	3
5*	1.45	0.44	−1.93	1.93	3
6	−1.93	1.98	−0.24	1.98	2
7	−0.24	1.98	−1.93	1.98	2
8	1.45	−2.65	1.45	2.65	1

5.4.4 Computer Analysis

StatXact has excellent capabilities for handling contingency tables. The data are entered in table format, and the analysis includes exact p-values and asymptotic p-values using the chi-square approximation. In the case of large contingency tables, the *p*-values may be approximated using random sampling of the permutations. Multiple comparisons are not available in StatXact. A printout of the analysis of the data in Table 5.4.2 is shown in Figure 5.4.1. This is accessed through the "Unordered R × C Table" option. StatXact also provides two other statistics, the likelihood ratio statistic and the Freeman–Halton statistic, which are not discussed here.

MINITAB and S-Plus implement the chi-square approximation from their menus of statistical procedures. Resampling Stats code for the analysis of the data in Table 5.4.2 is shown in Figures 5.4.2 and 5.4.3. In Figure 5.4.2, the observations are permuted between the two rows; in Figure 5.4.3, the observations are permuted among the three columns.

F I G U R E 5.4.1

StatXact Permutation Chi-Square Analysis of Data in Table 5.4.2

```
CHI-SQUARE TEST FOR INDEPENDENCE

Statistic based on the observed 2 by 3 table(x) :
    CH(X) : Pearson Chi-Square Statistic =       4.278

Asymptotic p-value: (based on Chi-Square distribution with 2 df )
    Pr { CH(X) .GE.        4.278 } =       0.1178

Exact p-value and point probability :
    Pr { CH(X) .GE.        4.278 } =       0.3143
    Pr { CH(X) .EQ.        4.278 } =       0.1714
```

F I G U R E 5.4.2

Resampling Stats Code for Chi-Square Permutation Test for Data in Table 5.4.2;
Observations Permuted between Rows

```
maxsize default 5000

'input data denoting columns to which observations belong
copy (1 1 2 2 2 3 3) dat

'input expected cell frequencies, row 1 listed first
'observed chi-square statistics is 4.278
copy (1.143 1.714 1.143 .857 1.286 .857 ) expect

'permute data between the two rows
repeat 5000
shuffle dat sdat
take sdat 1,4 row1
take sdat 5,7 row2
```

(continued on next page)

FIGURE 5.4.2

Resampling Stats Code for Chi-Square Permutation Test for Data in Table 5.4.2; Observations Permuted between Rows *(continued)*

```
'compute chi-square statistic
count row1 =1 obs_11
count row1 =2 obs_12
count row1 =3 obs_13
count row2 =1 obs_21
count row2 =2 obs_22
count row2 =3 obs_23
concat obs_11 obs_12 obs_13 obs_21 obs_22 obs_23 obs
subtract obs expect diff
square diff diff2
divide diff2 expect chi
sum chi chisq

'keep track of chi-square values
score chisq chidist
end


'get p-value
count chidist >=4.278 pval
divide pval 5000 pval
print pval


'print output here

PVAL      =       0.318
```

FIGURE 5.4.3

Resampling Stats Code for Chi-Square Permutation Test for Data in Table 5.4.2; Observations Permuted among Columns

```
maxsize default 5000

'input data denoting rows to which data below
copy (1 1 1 1 2 2 2) dat

'input expected cell frequencies, column 1, column 2, column 3.
'observed chi-square is 4.278
copy (1.143 .857 1.714 1.286 1.143 .857 ) expect

'permute data among the three columns
repeat 5000
```

(continued on next page)

FIGURE 5.4.3

Resampling Stats Code for Chi-Square Permutation Test for Data in Table 5.4.2;
Observations Permuted among Columns *(continued)*

```
shuffle dat sdat
take sdat 1,2 col_1
take sdat 3,5 col_2
take sdat 6,7 col_3

'compute chi-square statistic
count col_1 =1 obs_11
count col_1 =2 obs_21
count col_2 =1 obs_12
count col_2 =2 obs_22
count col_3 =1 obs_13
count col_3 =2 obs_23
concat obs_11 obs_21 obs_12 obs_22 obs_13 obs_23 obs
subtract obs expect diff
square diff diff2
divide diff2 expect chi
sum chi chisq

'keep track of chi-square values
score chisq chidist
end

'get p-value
count chidist >=4.278 pval
divide pval 5000 pval
print pval

'print output here

PVAL    =        0.32
```

5.5
Fisher's Exact Test for a 2 × 2 Contingency Table

The special case of the permutation test applied to a 2 × 2 contingency table is called *Fisher's exact test* for a 2 × 2 contingency table. The test is named for R. A. Fisher, the founder of modern statistics.

5.5.1 Probability Distribution Under the Null Hypothesis

When the column labels are permuted among the rows, as described in Section 5.4, both the row totals and the column totals remain fixed. The row totals remain fixed because the observations are permuted to maintain the same number originally in each row, and the column totals remain fixed because the same set of column labels are involved in each permutation. In the 2 × 2 table, knowledge of one cell entry along with knowledge of the row and column totals completely determines the entries in the table.

Consider the 2 × 2 contingency table in Table 5.5.1. Let X denote the number in row 1 and column A. A known value of X determines all other entries in the table. For instance, if $X = 3$, then the entry in row 1 and column B is 4, the entry in row 2 and column A is 1, and the entry in row 2 and column B is 2.

TABLE 5.5.1

A 2 × 2 Contingency Table with Fixed Marginal Totals

	Column A	Column B	Total
Row 1	X		7
Row 2			3
Total	4	6	10

Let us imagine that we have ten column labels: four letter A's for the four observations in column A and six letter B's for the six observations in column B. We wish to permute these column labels randomly in such a way that seven are in row 1 and three are in row 2. Since there are six letter B's, there must always be at least one letter A in row 1. Thus, the smallest possible value for X is 1 in this case. The largest possible value of X is 4, which occurs with four letter A's and three letter B's in row 1. Thus, X can take on the possible values 1, 2, 3, and 4 with positive probability. We compute the probability that X takes on each of these values.

If the null hypothesis of independence or homogeneity holds, then all possible $\binom{10}{7}$ assignments of the ten labels, seven to row 1 and three to row 2, are equally likely to occur. For illustration, we compute $P(X = 3)$. To have $X = 3$, we select three letter A's from the four to place in row 1 and column A, and we select four letter B's from the six to fill out row 1 and column B. Thus,

$$P(X = 3) = \frac{\binom{4}{3}\binom{6}{4}}{\binom{10}{7}} = .500$$

Similarly, we have

$$P(X = 1) = \frac{\binom{4}{1}\binom{6}{6}}{\binom{10}{7}} = .033$$

$$P(X = 2) = \frac{\binom{4}{2}\binom{6}{5}}{\binom{10}{7}} = .300$$

$$P(X = 4) = \frac{\binom{4}{4}\binom{6}{3}}{\binom{10}{7}} = .167$$

This distribution is a special case of the *hypergeometric distribution*.

Now let us consider how these computations may be used in determining a *p*-value for a permutation test. Suppose we have observed $X = 3$, and we wish to do an upper-tail test. That is, if there is a difference between row 1 and row 2, it would tend to produce large values of X. From the computations above, we see that $P(X \geq 3) = .667$. Suppose $X = 1$ and we do a lower-tail test. Then the value is statistically significant with $p = .033$.

EXAMPLE 5.5.1   A surgeon is interested in determining whether the administration of a certain drug can reduce the incidence of pulmonary embolism in patients undergoing high-risk surgery. Nineteen patients were selected for the study, with 11 getting the drug and 8 receiving the standard treatment. Data are shown in Table 5.5.2.

TABLE 5.5.2
Incidence of Pulmonary Embolism (P.E.)

	P.E.	No P.E.	Total
Drug	3	8	11
No Drug	4	4	8
Total	7	12	19

Let X denote the number in the Drug–P.E. category. Since a favorable outcome for the drug would be a low incidence of pulmonary embolism, a small value of X is of interest. In this case, the values 0, 1, . . . , 7 for X may all occur with positive probability. The p-value is

$$P(X \le 3) = \frac{\binom{7}{3}\binom{12}{8}}{\binom{19}{11}} + \frac{\binom{7}{2}\binom{12}{9}}{\binom{19}{11}} + \frac{\binom{7}{1}\binom{12}{10}}{\binom{19}{11}} + \frac{\binom{7}{0}\binom{12}{11}}{\binom{19}{11}} = .30$$

Thus, there is not sufficient indication in this study to conclude that the drug is effective in reducing the incidence of pulmonary embolism.   ∎

5.5.2 Computer Analysis

In StatXact, Fisher's exact test is accessed by selecting the "Unordered R × C Table" option and then choosing the Fisher–Freeman–Halton statistic. See Freeman and Halton (1951) for a discussion of this statistic. The exact one-sided p-value as computed in Example 5.5.1 is given in the output in Figure 5.5.1. S-Plus also implements this test under the "Compare Samples" and "Counts and Proportions" menu options.

FIGURE 5.5.1
Fisher's Exact Test for Data in Table 5.5.2 Using StatXact

```
FISHER'S EXACT TEST

Statistic based on the observed 2 by 2 table(x) :
    P(X)  = Hypergeometric Prob. of the table =      0.2292
    FI(X) = Fisher statistic             =       1.033

Asymptotic p-value: (based on Chi-Square distribution with 1 df )
     Two-sided:Pr{FI(X) .GE.         1.033} =       0.3094
     One-sided:0.5 * Two-sided        =       0.1547

Exact p-value and point probabilities :
 Two-sided:Pr{FI(X) .GE.       1.033}= Pr{P(X) .LE.       0.2292}=       0.3765
          Pr{FI(X) .EQ.       1.033}= Pr{P(X) .EQ.       0.2292}=       0.2292
 One-sided:Let y be the value in Row 1 and Column 1
    y =3 min(Y) =0 max(Y) =7 mean(Y) =      4.053 std(Y) =       1.067

          Pr { Y .LE. 3 } =       0.2966
          Pr { Y .EQ. 3 } =       0.2292
```

5.6
Contingency Tables with Ordered Categories

Although the most common statistic for analyzing $R \times C$ contingency tables is the chi-square statistic, other statistics may be preferred under certain circumstances. Specifically we consider alternatives to the chi-square statistic when there is an ordering among the categories of one or both factors in a contingency table.

A table in which the categories of one of the factors is ordered are called a *singly ordered* table. For instance, suppose we wish to compare the degrees of pain relief given by three different brands of pain relievers. If a patient's response is categorized as poor, fair, good, or excellent, then the degree of pain relief would be an ordered factor. If the categories of both factors are ordered, then the table is said to be *doubly ordered*. For instance, if patients are given one of three doses of a drug with the expectation that the higher doses will give a greater degree of relief, then both the drug treatment (dose 1, dose 2, dose 3) and the degree of pain relief (poor, fair, good, excellent) would be considered ordered.

5.6.1 Singly Ordered Tables

In the singly ordered table, the categories of the ordered factor are given numerical labels. The labels must be increasing according to the ordering of the categories but are otherwise arbitrary, since only the ranks of the labels will be used in the computations. For simplicity, we will use a label of 1 for the category designated as the smallest, a label of 2 for the next smallest category, and so on. In Examples 5.6.1 and 5.6.2, we show how the Wilcoxon rank-sum test with ties may be used when the unordered factor has just two categories, and the Kruskal–Wallis test with ties may be used when the unordered factor has more than two categories.

EXAMPLE **5.6.1**   Consider the pain relief data in Table 5.4.2. If there is a difference between the two methods of pain relief, then a researcher may hypothesize that the self-administered dose will be more satisfactory. Indeed, the data show that the proportion of satisfactory responses is higher in the self-administered case. To enable us to apply Wilcoxon's rank-sum test with ties, we designate the not-satisfied category as 1, the somewhat-satisfied category as 2, and the very-satisfied category as 3. The data are displayed in Table 5.6.1. The seven labels that come from combining the data from the two treatments are 1, 1, 2, 2, 2, 3, 3, and the ranks adjusted for ties are 1.5, 1.5, 4, 4, 4, 6.5, 6.5. The observations for the self-administered treatment are 2, 3, and 3 with rank sum $4 + 6.5 + 6.5 = 17$. From the Mann–Whitney–Wilcoxon option in StatXact, the exact p-value is .0857. By contrast, the p-value for the chi-square permutation test is .31. Neither test is significant at the 5% level in this case. However, in a small preliminary study, the researcher may be willing to adopt a 10% level of significance as an indication that further research is warranted, and this level of significance is shown by the Wilcoxon rank-sum test.

TABLE 5.6.1
Data from Table 5.4.2; Satisfaction with Pain Relief Displayed as an Ordered Factor

	Level of Satisfaction with Pain Relief
Physician-Prescribed	1, 1, 2, 2
Self-Administered	2, 3, 3

The next example shows the analysis of a singly ordered table with four ordered columns and three unordered rows.

EXAMPLE 5.6.2   Department heads of three colleges at Hypothetical State University were asked to rate the performance of the chief academic officer. The rating scale used in the study was poor, fair, good, very good, excellent. The data are displayed as a contingency table in Table 5.6.2. The chi-square test for association has $\chi^2 = 7.02$ with a *p*-value of .53. Now suppose we treat rating as an ordered factor and label the rating categories 1, 2, . . . , 5. The ratings for the individuals are shown in Table 5.6.3. Applying the Kruskal–Wallis test with ties to these data, we find $KW_{ties} = 6.43$ with a *p*-value of .04. Thus, the Kruskal–Wallis statistic is able to detect the effect of college upon the ratings, whereas the chi-square statistic is not.

TABLE 5.6.2
Ratings of the Chief Academic Officer at Hypothetical State University with Data Displayed as a Contingency Table

	Poor (1)	Fair (2)	Good (3)	Very Good (4)	Excellent (5)
College A	6	4	3	2	0
College B	2	2	3	2	1
College C	1	2	3	4	2

TABLE 5.6.3
Data from Table 5.6.2 with Rating Displayed as an Ordered Factor

College A	1, 1, 1, 1, 1, 1, 2, 2, 2, 2, 3, 3, 3, 4, 4
College B	1, 1, 2, 2, 3, 3, 3, 4, 4, 5
College C	1, 2, 2, 3, 3, 3, 4, 4, 4, 4, 5, 5

The large difference between the *p*-values for the Kruskal–Wallis statistic and the chi-square statistic in the preceding example is due to the types of alternatives that the two statistics are designed to detect. The chi-square statistic is an omnibus statistic that is designed to detect any association, whereas the Kruskal–Wallis statistic is designed specifically to detect an ordering effect. If it should turn out that the effect is something other than an ordering effect, then the Kruskal–Wallis statistic would be inappropriate but the chi-square statistic could still be used.

5.6.2 Doubly Ordered Tables

To analyze doubly ordered contingency tables, we may apply the Jonckheere–Terpstra statistic with ties. This procedure is shown in Example 5.6.3.

EXAMPLE **5.6.3**   An entomologist was interested in comparing the effectiveness of three methods of controlling insect damage to alfalfa plants. Plants were infected with insects and treated with product A, B, or C. The damage done by the insects to each plant was classified as severe, moderate, slight, or none. It was anticipated that the plants with treatment A would have the most damage, followed by those with treatment B, and then C. The data are shown in Table 5.6.4.

TABLE 5.6.4

Level of Insect Damage for Each of Three Treatments; Doubly Ordered Table

Response/ Treatment	Severe	Moderate	Slight	None
A	10	12	17	30
B	9	9	11	35
C	7	8	12	43

Suppose we label the responses as severe = 1, moderate = 2, slight = 3, and none = 4. If the assumption about the ordering of the treatments is correct, then those with A would tend to have the lower labels, those with B the intermediate labels, and those with C the highest labels. Application of StatXact shows a significant difference among the treatments with $p = .02$ (based on 10,000 randomly selected permutations of the data). On the other hand, the Kruskal–Wallis test, which does not consider an ordering among the treatments, has $p = .14$. The chi-square test for association has $p = .55$.   ∎

Ordered categorical responses occur in many applied situations. The potential gain in power in using techniques for singly ordered or doubly ordered tables over what can be achieved with the chi-square test for association can be substantial.

5.6.3 Computer Analysis

Any computer package that will do the Wilcoxon rank-sum test, the Kruskal–Wallis test, or the Jonckheere–Terpstra with ties may be used for contingency tables with ordered categories. StatXact allows data to be entered in contingency table format and provides options for single ordered and doubly ordered tables.

5.7
Mantel–Haenszel Test

Suppose a study is done at three medical centers to compare two procedures, I and II, for treating a particular condition. Each patient's outcome is classified as either "complications occur" (C) or "no complications occur" (NC). Data may be taken in one of two ways. We may select patients randomly from a target population at each center and cross-classify the data according to the treatment received and the outcome, or we may have a fixed number of individuals predetermined for each treatment at each center and classify their outcomes as C or NC.

Because of differences among patients and other factors, it is expected that there will be differences in the rates of complications among centers. However, it can be assumed that the same procedure will be better at all centers if in fact the procedures differ. Data for this hypothetical study are given in Table 5.7.1. The centers are termed *strata*, and it is assumed that the responses are independent from stratum to stratum.

5.7.1 Hypotheses to Be Tested

Let us assume we have s strata. Let $p_{ij}(k)$ denote the probability that an observation is in row i and column j given stratum k. All probabilities will be assumed to be conditioned on the strata. Let $p_{j|i}(k)$ denote the conditional probability that an observation is classified in column j given that it is in row i and stratum k, $i = 1, 2$, $j = 1, 2, k = 1, 2, \ldots, s$. We state the null and alternative hypotheses of interest in three equivalent ways, each of which may be useful for interpretation.

First we express the hypotheses in terms of conditional probabilities. The null hypothesis is

$$H_0: p_{1|1}(k) = p_{1|2}(k), \quad k = 1, 2, \ldots, s$$

In our example, H_0 true would imply that the rate of complications for procedure I is the same as the rate for procedure II at each center. The alternative hypothesis is

$$H_a: p_{1|1}(k) \geq p_{1|2}(k), \quad k = 1, 2, \ldots, s \text{ or}$$

$$p_{1|1}(k) \leq p_{1|2}(k), \quad k = 1, 2, \ldots, s$$

TABLE 5.7.1

Comparison of the Effect of Procedures I and II on the Incidence of Complications (C = complications, NC = no complications); Entries Are Number of Cases in Each Category

Center 1

Procedure	C	NC	Row Totals
I	2	8	10
II	7	8	15
Column Totals	9	16	25

Center 2

Procedure	C	NC	Row Totals
I	3	9	12
II	14	28	42
Column Totals	17	37	54

Center 3

Procedure	C	NC	Row Totals
I	1	9	10
II	13	17	30
Column Totals	14	26	40

with strict inequality holding for at least one k. In our example, H_a true would imply that the rate of complications for procedure I is either greater than or equal to the rate for procedure II at all centers or less than or equal to the rate for procedure II at all centers, with strict inequality holding for at least one center.

The second way of stating the hypotheses is in terms of the independence of each of the 2×2 tables. The null hypothesis is

$$H_0: \; p_{11}(k) = p_{1.}(k)p_{.1}(k), \quad k = 1, 2, \ldots, s$$

where $p_{1.}(k)$ and $p_{.1}(k)$ are the marginal probabilities for row 1 and column 1, respectively, of the kth stratum. The alternative is

$$H_a: \; p_{11}(k) \geq p_{1.}(k)p_{.1}(k), \quad k = 1, 2, \ldots, s \; \text{ or }$$

$$p_{11}(k) \leq p_{1.}(k)p_{.1}(k), \quad k = 1, 2, \ldots, s$$

The alternative hypothesis expressed this way states that the probability that an observation is in row 1 and column 1 is greater than or equal to what one would expect under the independence assumption for all strata, or the probability is less than or equal to what one would expect under the independence assumption for all strata, with strict inequality holding for at least one stratum.

The third way of expressing the hypotheses is in terms of *odds ratios* defined by

$$\theta(k) = \frac{P_{11}(k)p_{22}(k)}{p_{12}(k)p_{21}(k)}$$

To see where this expression comes from, we note that $p_{11}(k)/p_{12}(k)$ is the odds that an observation from row 1 of the kth stratum falls in column 1, and $p_{21}(k)/p_{22}(k)$ is the odds that an observation from row 2 of the kth stratum falls in column 1. The ratio of these two quantities is $\theta(k)$. The null and alternative hypotheses in terms of the odds ratios are

$$H_0: \theta(k) = 1, \quad k = 1, 2, \ldots, s$$

$$H_a: \theta(k) \geq 1, \quad k = 1, 2, \ldots, s \quad \text{or}$$

$$\theta(k) \leq 1, \quad k = 1, 2, \ldots, s$$

To illustrate the interpretation, suppose the kth stratum has the probabilities as given in Table 5.7.2. The odds that an observation from row 1 is in column 1 is 3/4, and the odds that an observation from row 2 is in column 1 is 1/2. The odds ratio is 3/2. This value indicates that the odds that an observation is in column 1 are 1.5 times greater in row 1 than in row 2. For instance, if column 1 represents a medical complication and row 1 and row 2 represent two medical procedures, then the odds that procedure 1 has a complication is 1.5 times greater than the odds that procedure 2 has a complication.

TABLE 5.7.2
Entries $p_{ij}(k)$ in a 2×2 Contingency Table

.3	.4
.1	.2

5.7.2 Testing Hypotheses for Stratified Contingency Tables

Now let us turn to testing these hypotheses. Let X_k denote the number of observations classified in row 1 and column 1 of the kth stratum. A permutation test and a chi-square approximation will be based on ΣX_k. Let N_k denote the number observations in the kth stratum. Let r_{1k} and r_{2k} denote the number of observations in rows 1 and 2, respectively, of the kth stratum, and let c_{1k} and c_{2k} denote the number of observations in columns 1 and 2 of the kth stratum.

To conduct a permutation test, we permute the observations within each 2×2 contingency table as described in Section 5.4.2. The number of possible permutations is

$$\prod_{k=1}^{s} \binom{N_k}{r_{1k}}$$

If the null hypothesis is true, then all permutations are equally likely to occur. The expected value and variance of X_k, based on the permutation distribution, are

$$E\left(X_k\right) = \frac{r_{1k}c_{1k}}{N_k}$$

$$\mathrm{var}\left(X_k\right) = \frac{r_{1k}r_{2k}c_{1k}c_{2k}}{N_k^2\left(N_k - 1\right)}$$

The Mantel–Haenszel statistic is defined as

$$\mathrm{MH} = \frac{\left(\sum_{k=1}^{s}\left[X_k - E\left(X_k\right)\right]\right)^2}{\sum_{k=1}^{s} \mathrm{var}\left(X_k\right)}$$

The numerator of MH is a measure of how much the total number of observations in row 1 and column 1 deviates from what one would expect under the assumption of independence. It is designed to pick up treatment effects if all the X_k's tend to be either larger than one would expect under the independence assumption or smaller than one would expect under the independence assumption. It would be inappropriate if X_k tended to be larger than $E(X_k)$ in some strata and smaller than $E(X_k)$ in other strata. For large samples, the distribution under the null hypothesis is approximately chi-square with 1 degree of freedom. For small samples, we may obtain the permutation distribution of MH.

EXAMPLE 5.7.1   The computation of MH for the data in Table 5.7.1 is given here. For the data from center 1,

$$X_1 = 2, \quad E\left(X_1\right) = \frac{(10)(9)}{25} = 3.6, \quad \mathrm{var}\left(X_1\right) = \frac{(10)(15)(9)(16)}{(25)^2(24)} = 1.44$$

Similarly, $X_2 = 3$, $E(X_2) = 3.78$, var$(X_2) = 2.05$, $X_3 = 1$, $E(X_3) = 3.5$, and var$(X_3) = 1.75$. Thus,

$$\mathrm{MH} = \frac{\left[2 + 3 + 1 - (3.6 + 3.78 + 3.5)\right]^2}{1.44 + 2.05 + 1.75} = 4.54$$

The *p*-value from a chi-square distribution with 1 degree of freedom is .033, which indicates a difference between the two procedures in the rates of complications. Moreover, procedure I has a smaller proportion of complications across strata than procedure II, so it would be considered the better of the two procedures in terms of the rate of complications. StatXact obtains the permutation distribution of $S = \Sigma X_k$ where X_k is defined using cell (1, 2). This test gives a one-sided *p*-value of .0248 or a two-sided *p*-value of .0496. ∎

5.7.3 Estimation of the Odds Ratio

If there is a common odds ratio—that is, if $\theta(1) = \theta(2) = \cdots = \theta(s)$—then it is possible to obtain an estimate and a confidence interval for the common value θ. Let n_{ijk} denote the observation in row i and column j of stratum k, $i = 1, 2, j = 1, 2, k = 1, 2, \ldots, s$. Define quantities A and B as

$$A = \sum_{k=1}^{s} \frac{n_{11k}n_{22k}}{N_k}$$

$$B = \sum_{k=1}^{s} \frac{n_{12k}n_{21k}}{N_k}$$

The Mantel–Haenszel estimate of the common odds ratio is

$$\hat{\theta} = \frac{A}{B}$$

The variable $\log(\hat{\theta})$ has an approximate normal distribution with a mean of $\log(\theta)$ and a variance given by

$$\text{var}\left(\log \hat{\theta}\right) = \frac{\sum_{k=1}^{s} \left(n_{11k} + n_{22k}\right)\left(n_{11k}n_{22k}\right)/N_k^2}{2A^2}$$

$$+ \frac{\sum_{k=1}^{s} \left(n_{11k} + n_{22k}\right)\left(n_{12k}n_{21k}\right) + \left(n_{12k} + n_{21k}\right)\left(n_{11k}n_{22k}\right)/N_k^2}{2AB}$$

$$+ \frac{\sum_{k=1}^{s} \left(n_{12k} + n_{21k}\right)\left(n_{12k}n_{21k}\right)/N_k^2}{2B^2}$$

The normal approximation can be used to make a confidence interval for $\log(\theta)$ and then exponentiated to obtain a confidence interval for θ.

EXAMPLE 5.7.2   Assuming a common odds ratio for the data in Table 5.7.1, we find the Mantel–Haenszel estimate to be

$$\hat{\theta} = \frac{16/25 + 84/54 + 17/40}{56/25 + 126/54 + 117/40} = \frac{2.62}{7.50} = 0.35$$

The estimate of the variance of $\log(\hat{\theta})$ is

$$\frac{(10)(16)/25^2 + (31)(84)/54^2 + (18)(17)/40^2}{(2)(2.62)^2}$$

$$+ \frac{\left[(10)(56) + (15)(16)\right]/25^2 + \left[(31)(126) + (23)(84)\right]/54^2 + \left[(18)(117) + (22)(17)\right]/40^2}{(2)(2.62)(7.50)}$$

$$+ \frac{(15)(56)/25^2 + (23)(126)/54^2 + (22)(117)/40^2}{(2)(7.50)^2} = 0.26$$

An approximate 95% confidence interval for $\log(\theta)$ is

$$\log(0.35) \pm 1.96\sqrt{(0.26)} \text{ or } (-2.05, -0.05)$$

Exponentiation gives the interval $(0.13, 0.95)$ for θ.   ∎

The MH test is based on the counts in row 1 and column 1. This is an arbitrary choice. The test could just as well have been developed using the counts in any other cell. However, if one does so, one must be cautious of the conclusions. For instance, if the test were based on counts in row 1 and column 2, then the direction of a one-sided test would be reversed. That is, a tendency for a smaller than expected number of observations in cell (1, 1) would imply a tendency for a larger than expected number of observations in cell (1, 2).

In defining the odds ratio, one could use the reciprocal of the ratio defined above. The reciprocal would define the odds ratio of observations falling in column 2 rather than the odds ratio of observations falling in column 1. In this case, the confidence interval for the logarithm of the odds ratio would be the negative of the interval given above.

The derivation of the variance of $\log(\hat{\theta})$ is due to Robins, Breslow, and Greenland (1986). A useful source for the analysis of contingency tables and categorical data, including the MH test, is Agresti (1990).

5.7.4 Computer Analysis

Procedures for stratified 2×2 tables are implemented in StatXact. Cell (1, 2) is used in defining the MH statistic, and the inverse of the odds ratio defined above is computed. StatXact includes the MH test and an exact confidence interval for the odds ratio, which are accessed from the "Stratified 2 × 2 Tables" menu option along with the CI on Common Odds-Ratio procedure. The Homogeneity of Odds-Ratios procedure may be used to check whether or not the assumption of a common odds ratio is reasonable. Output is shown in Figure 5.7.1. Italicized comments are included to show the interpretation of the results.

S-Plus implements the chi-square approximation for the MH test under the "Compare Samples" and "Counts and Proportions" menu options.

FIGURE 5.7.1
StatXact CI for Common Odds-Ratio, and Homogeneity of Odds-Ratios Procedure Applied to Data in Table 5.7.1

```
ESTIMATION AND TESTING OF COMMON ODDS RATIO
[ 3 2x2 informative tables  ]

Summary of Exact Distribution of S [sum of counts in cell(1,2)]:
   Min          Max          Mean         Std-dev       Observed       Standardized
      1.000         32.00         21.12          2.289          26.00            2.131

Exact p-values for testing that the Common odds Ratio is 1:
   One-sided: Pr { S .GE.        26.00 }                    =        0.0248
```

The line above is the exact MH p-value based on the sum of the counts in cell (1, 2).

```
                  Pr { S .EQ.        26.00 }                =        0.0174
   Two-sided: Method 1: 2 * One-sided                       =        0.0496
              Method 2: (Sum of Probs .LE.       0.0174)    =        0.0473
              Method 3: Pr{|S-Mean| .GE. |       26.00-Mean|} =      0.0473

Exact Estimation of Common Odds Ratio:
       Conditional maximum likelihood estimate:        2.856
       95.00 % Conf. Intervals:
          Exact          : (        1.002 ,       9.480)
```

The endpoints of the exact interval for the common odds ratio for cell (1, 1) are the reciprocals of the endpoints of the interval above.

```
          Mid-p corrected  : (        1.088 ,       8.391)

Mantel-Haenszel Inference:
       Common Odds Ratio estimate:         2.861
       Two-sided p-value:     0.0376 (with RBG variance)
                             0.0331 (M-H variance)
```
(continued on next page)

FIGURE 5.7.1
StatXact CI for Common Odds-Ratio, and Homogeneity of Odds-Ratios Procedure
Applied to Data in Table 5.7.1 *(continued)*

```
The p-value above is based on the chi-square approximation for the MH statistic
as shown in Example 5.7.1 using M-H variance.

      95.00% CI with RBG variance: (      1.062 ,      7.708)

Except for rounding, the endpoints of the interval above are the reciprocals of the
endpoints in Example 5.7.2, which were obtained from the normal approximation.


TEST FOR HOMOGENEITY OF ODDS RATIOS
[ 3   2x2 informative tables ]

Observed Statistics:
      BD:  Breslow and Day Statistic =      1.470
      ZE:  Zelen Statistic           =      0.1034

Asymptotic p-value: (based on Chi-Square distribution with 2 df )
      Pr { BD .GE.      1.470 }   =      0.4795

Exact p-value:
      Pr { ZE .LE.      0.1034 }   =      0.5718

This test shows that the assumption of a common odds ratio is reasonable for this
data set.
```

5.8
McNemar's Test

McNemar's test is used in paired-comparison studies when the responses are dichotomous. For example, suppose members of a focus group view a political debate between two candidates. They are asked before the debate whom they would choose as a candidate, and they are asked after the debate whom they would choose. Suppose we denote the candidates A and B. We are interested in knowing whether the proportion who favor A before the debate is the same as or different from the proportion who favor A after the debate.

Here the data are paired. The possible responses for each member of the focus group are (A, A), (A, B), (B, A), and (B, B), where the first letter is the person's choice before the debate and the second letter is the choice after the debate. Table 5.8.1 shows a hypothetical outcome displayed in contingency table format.

TABLE 5.8.1
Choice of Candidate Before and After Debate

	Chose A After Debate	Chose B After Debate
Chose A Before Debate	3	9
Chose B Before Debate	2	6

In Table 5.8.1, 12 of the 20 members of the focus group chose A before the debate, but only 5 of the 20 chose A after the debate. Thus, members of the focus group appear to favor B's presentation. We are interested in determining whether the switch from A to B is statistically significant.

5.8.1 Notation and Hypotheses

Suppose we have paired responses labeled (A, A), (A, B), (B, A), and (B, B). Assume a random sample of N individuals has been selected and the paired responses have been arranged in a 2×2 contingency table as shown in Table 5.8.2. Let P_{AA}, P_{AB}, P_{BA}, and P_{BB} denote the probabilities that an individual responds (A, A), (A, B), (B, A), and (B, B), respectively. The probability that A is the first response is $P_{AA} + P_{AB}$, and the probability that A is the second response is $P_{AA} + P_{BA}$. We are interested in testing the null hypotheses $P_{AA} + P_{AB} = P_{AA} + P_{BA}$ or, equivalently,

$$H_0: P_{AB} = P_{BA}$$

against one-sided or two-sided alternatives.

TABLE 5.8.2
Two-Way Table for Paired Data

	A	B
A	X_{AA}	X_{AB}
B	X_{BA}	X_{BB}

5.8.2 Test Statistic and Permutation Distribution

We consider four test statistics. These statistics are based on X_{AB} and X_{BA}, the number of switches from A to B and the number of switches from B to A, respectively. The statistics are

$$T_1 = X_{AB}$$

$$T_2 = X_{AB} - X_{BA}$$

$$T_3 = \frac{X_{AB} - X_{BA}}{\sqrt{X_{AB} + X_{BA}}}$$

$$T_4 = T_3^2$$

The conditional distribution of T_1 will be obtained given that the number of individuals who switch is fixed. Now the number who switch is $n = X_{AB} + X_{BA}$. If n is fixed and if $P_{AB} = P_{BA}$, then among the individuals who switch, there is an equal chance that the switch is from A to B or B to A. If we assign a plus sign to individuals who are classified (A, B) and a minus sign to those who are classified (B, A), then X_{AB} is the number of plus signs. The outcome of whether an individual is (A, B) or (B, A) under the null hypotheses is equivalent to assigning pluses and minuses with probability $p = .5$ to the individuals who switch. The conditional distribution of X_{AB}, given n, is a binomial distribution with mean $.5n$ and variance $.25n$ under the null hypotheses. It follows that the test based on X_{AB} is just the sign test as discussed in Section 4.3.1.

Again, for a given value of n, we have

$$X_{AB} - X_{BA} = X_{AB} - (n - X_{AB}) = 2X_{AB} - n$$

Thus, T_1 and T_2 are equivalent statistics. The statistic T_3 can be expressed as

$$T_3 = \frac{X_{AB} - .5n}{\sqrt{.25n}}$$

This is just the standardized version of the sign statistic, and it also is equivalent to the statistic T_1. The statistic T_3 has an approximate standard normal distribution under H_0. Finally, using T_4 is equivalent to doing a two-tail test with T_3. It has an approximate chi-square distribution with 1 degree of freedom. Using the chi-square approximation with statistic T_4 is typically called *McNemar's test*.

EXAMPLE 5.8.1   For the data in Table 5.8.1, 9 of the 11 individuals who switched did so from A to B. From Table A10 in the Appendix the one-sided p-value is .033. For the normal approximation, we see that

$$T_3 = \frac{9 - 5.5}{\sqrt{2.75}} = 2.11$$

From the standard normal distribution, $p = P(Z \geq 2.11) = .0174$.   ∎

5.8.3 Computer Analysis

StatXact implements McNemar's test. A printout of the analysis of the data in Table 5.8.1 is shown in Figure 5.8.1. StatXact computes the statistic T_2. S-Plus has the chi-square approximation for T_4 available from its menu of statistical procedures.

FIGURE 5.8.1

McNemar's Test in StatXact for Data in Table 5.8.1

```
MCNEMAR'S TEST

Statistic based on the observed 2 by 2 table(x) :

        Min          Max        Mean      Std-dev     Observed Standardized
     -11.00        11.00      0.0000        3.317        7.000         2.111

Asymptotic Inference:
    One-sided p-value: Pr { Test Statistic .GE. Observed }  =      0.0174
    Two-sided p-value: 2 * One-sided                        =      0.0348

Exact Inference:
    One-sided p-value: Pr {  Test Statistic .GE. Observed } =      0.0327
                       Pr {  Test Statistic .EQ. Observed } =      0.0269
    Two-sided p-value: 2*One-Sided                          =      0.0654
```

Exercises

1 Find the permutation distribution of the slope of the least squares line for the height and weight data in the table.

Height	68	70	74
Weight	145	155	160

2 Generate the permutation distributions of Spearman's r_s (see Section 5.2) and Kendall's τ (see Section 5.3) for the data in Exercise 1.

3 The data in the table are the ages (in days) of concrete cylinders and the compressive strengths of the cylinders.

Age	3	7	15	24	85	180	360
Strength	2500	3200	4300	5300	5900	6700	6900

a Plot the data to show a nonlinear relationship. Compute Pearson's correlation, Spearman's correlation, and Kendall's tau.

b Test for significant association using each of the measures of association in part a.

4 The data are the mothers' heights and daughters' heights (in inches) of 11 mother–daughter pairs. Test for significant association using Pearson's correlation, Spearman's correlation, and Kendall's tau.

Mother	70	69	65	64	66	65	64	66	60	70	66
Daughter	67	64	62	64	69	70	65	66	63	74	60

5   The data obtained from graphs in Matlack (2001) are mean soil moisture readings (in g/m²) versus the relative abundance of shrews (in number of shrews per trap line). Test for significant association between the two variables using Spearman's correlation and Kendall's tau adjusted for ties.

Moisture	355	370	380	380	380	400	400	415	415	415	415	430	440	440	470
Shrew Abundance	0.5	2	2	0.5	3	0	4	0.5	2	4	4	1	2	5	7

6   Find the permutation distribution of the chi-square statistic for the 2×2 table.

	Column 1	Column 2
Row 1	2	0
Row 2	1	2

7   In a study of contamination of farm wells, contaminated water was classified as either "low contamination" or "high contamination," and the distance of the well from a potential source of organic contamination was classified as either "nearby" or "not nearby." Results are shown in the table. Test for significant association between contamination and distance using Fisher's exact test.

	Nearby	Not Nearby
Low	4	3
High	9	0

8   Test for association between the English and mathematics grades of the 56 college applicants.

Math (column)/English (row)	A	B	C
A	6	8	8
B	4	12	12
C	1	3	2

9   The data are the ACT mathematics scores and high school mathematics grades of 56 college applicants.

Grades (column)/ACT scores (row)	A	B	C
30 or greater	4	5	1
24–29	5	8	4
18–23	1	6	10
17 or less	3	3	6

a Test for association between the two factors using a permutation chi-square test.

b Test for association using with the Kruskal–Wallis test with ties, treating grades as the ordered factor (A = 1, B = 2, C = 3). Compare the *p*-value with that in part a.

c Test for association using the Jonckheere–Terpstra statistic with ties, treating both grades and ACT scores as ordered factors. Compare the *p*-value with those in parts a and b.

10 Refer to the data in Exercise 9. Suppose we would like to compare the proportion of A's among the four ACT categories. Obtain the permutation distribution of $Q^* = \max_i |Z_i|$, where the maximum is taken over the six possible pairs of comparisons of proportions (proportion of A's in the 30-or-greater group versus proportion of A's in the 24–29 group, etc.). Determine where significant differences exist.

11 Data on the presence or absence of a certain insect on plants were taken at three locations. The study used two types of plants: those that are not resistant to the insect and those that were bred to be resistant.

	Location 1	
	Insect Present	Insect Not Present
Resistant	4	16
Not Resistant	9	11

	Location 2	
	Insect Present	Insect Not Present
Resistant	1	18
Not Resistant	6	14

	Location 3	
	Insect Present	Insect Not Present
Resistant	7	11
Not Resistant	10	9

a Based on the data, is there a significant association between the type of plant and the presence of the insect?

b   Assuming a common odds ratio, compute a 95% confidence interval for the odds ratio and interpret the result.

12   Thirty recreational basketball players were asked to shoot two free throws. Data on whether they made or missed their shots are shown in the table. The question of interest is whether the probability of making a shot on the first attempt is different from the probability of making a shot on the second attempt.

	Missed Second Attempt	Made Second Attempt
Missed First Attempt	4	14
Made First Attempt	5	7

a   Use McNemar's test to answer this question.

b   Suppose the pairings of the observations were ignored, and we analyzed the data using a permutation chi-square test as discussed in Section 5.4. Compare this *p*-value with the *p*-value in part a.

Theory and Complements

13   **A Computational Formula.** Let

$$D = \sum_{i=1}^{n} \left[R(X_i) - R(Y_i) \right]^2$$

a   It can be shown that

$$r_s = 1 - \frac{6D}{n(n^2 - 1)}$$

when the data have no ties. Apply this formula to the data in Table 5.2.1 to show by example that this formula gives r_s.

b   Verify the formula using the following steps. First note that the Pearson correlation can be expressed as

$$r = \frac{\sum_{i=1}^{n} X_i Y_i - n\overline{X}\,\overline{Y}}{n\sigma_x \sigma_y}$$

where σ_x and σ_y are the population standard deviations of the sets $\{X_1, X_2, \ldots, X_n\}$ and $\{Y_1, Y_2, \ldots, Y_n\}$, respectively. Use the fact that $R(X)$ and $R(Y)$ consist of the integers 1 to n to show that

$$\sum_{i=1}^{n} R(X_i)R(Y_i) = \frac{n(n+1)(2n+1)}{12} - \frac{D}{2}$$

and

$$r_s = \frac{\sum_{i=1}^{n} R(X_i)R(Y_i) - A}{B}$$

where

$$A = \frac{n(n+1)^2}{4}, \quad B = \frac{n(n-1)(n+1)}{12}$$

Substitute for $\sum R(X_i)R(Y_i)$ in r_s in terms of D and simplify to obtain the formula.

14 The following procedures may be used to estimate the slope of a regression line and to obtain a confidence interval for the slope: Suppose we have X and Y related by the equation $Y_i = \beta_0 + \beta_1 X_i + \varepsilon_i$ as in Section 5.1. Suppose the X_i's are distinct and $X_1 < X_2 < \ldots < X_n$.

The estimator β_1 is the median of the $N = \binom{n}{2}$ slopes of the form $(Y_j - Y_i)/(X_j - X_i)$, where $i < j$. If we denote the slopes in order from smallest to largest as $m_{(1)} < m_{(2)} < \ldots < m_{(N)}$, then the confidence interval is of the form $m_{(L)} < \beta_1 < m_{(U)}$, where the limits are chosen to include β_1 with a desired level of confidence. We note that $\beta_1 < m_{(U)}$ if and only if at most $U - 1$ of the slopes are less than or equal to β_1 or, equivalently, at least $N - U + 1$ of the slopes are greater than β_1. Since $(Y_j - Y_i)/(X_j - X_i) > \beta_1$ if and only if $Y_j - (\beta_0 + \beta_1 X_j) > Y_i - (\beta_0 + \beta_1 X_i)$ and since $\varepsilon_i = Y_i - (\beta_0 + \beta_1 X_i)$, it follows that the probability that there are at least $N - U + 1$ slopes greater than β_1 is the same as the probability that there are at least $N - U + 1$ concordant pairs among the pairs (i, ε_i), (j, ε_j), $i < j$. The distribution of the number of concordant pairs may be determined from the distribution of Kendall's tau using the relationships developed in Section 5.3. Hence, it is possible to determine $P(\beta_1 < m_{(U)})$ from the distribution of Kendall's tau. A similar procedure applies to the lower end of the confidence interval. Details are left to the reader. This procedure was proposed by Theil (1950).

15 Show that the fraction of permutations of the data that give rise to the configuration in Table 5.4.1 is

$$\frac{\prod_{i=1}^{r}(n_{i.})!\prod_{j=1}^{c}(n_{.j})!}{(n)!\prod_{i=1}^{r}\prod_{j=1}^{c}(n_{ij})!}$$

This formula applies whether we permute observations among rows or permute them among columns.

16 The following is a derivation of the formula for the variance in the Mantel–Haenszel test.

  a Using the results in Section 2.10, show that if a finite population of size N contains a proportion p of 1's and a proportion $1 - p$ of 0's, and if n items are selected at random from the population without replacement, then the number of 1's in the sample has a mean of np and a variance of

$$\frac{N-n}{N-1}np(1-p)$$

b   Suppose we have this 2×2 contingency table:

	Column 1	Column 2	Row Total
Row 1	X		r_1
Row 2			r_2
Column Totals	c_1	c_2	N

If there is no association between the row and column factors, then X can be regarded as being the number of 1's selected at random without replacement in a sample of size r_1 from a finite population of size N, where the proportion of 1's is c_1/N and the proportion of 0's is c_2/N. Apply part a to show that the mean of X is r_1c_1/N and the variance of X is

$$\frac{r_2}{N-1}r_1\frac{c_1}{N}\frac{c_2}{N} = \frac{r_1r_2c_1c_2}{N^2(N-1)}$$

c   To obtain the mean and variance of the Mantel–Haenszel statistic, apply the results of part b to each stratum.

6

Multivariate Tests

A Look Ahead   Seldom in research is a single variable measured. In studies on animals, for instance, we may measure weight gains, body fat percentages, and feed consumption, and we may measure these amounts at several times during a study. A univariate analysis would consider each variable separately, determining how each is affected by the treatments. However, it may be advantageous to consider the variables simultaneously. This perhaps will give a clearer picture of the effect of the treatments. Multivariate techniques are designed to take into account several variables at once. In this chapter, we present some simple multivariate techniques for comparing two treatments.

6.1
Two-Sample Multivariate Permutation Tests

We now consider the problem of comparing two treatments in experiments in which multiple response variables are measured on each experimental unit. A grain scientist who is comparing breads made from two types of flour might be interested in several sensory characteristics, such as aroma, sweetness, and moistness. An anthropologist might compare the skeletal remains of two groups of primates according to measurements of lengths or diameters of hand, arm, leg, and cranial bones. Measurements of a single characteristic taken at several different times also represent multiple responses. For instance, a physician might monitor cholesterol levels on a group of patients each week for a five-week period as a means of determining which of two treatments is more effective at reducing cholesterol levels.

Statistical analysis of such data may be done one variable at a time, or it may be done by incorporating all information into a multivariate analysis. There are two reasons for doing multivariate analysis. (1) The researcher can control the experiment-wise error rate, which is the probability of incorrectly declaring that the two treatments differ on at least one response variable when in fact they do not differ on any. (2) An appropriately chosen test procedure that uses all of the response variables may have greater power to detect differences between treatments than an analysis on any single variable.

6.1.1 Notation and Assumptions

Assume we have m experimental units for treatment 1 and n for treatment 2, and each experimental unit has k response variables. The units are assumed to be selected at random from two populations. The vector of observations for the two treatments are denoted as

$$\text{Treatment 1: } X' = \left(X_{i1}, X_{i2}, \ldots, X_{ik}\right), \quad i = 1, \ldots, m$$

$$\text{Treatment 2: } Y' = \left(Y_{i1}, Y_{i2}, \ldots, Y_{ik}\right), \quad i = 1, \ldots, n$$

The vectors of sample means are given by

$$\text{Treatment 1: } \overline{X}' = \left(\overline{X}_1, \overline{X}_2, \ldots, \overline{X}_k\right)$$

$$\text{Treatment 2: } \overline{Y}' = \left(\overline{Y}_1, \overline{Y}_2, \ldots, \overline{Y}_k\right)$$

The covariance between the uth and vth response variables on treatment 1 is

$$C_{Xuv} = \frac{\sum_{i=1}^{m}\left(X_{iu} - \overline{X}_u\right)\left(X_{iv} - \overline{X}_v\right)}{m - 1}$$

The covariance C_{Yuv} between the uth and vth response variables on treatment 2 is defined in the analogous way. The pooled covariance is

$$C_{uv} = \frac{(m-1)C_{Xuv} + (n-1)C_{Yuv}}{m + n - 2}$$

Finally, we let $C = [C_{uv}]$ denote the $k \times k$ matrix of pooled covariances.

A common multivariate statistic for comparing two treatments is *Hotelling's T^2* defined by

$$T^2 = \frac{mn}{m+n}\left(\overline{\mathbf{X}} - \overline{\mathbf{Y}}\right)' \mathbf{C}^{-1}\left(\overline{\mathbf{X}} - \overline{\mathbf{Y}}\right)$$

If the observations are selected randomly from multivariate normal populations in which the covariance matrices are the same for the two populations, and if there are no differences between the means of the distributions, then the statistic

$$F = \frac{m + n - k - 1}{(m + n - 2)k} T^2$$

has an F-distribution with degrees of freedom k and $m + n - k - 1$. The null hypothesis being tested is

$$H_0: \mu_{Xj} = \mu_{Yj}, \quad j = 1, \ldots, k$$

where μ_{xj} and μ_{yj} are the population means for the *j*th response in treatments 1 and 2, respectively. The alternative hypothesis is that at least one population mean for treatment 1 is not the same as the corresponding population mean for treatment 2. If the assumption of normality is not acceptable, then one may carry out a permutation test using T^2 or F.

6.1.2 The Permutation Version of Hotelling's T^2

In a two-sample multivariate permutation test, multivariate vectors are permuted between treatments in the same way that individual observations are permuted between treatments in a univariate two-sample permutation test. This is illustrated in the following example.

EXAMPLE **6.1.1**   Table 6.1.2 is a display of the ten possible permutations of the bivariate vectors in Table 6.1.1 and the corresponding F statistics. The F statistic for the original data is 5.00. Two values of F are this large or larger among the ten possible permuted data sets. Thus, $p = .20$. If we assume that the observations are from a normal distribution, then the F statistic has an F-distribution with degrees of freedom 2 and 2. The p-value based on this F-distribution is .17.

TABLE **6.1.1**

Data to Illustrate a Multivariate Permutation Test

Treatment 1	(50, 5) (60, 4)
Treatment 2	(30, 7) (34, 8) (40, 6)

TABLE **6.1.2**

Permutation F-Test for Data in Table 6.1.1; $F = (m + n - k - 1)T^2/(m + n - 2)k$

Permutation	Treatment 1	Treatment 2	F
1*	(50,5) (60,4)	(30,7) (34,8) (40,6)	5.0
2	(50,5) (30,7)	(60,4) (34,8) (40,6)	0.5
3	(50,5) (34,8)	(60,4) (30,7) (40,6)	0.7
4	(50,5) (40,6)	(60,4) (30,7) (34,8)	0.2
5	(60,4) (30,7)	(50,5) (34,8) (40,6)	0.2
6	(60,4) (34,8)	(50,5) (30,7) (40,6)	3.0
7	(60,4) (40,6)	(50,5) (30,7) (34,8)	0.5
8	(30,7) (34,8)	(50,5) (60,4) (40,6)	3.0
9	(30,7) (40,6)	(50,5) (60,4) (34,8)	5.0
10	(34,8) (40,6)	(50,5) (60,4) (30,7)	0.7

6.1.3 Other Multivariate Statistics

Other statistics besides Hotelling's T^2 are appropriate for multivariate data. We consider two permutation tests that are based on the computation of the t statistic for each of the response variables. Let

$$ t_j = \frac{\overline{X}_j - \overline{Y}_j}{\sqrt{C_{jj}\left(\dfrac{1}{m} + \dfrac{1}{n}\right)}} $$

denote the two-sample t statistic for testing for a difference between treatments 1 and 2 on response variable j, $j = 1, 2, \ldots, k$. For instance, if height and weight are the two response variables, then we would have a t statistic for comparing the heights of the two groups and a t statistic for comparing the weights. Note that C_{jj} is the usual pooled sample variance obtained on the data from the jth response variable. The statistics that we consider are (1) the maximum of the absolute values of the t statistics and (2) the maximum of the one-sided t statistics. We denote these as

$$ t_{\text{max abs}} = \max\left(|t_1|, |t_2|, \ldots, |t_k|\right) $$

$$ t_{\text{max}} = \max\left(t_1, t_2, \ldots, t_k\right) $$

The $t_{\text{max abs}}$ statistic is a two-sided test statistic in which the direction of the differences between means is not specified. The t_{max} statistic is for an upper-tail, one-sided test. It would be appropriate if we wish to detect whether or not the means of the responses for treatment 1 are greater than the means of the corresponding responses for treatment 2. Multivariate one-sided tests were proposed by Boyett and Shuster (1977) in the context of medical applications. Blair and colleagues (1994) showed that one-sided multivariate tests enjoy substantial power advantages over Hotelling's T^2 test under certain conditions.

Let $t(\alpha)_{\text{max abs}}$ denote the critical value of $t_{\text{max abs}}$ for a test at level of significance α obtained from the permutation distribution. If the observed value of $t_{\text{max abs}}$ is greater than or equal to $t(\alpha)_{\text{max abs}}$, then we can assert at level of significance α that the treatments are different. We may also obtain the p-value as the fraction of the permutation distribution that is greater than or equal to the observed value of $t_{\text{max abs}}$. To determine which variables differ between treatments, we can assert that the treatments differ significantly on the jth response variable if $|t_j| \geq t(\alpha)_{\text{max abs}}$. This procedure controls the experiment-wise error rate in the sense that if there are no differences between the means of the responses for the two treatments, then

$$ P\left[\text{at least one } |t_j| \geq t(\alpha)_{\text{max abs}}\right] = P\left[t_{\text{max abs}} \geq t(\alpha)_{\text{max abs}}\right] = \alpha $$

We may also obtain a p-value for the jth variable as the fraction of the permutation distribution that is greater than $|t_j|$. Similar considerations apply to the test statistic t_{max}.

EXAMPLE 6.1.2   Table 6.1.3 shows $t_{max\ abs}$ and t_{max} for the ten permutations of the data in Table 6.1.2. The largest value for both $t_{max\ abs}$ and t_{max} is 3.85, which occurs with the original observations. Thus, $p = .10$ in testing for differences between treatments using either the two-sided test or the upper-tail, one-sided test. Since $t_1 = 3.85$ and $t_2 = -3.00$, both the two-sided test and the upper-tail, one-sided test show that treatment 1 differs from treatment 2 on the first response variable at the 10% level of significance but not on the second variable.

TABLE 6.1.3

Permutation Distributions for $t_{max\ abs}$ and t_{max} Based on Permutations in Table 6.1.2

Permutation	t_1	t_2	$t_{max\ abs}$	t_{max}
1*	3.85	−3.00	3.85	3.85
2	−0.37	0.00	0.37	0.00
3	−0.10	0.52	0.52	0.52
4	0.29	−0.52	0.52	0.29
5	0.29	−0.52	0.52	0.29
6	0.57	0.00	0.57	0.57
7	1.11	−1.22	1.22	1.11
8	−2.37	3.00	3.00	3.00
9	−1.24	0.52	1.24	0.52
10	−0.83	1.22	1.22	1.22

∎

EXAMPLE 6.1.3   An animal scientist obtained pH measurements on beef carcasses that had been subjected to one of two treatments. Six carcasses received each treatment. For each carcass, the pH measurements were made six times, with the objective being to compare the treatments across time. The data are shown in Table 6.1.4. The t statistics for the six times are $t_1 = 0.25$, $t_2 = 1.37$, $t_3 = 6.66$, $t_4 = 3.31$, $t_5 = 2.39$, $t_6 = -0.87$. It was expected that treatment 1 would have higher pH levels than treatment 2, so a one-sided test is appropriate. The estimated p-value for the observed value of $t_{max} = 6.66$ is .002 based on 5000 randomly selected permutations. Table 6.1.5 displays critical values for t_{max} based on 5000 randomly selected permutations. Comparing the observed t statistics with the critical values of t_{max} in this table, we find that $p < .01$ for time 3, $p < .025$ for time 4, and $p < .10$ for time 5.

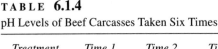

TABLE 6.1.4

pH Levels of Beef Carcasses Taken Six Times

Treatment	Time 1	Time 2	Time 3	Time 4	Time 5	Time 6
1	6.81	6.16	5.92	5.86	5.80	5.39
1	6.68	6.30	6.12	5.71	6.09	5.28
1	6.34	6.22	5.90	5.38	5.20	5.46
1	6.68	5.24	5.83	5.49	5.37	5.43
1	6.79	6.28	6.23	5.85	5.56	5.38
1	6.85	6.51	5.95	6.06	6.31	5.39
2	6.64	5.91	5.59	5.41	5.24	5.23
2	6.57	5.89	5.32	5.41	5.32	5.30
2	6.84	6.01	5.34	5.31	5.38	5.45
2	6.71	5.60	5.29	5.37	5.26	5.41
2	6.58	5.63	5.38	5.44	5.17	6.62
2	6.68	6.04	5.62	5.31	5.41	5.44

TABLE 6.1.5

Critical Values for t_{max} Based on 5000 Randomly Selected Permutations
of the Response Vectors

Statistic/Level	.10	.05	.025	.01
t_{max}	2.14	2.69	3.24	3.64

6.1.4 Computer Analysis

Multivariate permutation tests are not available in StatXact or MINITAB, and they are not available from the menu in S-Plus, although one may use the capabilities of the S-Plus programming language to carry out such tests. Multivariate permutation tests may be carried out using SAS® PROC MULTTEST. See Westfall and colleagues (1999). With the Resampling Stats add-in to the Excel spreadsheet (Blank, Seiter, and Bruce, 1999), we may take advantage of the computational features of the spreadsheet and invoke the option for permuting vectors (rows in Excel) to obtain the permutation distribution. We used this procedure in Example 6.1.3.

Figure 6.1.1 shows standard Resampling Stats code for a multivariate permutation test applied to the data in Table 6.1.1. Permutations of vectors are done as follows: We create an index vector $(1, 2, \ldots, m + n)$ to identify the experimental units. We shuffle the index vector and then select the data for each variable corresponding to the index that was selected.

FIGURE 6.1.1

Resampling Stats Code for Two-Sample Multivariate Permutation Test with t_{max} Statistic; Data in Table 6.1.1

```
maxsize default 5000

'create data vector for each variable
'first 2 elements are treatment 1
'last 4 are treatment 2
copy (50 60 30 34 40) v1
copy (5 4 7 8 6) v2

'create index
copy (1 2 3 4 5) index

'permute vectors

repeat 5000
shuffle index s_index

'variable 1 computations
take v1 s_index sv1
take sv1 1,2 trt1_v1
take sv1 3,5 trt2_v1

'difference of means
mean trt1_v1 m11
mean trt2_v1 m21
subtract m11 m21 diff_v1

'pooled standard deviation
sumsqrdev trt1_v1 m11 ss1_v1
sumsqrdev trt2_v1 m21 ss2_v1
add ss1_v1 ss2_v1 ss_v1
divide ss_v1 3 var_v1
sqrt var_v1 sp_v1

'compute t-stat variable 1
'note sqrt(1/2 + 1/3) = .91287
multiply sp_v1 .91287 denom_v1
divide diff_v1 denom_v1 t_v1




(Continued next column)
```

```
'variable 2 computations

take v2 s_index sv2
take sv2 1,2 trt1_v2
take sv2 3,5 trt2_v2

'difference of means
mean trt1_v2 m12
mean trt2_v2 m22
subtract m12 m22 diff_v2

'pooled standard deviation
sumsqrdev trt1_v2 m12 ss1_v2
sumsqrdev trt2_v2 m22 ss2_v2
add ss1_v2 ss2_v2 ss_v2
divide ss_v2 3 var_v2
sqrt var_v2 sp_v2

'compute t-stat variable 2
'note sqrt(1/2 + 1/3) = .91287
multiply sp_v2 .91287 denom_v2
divide diff_v2 denom_v2 t_v2

'get tmax statistic
concat t_v1 t_v2 t_all
max t_all t_max

'keep track of t_max
score t_max dist

end

'get pvalue for observed t_max = 3.845
count dist >=3.845 pval
divide pval 5000 pval
print pval

'print pvalue here

'print pvalue here

PVAL    =    0.1022
```

6.2
Two-Sample Multivariate Rank Tests

Assume we have taken multivariate observations as in Section 6.1. To construct a multivariate rank test, we first obtain separate rankings for each variable. That is, for each response variable, we combine the data and rank the observations from smallest to largest using average ranks for tied observations. We then replace the original variables by their ranks. A *multivariate rank test* is a multivariate permutation test performed on these ranked observations.

We may compute a statistic analogous to Hotelling's T^2 simply by substituting ranks for the actual observations, or we may compute Wilcoxon rank-sum statistics analogous to $t_{max\ abs}$ or t_{max}. Specifically, for the k response variables let $W_1, W_2, \ldots, W_k$ denote the Wilcoxon rank-sum statistics, where each statistic is computed as the sum of the ranks for treatment 1. For the jth response variable, let

$$Z_j = \frac{W_j - E(W_j)}{\sqrt{\text{var}(W_j)}}, \quad j = 1, \ldots, k$$

where $\text{var}(W_j)$ is adjusted for ties if necessary. Two statistics of interest are

$$Z_{max\ abs} = \max\left(|Z_1|, |Z_2|, \ldots, |Z_k|\right)$$

$$Z_{max} = \max\left(Z_1, Z_2, \ldots, Z_k\right)$$

We carry out multivariate permutation tests using these statistics in the same way we carry out tests using $t_{max\ abs}$ and t_{max}.

EXAMPLE 6.2.1   Table 6.2.1 gives the ranks of the pH data in Table 6.1.4. The Z statistics corresponding to the Wilcoxon rank-sum statistics of observed data are $Z_1 = 0.81$, $Z_2 = 1.92$, $Z_3 = 2.88$, $Z_4 = 2.41$, $Z_5 = 1.76$, and $Z_6 = -0.48$. Since the researcher expects the pH responses to be the same or higher for treatment 1 than treatment 2, the one-sided statistic Z_{max} is appropriate in this case. Critical values for Z_{max} based on 5000 randomly selected permutations are shown in Table 6.2.2. Comparing the Z_i's with these critical values, we find $p < .01$ at time 3 and $p < .05$ at time 4. The significance levels are .673, .140, .005, .037, .194, and .994, respectively.   ∎

Other scoring systems may be used instead of ranks in multivariate tests—for instance, exponential scores. The choice should be guided by the same considerations as in the case of one response variable.

TABLE **6.2.1**

Ranks of pH Levels in Beef Carcasses Taken Six Times

Treatment	Time 1	Time 2	Time 3	Time 4	Time 5	Time 6
1	10	8	9	11	10	5.5
1	6	11	11	9	11	2
1	1	9	8	4	2	11
1	6	1	7	8	6	8
1	9	10	12	10	9	4
1	12	12	10	12	12	5.5
2	4	5	5	5.5	3	1
2	2	4	2	5.5	5	3
2	11	6	3	1.5	7	10
2	8	2	1	3	4	7
2	3	3	4	7	1	12
2	6	7	6	1.5	8	9

TABLE **6.2.2**

Critical Values for Z_{max}

Statistic/Level	.10	.05	.025	.01
Z_{max}	1.94	2.25	2.49	2.72

6.2.1 A Sum of Wilcoxon Statistics

Suppose the anticipated effect of the treatments is to cause the observations from treatment 1 to be on average larger than the observations from treatment 2 for all of the variables. There may not be sufficient power in a one-sided Wilcoxon rank-sum test on any individual response variable to detect a difference between treatments, but a sum of such statistics may reveal a difference. This is much like the situation in testing for ordered alternatives with the Jonckheere–Terpstra statistic, in which a sum of Wilcoxon statistics is used to test for the effect of interest.

Sum Statistic with No Ties

Assume the data contain no ties. The sum statistic is

$$W_{sum} = W_1 + W_2 + \cdots + W_k$$

where W_j is the rank sum for treatment 1 on response variable j, $j = 1, \ldots, k$. We may find the permutation distribution of this statistic. In addition, a simple normal approximation is available.

The expected value is given by

$$E\left(W_{\text{sum}}\right) = \frac{km(m+n+1)}{2}$$

Let W_{ij} denote the rank of the *i*th experimental unit on the *j*th variable. Let

$$S_j = \sum_{j=1}^{k} W_{ij}$$

and let σ_S^2 denote the population of the S_j's. The variance of W_{sum} is given by

$$\text{var}\left(W_{\text{sum}}\right) = \frac{mn}{m+n-1}\sigma_S^2$$

We refer the statistic

$$Z = \frac{W_{\text{sum}} - E\left(W_{\text{sum}}\right)}{\sqrt{\text{var}\left(W_{\text{sum}}\right)}}$$

to the standard normal distribution.

Sum Statistic with Ties

If the data contain only a few ties, then the procedure is to compute W_j using average ranks but make no other adjustment. If the number of ties is substantial, then we standardize the W_j's before summing; that is,

$$Z_{\text{sum}} = \sum_{j=1}^{k} \frac{W_j - E\left(W_j\right)}{\sqrt{\text{var}\left(W_j\right)}}$$

where W_j is computed using average ranks and var(W_j) is adjusted for ties. Again we may either perform a permutation test on Z_{sum} or use a normal approximation.

For the normal approximation, we have $E(Z_{\text{sum}}) = 0$. We standardize the ranks for each variable using

$$Z_{ij} = \frac{W_{ij} - E\left(W_j\right)}{\sqrt{\text{var}\left(W_j\right)}}$$

Let

$$SZ_j = \sum_{j=1}^{k} Z_{ij}$$

and let σ_Z^2 denote the population variance of the SZ_j's. The variance of Z_{sum} is given by

$$\text{var}(Z_{\text{sum}}) = \frac{mn}{m+n-1}\sigma_Z^2$$

We refer the statistic

$$Z = \frac{Z_{\text{sum}}}{\sqrt{\text{var}(Z_{\text{sum}})}}$$

to the standard normal distribution.

EXAMPLE 6.2.2   For the ranked data in Table 6.2.1, we have $W_1 = 44$, $W_2 = 51$, $W_3 = 57$, $W_4 = 54$, $W_5 = 50$, $W_6 = 36$, and $W_{\text{sum}} = 292$. The S_j's are the sums of the 12 rows in Table 6.2.1, which are 53.5, 50, 35, 36, 54, 63.5, 23.5, 21.5, 38.5, 25, 30, and 37.5. The population variance of these numbers is 167.3. Thus,

$$\text{var}(W_{\text{sum}}) = \frac{(6)(6)}{11}167.3 = 547.5$$

Moreover, $E(W_{\text{sum}}) = 234$. Thus,

$$Z = \frac{292 - 234}{\sqrt{547.5}} = 2.48$$

The upper-tail p-value is .007, indicating a significant difference between the treatments. Adjustment for ties gives virtually the same level of significance. We may obtain an exact test by applying a two-sample permutation test to the 12 scores, where the first six belong to treatment 1 and the second six belong to treatment 2. From StatXact, this p-value is .0065. ∎

6.2.2 Computer Analysis

Resampling Stats code for the multivariate permutation tests in Section 6.1 may be appropriately modified to obtain p-values for multivariate rank tests. Both MINITAB and S-Plus have menu options for ranking data, and after ranks are obtained, a sum of Wilcoxon statistics, its variance, and the normal approximation may be computed. In addition, we may apply a two-sample permutation test to the S_j's or the SZ_j's using StatXact to obtain exact levels of significance for the sum of the Wilcoxon statistics.

6.3
Multivariate Paired Comparisons

We extend the notion of paired comparisons to the multivariate setting. Vectors of observations are paired according to some characteristic. For instance, the pairs could represent science, mathematics, and language scores taken before and after students have undergone intensive preparation for a college entrance examination. The data consist of pairs of vectors with k response variables on each vector. The vectors are denoted as follows:

$$\text{Vector 1 of }i\text{th pair: } \mathbf{X}'_i = \left(X_{i1},\ X_{i2},\ \ldots,\ X_{ik}\right),\ \ i = 1,\ \ldots,\ n$$

$$\text{Vector 2 of }i\text{th pair: } \mathbf{Y}'_i = \left(Y_{i1},\ Y_{i2},\ \ldots,\ Y_{ik}\right),\ \ i = 1,\ \ldots,\ n$$

Let $D_{ij} = X_{ij} - Y_{ij}$ denote the difference between the jth response variable on the ith pair, $i = 1, 2, \ldots, n, j = 1, \ldots, k$. The vector of the differences is denoted

$$\mathbf{D}'_i = \left(D_{i1},\ D_{i2},\ \ldots,\ D_{ik}\right),\ \ i = 1,\ \ldots,\ n$$

and the vector of the sample means of the differences is denoted

$$\overline{\mathbf{D}}' = \left(\overline{D}_1,\ \overline{D}_2,\ \ldots,\ \overline{D}_k\right)$$

Let Δ_j denote the mean (or median) of the distribution of the jth response variable. The null hypothesis to be tested is $H_0\colon \Delta_j = 0, j = 1, 2, \ldots, k$. The alternative hypothesis may be either two-sided or one-sided. The two-sided alternative is $H_a\colon \Delta_j \neq 0$ for at least one j, $j = 1, 2, \ldots, k$. The upper-tail, one-sided alternative is $H_a\colon \Delta_j > 0$ for at least one j, $j = 1, 2, \ldots, k$. Similarly, the lower-tail hypothesis is $H_a\colon \Delta_j < 0$ for at least one j.

Various multivariate tests statistics may be applied to the vector of differences. The Hotelling's T^2 statistic for paired comparisons, which tests the two-sided alternative, is

$$T^2 = n\overline{\mathbf{D}}'\mathbf{C}^{-1}\overline{\mathbf{D}}$$

where $\mathbf{C}$ is the $k \times k$ matrix of covariance with this uvth entry:

$$C_{uv} = \frac{\sum\limits_{i=1}^{n}\left(D_{iu} - \overline{D}_u\right)\left(D_{iv} - \overline{D}_v\right)}{n - 1}$$

If the vector $\mathbf{D}$ has a multivariate normal distribution, then

$$F = \frac{n - k}{(n - 1)k}T^2$$

has an F-distribution with k degrees of freedom for the numerator and $n - k$ degrees of freedom for the denominator.

6.3.1 A Permutation Test Based on T^2

If the assumption of normality is not acceptable, then the permutation distribution of T^2 or F may be obtained. This procedure is carried out as follows: Under the null hypothesis, either vector for the ith pair is as likely to be vector 1 as vector 2. Thus, the vector of differences is equally likely to be either the observed vector of differences itself or the negative of the observed vector of differences. For instance, if the vectors are $X_i' = (10, 5, 8)$ and $Y_i' = (5, 6, 11)$, then the vector of differences under the null hypothesis is equally likely to be either $D_i' = (5, -1, -3)$ or $D_i' = (-5, 1, 3)$. If there are n pairs of vectors, there are 2^n possible permutations of the pairs and thus 2^n possible sets of difference vectors. We may compute Hotelling's T^2, or some other appropriate statistic, on each set of difference vectors and obtain the permutation distribution of the statistic in question. From there a p-value for the observed statistic is obtained and a decision to accept or reject the null hypothesis reached.

EXAMPLE 6.3.1 Suppose four students are given mathematics and English tests in a quiet environment and in an environment in which music is playing in the background. The test scores and differences are shown in Table 6.3.1. Table 6.3.2 lists the 16 possible sets of difference vectors and the corresponding Hotelling's T^2. The test statistic can take on eight distinct values (0.11, 0.70, 0.97, 3.25, 5.32, 5.61, 17.51, 19.55), each with probability 1/8 under the null hypothesis. The observed value (set 1) has $T^2 = 5.32$, which is the fourth largest value with $p = .50$.

TABLE 6.3.1
Math and English Test Scores with and without Music

Student	With Music		Without Music		Difference	
	Math	English	Math	English	Math	English
1	82	60	72	62	10	-2
2	75	71	70	68	5	3
3	85	59	87	64	-2	-5
4	90	77	87	78	3	-1

■

6.3.2 Random Sampling the Permutations

Suppose n is large enough so that enumeration of all possible permutations is not feasible. We now consider how to randomly sample the permutations. Let M denote the matrix whose rows are the difference vectors D_i'. For the data in Example 6.3.1, we have

TABLE 6.3.2

Sets of Difference Vectors and T^2 Obtained from Permutations of the Data in Table 6.3.1

10	−2	10	−2	10	−2	10	−2
5	3	5	3	5	3	−5	−3
−2	−5	−2	−5	2	5	−2	−5
3	−1	−3	1	3	−1	3	−1
$T^2 = 5.32\ (1^*)$		$T^2 = 0.97\ (2)$		$T^2 = 19.55\ (3)$		$T^2 = 17.51\ (4)$	
−10	2	10	−2	10	−2	10	−2
5	3	5	3	−5	−3	−5	−3
−2	−5	2	5	−2	−5	2	5
3	−1	−3	1	−3	1	3	−1
$T^2 = 0.11\ (5)$		$T^2 = 5.61\ (6)$		$T^2 = 3.25\ (7)$		$T^2 = 0.70\ (8)$	
−10	2	−10	2	−10	2	−10	2
−5	−3	5	3	5	3	−5	−3
−2	−5	2	5	−2	−5	2	5
3	−1	3	−1	−3	1	3	−1
$T^2 = 5.61\ (9)$		$T^2 = 3.25\ (10)$		$T^2 = 0.70\ (11)$		$T^2 = 0.97\ (12)$	
−10	2	−10	2	10	−2	−10	2
−5	−3	5	3	−5	−3	−5	−3
−2	−5	2	5	2	5	2	5
−3	1	−3	1	−3	1	−3	1
$T^2 = 19.55\ (13)$		$T^2 = 17.51\ (14)$		$T^2 = 0.11\ (15)$		$T^2 = 5.32\ (16)$	

$$M = \begin{pmatrix} 10 & -2 \\ 5 & 3 \\ -2 & -5 \\ 3 & -1 \end{pmatrix}$$

Let U' denote a k-dimensional vector of the form $(\pm 1, \pm 1, \ldots, \pm 1)$. A permutation of the original data results in a difference vector of the form $U'M$. For instance, permutation 8 in Table 6.3.2 may be expressed as

$$(1, \; -1, \; -1, \; 1) \begin{pmatrix} 10 & -2 \\ 5 & 3 \\ -2 & -5 \\ 3 & -1 \end{pmatrix} = \begin{pmatrix} 10 & -2 \\ -5 & -3 \\ 2 & 5 \\ 3 & -1 \end{pmatrix}$$

We can randomly generate a vector U' by independently generating each component of the vector to be either +1 or –1 with probability 1/2. It follows that $U'M$ is a random permutation of the differences.

We may base the permutation test on randomly selected differences if enumeration of all possible sets of differences is not feasible. In implementing this procedure in a computer program, we must take care not to inadvertently assign plus and minus signs independently to each of the variables. The signs of the elements of a difference vector must be either all the same as those of the original vector or all their negatives.

6.3.3 Other Multivariate Statistics

In addition to a permutation test based on the Hotelling T^2 statistic, we may obtain two-sided and one-sided tests based on paired t statistics computed on each of the variables. Let S_j denote the sample standard deviation of the differences on the jth response variable D_{ij}, $i = 1, 2, \ldots, n$. The paired t statistic for the jth variable is

$$t_j = \frac{\overline{D}_j}{S_j / \sqrt{n}}$$

The two-sided and upper-tail, one-sided test statistics are

$$t_{\text{max abs}} = \max\left(|t_1|, \; |t_2|, \; \ldots, \; |t_k|\right)$$

$$t_{\text{max}} = \max\left(t_1, \; t_2, \; \ldots, \; t_k\right)$$

EXAMPLE 6.3.2   The statistics $t_{\text{max abs}}$ and t_{max} for the 16 permutations in Table 6.3.2 are shown in Table 6.3.3. For the observed data (set 1), we find $t_{\text{max abs}} = 1.61$. Since 8 of the 16 values of $t_{\text{max abs}}$ are greater than or equal to 1.61, the p-value for this statistic is .50. Also we find $t_{\text{max}} = 1.61$, but only four values of t_{max} are greater than or equal to 1.61, so the p-value for this statistic is .25.

TABLE 6.3.3

Values of $t_{max\ abs}$ and t_{max} for Permutations in Table 6.3.2

Set	t_1	t_2	$t_{max\ abs}$	t_{max}
1*	1.61	−0.76	1.61	1.61
2	0.81	−0.43	0.81	0.81
3	2.81	0.76	2.81	2.81
4	0.46	−3.22	3.22	0.46
5	−0.30	−0.14	0.30	−0.14
6	1.29	1.17	1.29	1.29
7	0.00	−1.80	1.80	0.00
8	0.81	−0.14	0.81	0.81
9	−1.29	−1.17	1.29	−1.17
10	0.00	1.80	1.80	1.80
11	−0.81	0.14	0.81	0.14
12	−0.81	0.43	0.81	0.43
13	−2.81	−0.76	2.81	−0.76
14	−0.46	3.22	3.22	3.22
15	0.30	0.14	0.30	0.30
16	−1.61	0.76	1.61	0.76

6.3.4 Computer Analysis

Resampling Stats code for a multivariate paired-comparison test using the t_{max} statistic is shown in Figure 6.3.1. Multivariate paired-comparison tests are not available in StatXact, MINITAB, or the S-Plus statistics menu.

6.4
Multivariate Rank Tests for Paired Comparisons

Rank tests for paired comparisons follow the same pattern as the permutation tests for paired comparisons in Section 6.3. We obtain signed ranks for the differences in responses on each variable and obtain the permutation distribution of an appropriate statistic based on signed ranks. For instance, a rank test may be obtained by applying the permutation version of Hotelling's T^2 to the signed ranks. We may also base multivariate statistics on the Wilcoxon statistic. Specifically, let

FIGURE 6.3.1

Resampling Stats Code for Multivariate Paired Permutation Test Applied to Data in Table 6.3.1

```
maxsize default 5000

'input differences for first variable
copy (10 5 -2 3) v1
'input differences for second variable
copy (-2 3 -5 -1) v2

'observed t-statistics are tmax = 1.61, t1 = 1.61, t2 = -.76

'randomly select 5000 permutations
repeat 5000
sample 4 (-1 1) rannum
multiply rannum v1 sv1
multiply rannum v2 sv2

'compute t-statistics
mean sv1 mv1
mean sv2 mv2
variance sv1 var_v1
variance sv2 var_v2
divide var_v1 4 vn1
divide var_v2 4 vn2
sqrt vn1 se1
sqrt vn2 se2
divide mv1 se1 t1
divide mv2 se2 t2

'get tmax and keep track of values
concat t1 t2 t
max t tmax
score tmax tmaxdist

end

count tmaxdist >= 1.61 pvalmax
divide pvalmax 5000 pvalmax
count tmaxdist >= 1.61 pvalt1
divide pvalt1 5000 pvalt1
count tmaxdist >= -.76 pvalt2
divide pvalt2 5000 pvalt2
print pvalmax pvalt1 pvalt2

'print output here

PVALMAX  =      0.2534
PVALT1   =      0.2534
PVALT2   =      0.9338
```

$$Z_j = \frac{\mathrm{SR}_{+j} - E(S_{+j})}{\sqrt{\mathrm{var}(S_{+j})}}, \quad j = 1, \ldots, k$$

where SR_{+j} is the sum of the positive signed ranks for the jth variable, and the expected value and variance have been adjusted for ties if necessary. Two statistics of interest are

$$Z_{\mathrm{max\ abs}} = \max(|Z_1|, |Z_2|, \ldots, |Z_k|)$$

$$Z_{\mathrm{max}} = \max(Z_1, Z_2, \ldots, Z_k)$$

which test two-sided and upper-tail, one-sided alternatives, respectively.

EXAMPLE 6.4.1   For the differences of the data in Table 6.3.1, the matrix of signed ranks of the four students is

$$\begin{pmatrix} 4 & -2 \\ 3 & 3 \\ -1 & -4 \\ 2 & -1 \end{pmatrix}$$

The expected value and variance of the signed-rank statistic without ties are $E(\mathrm{SR}_+) = 5$ and $\mathrm{var}(\mathrm{SR}_+) = 7.5$, respectively. The Z statistics computed on the signed ranks are $Z_1 = (9 - 5)/\sqrt{(7.5)} = 1.46$ and $Z_2 = (3 - 5)/\sqrt{(7.5)} = -0.73$. Thus, $Z_{\mathrm{max\ abs}} = Z_{\mathrm{max}} = 1.46$. Table 6.4.1 contains the values of $Z_{\mathrm{max\ abs}}$ and Z_{max} for the 16 sets of differences in Table 6.3.2. Eight of the 16 values of $Z_{\mathrm{max\ abs}}$ are greater than or equal to 1.46, so the p-value for $Z_{\mathrm{max\ abs}}$ is .50. Similarly, the p-value for Z_{max} is .25.   ∎

6.4.1 Testing Medians of a Symmetric Distribution

The methodology here and in Section 6.3 may also apply to certain situations in which the data do not arise as differences in a paired-comparison experiment. Suppose the vectors $D_i = (D_{i1}, D_{i2}, \ldots, D_{ik})$, $i = 1, 2, \ldots, n$, are randomly selected from a multivariate population whose distribution is symmetric about a median or mean vector $\Delta = (\Delta_1, \Delta_2, \ldots, \Delta_k)$ in the sense that the distribution of $D_{ij} - \Delta_j$ is the same as that of $-(D_{ij} - \Delta_j)$, $i = 1, \ldots, n, j = 1, \ldots, k$. If we wish to test $\Delta_j = 0, j = 1, \ldots, k$, against either two-sided or one-sided alternatives, we may use the methodology for paired comparisons, since the difference vectors D_i and $-D_i$ have identical distributions under the null hypothesis as in the paired-comparison situation.

TABLE 6.4.1

Values of $Z_{\text{max abs}}$ and Z_{max} for Sets of Differences in Table 6.3.2

Set	Z_1	Z_2	$Z_{\text{max abs}}$	Z_{max}
1	1.46	−0.73	1.46	1.46
2	0.73	−0.37	0.73	0.73
3	1.83	0.73	1.83	1.83
4	0.37	−1.83	1.83	0.37
5	0.00	0.00	0.00	0.00
6	1.10	1.10	1.10	1.10
7	−0.37	−1.46	1.46	−0.37
8	0.73	−0.37	0.73	0.73
9	−1.10	−1.10	1.10	−1.10
10	0.37	1.46	1.46	1.46
11	−0.73	0.37	0.73	0.37
12	−0.73	0.37	0.73	0.37
13	−1.83	−0.73	1.83	−0.73
14	−0.37	1.83	1.83	1.83
15	0.00	0.00	0.00	0.00
16	−1.46	0.73	1.46	0.73

EXAMPLE 6.4.2   Suppose we wish to test for changes between adjacent time periods using the pH data in Table 6.1.4. Here, if a change over time occurs, it is expected that pH levels will decline for both treatments. Thus, for the purpose of testing for this change, we combine the data from the two treatments. Combining data like this would not be appropriate if the pH levels would tend to increase for one treatment and decrease for another, or if it is of interest to compare the rates of decline between the two treatments. Let T_{ij} denote the pH measurement on the ith carcass taken at the jth time period, $i = 1, 2, \ldots, 12, j = 1, 2, \ldots, 6$. Differences of interest are of the form $D_{ij} = T_{ij} - T_{i(j+1)}, j = 1, 2, \ldots, 5$, where we assume that the difference vector $\boldsymbol{D}_i = (D_{i1}, D_{i2}, \ldots, D_{i5})$ is distributed symmetrically about $(0, 0, \ldots, 0)$ under the null hypothesis. Table 6.4.2 lists these differences, their signed ranks, and the sums of the positive signed ranks with their Z statistics.

Signed ranks corresponding to a random permutation of the data may be obtained as $\boldsymbol{U'M}$, where $\boldsymbol{U'}$ is a 1×12 vector whose elements are independently generated to be either 1 or −1 with probability 1/2, and $\boldsymbol{M}$ is the 12×5 matrix of signed ranks in Table 6.4.2. Table 6.4.3 contains simulated percentiles for the permutation distribution of Z_{max} based on 5000 randomly selected permutations. Since $Z_{\text{max}} = $

3.06 for the original data, there is a significant change over time ($p < .01$). The upper-tail test shows that the mean changes in pH are statistically significantly different from 0 ($p < .05$) for the first two differences.

TABLE 6.4.2

Differences and Signed Ranks for pH Measurements in Table 6.1.4

Carcass	Time 1 – Time 2	Time 2 – Time 3	Time 3 – Time 4	Time 4 – Time 5	Time 5 – Time 6
1	0.65 (6)	0.24 (3)	0.06 (2.5)	0.06 (1)	0.41 (9)
2	0.38 (3)	0.18 (2)	0.41 (11)	−0.38 (−12)	0.81 (10)
3	0.12 (1)	0.32 (6.5)	0.52 (12)	0.18 (8)	−0.26 (−8)
4	1.44 (12)	−0.59 (−11)	0.34 (9)	0.12 (6)	−0.06 (−4)
5	0.51 (4)	0.05 (1)	0.38 (10)	0.29 (11)	0.18 (7)
6	0.34 (2)	0.56 (9)	−0.11 (−6)	−0.25 (−9)	0.92 (11)
7	0.73 (8)	0.32 (6.5)	0.18 (7)	0.17 (7)	0.01 (1)
8	0.68 (7)	0.57 (10)	−0.09 (−5)	0.09 (3)	0.02 (2)
9	0.83 (9)	0.67 (12)	0.03 (1)	−0.07 (−2)	−0.07 (−5)
10	1.11 (11)	0.31 (5)	−0.08 (−4)	0.11 (5)	−0.15 (−6)
11	0.95 (10)	0.25 (4)	−0.06 (−2.5)	0.27 (10)	−1.45 (−12)
12	0.64 (5)	0.42 (8)	0.31 (8)	−0.10 (−4)	−0.03 (−3)
SR_+	78	67	60.5	51	40
Z	3.06	2.20	1.69	0.94	0.08

TABLE 6.4.3

Simulated Percentiles for Z_{max}

Statistic/Level	90	95	97.5	99
Z_{max}	1.77	2.04	2.28	2.59

6.4.2 A Sum of Signed-Rank Statistics

As in the case of two samples, a sum of Wilcoxon statistics may be useful when differences in each variable are expected to be small but in the same direction. For instance, this would occur if a modification of a person's diet produced small reductions in blood pressure and cholesterol levels. Suppose we have paired differences D_{ij}, $i = 1, 2, \ldots, n, j = 1, 2, \ldots, k$, obtained on n experimental units. We rank the absolute values of $D_{1j}, D_{2j}, \ldots, D_{nj}$, and we let $S_{1j}, S_{2j}, \ldots, S_{nj}$ denote their signed ranks, $j = 1, 2, \ldots, k$. In the case of no ties in the data, the statistic that we compute is the sum of the signed ranks

$$\mathrm{SR}_{\mathrm{sum}} = \sum_{i=1}^{n} \sum_{j=1}^{k} S_{ij}$$

If SR_{+j} is the sum of the positive signed ranks on each variable, then

$$2\sum_{j=1}^{k} \mathrm{SR}_{+j} = \mathrm{SR}_{\mathrm{sum}} - \frac{kn(n-1)}{2}$$

Thus, $\mathrm{SR}_{\mathrm{sum}}$ is equivalent to a sum of the Wilcoxon signed-rank statistics. We may obtain the permutation distribution of this statistic, or we may use a normal approximation.

A randomly selected value of $\mathrm{SR}_{\mathrm{sum}}$ under the null hypothesis of no difference between the treatments may be represented as

$$\mathrm{SR}_{\mathrm{sum}} = \sum_{i=1}^{n} \sum_{j=1}^{k} U_i S_{ij}$$

where the U_i's are independent and $U_i = 1$ or -1 with probability .5 each. It follows that $E(\mathrm{SR}_{\mathrm{sum}}) = 0$ and

$$\mathrm{var}(\mathrm{SR}_{\mathrm{sum}}) = \sum_{i=1}^{n} \left(\sum_{j=1}^{k} S_{ij} \right)^2$$

The statistic

$$Z = \frac{\mathrm{SR}_{\mathrm{sum}}}{\sqrt{\mathrm{var}(\mathrm{SR}_{\mathrm{sum}})}}$$

is referred to the standard normal distribution.

With a small number of ties, we may apply the procedure without ties to the adjusted signed ranks. Otherwise, replace the signed ranks with

$$Z_{ij} = \frac{S_{ij}}{\sqrt{\sum_{i=1}^{n} S_{ij}^2}}$$

The Z_{ij}'s are the standardized signed-rank statistics for the jth variable. The statistic is

$$Z_{\mathrm{sum}} = \sum_{i=1}^{n} \sum_{j=1}^{k} Z_{ij}$$

We have

$$E(Z_{\mathrm{sum}}) = 0, \quad \mathrm{var}(Z_{\mathrm{sum}}) = \sum_{i=1}^{n} \left(\sum_{j=1}^{k} Z_{ij} \right)^2$$

EXAMPLE 6.4.3 We may apply the sum of signed-rank statistics to the data in Table 6.4.2. From Table 6.4.2, we find that the sum of the signed ranks is $SR_{sum} = 203$. The sums of the ranks for each of the 12 carcasses are 21.5, 14, 19.5, 12, 33, 7, 29.5, 17, 15, 11, 9.5, and 14. Thus,

$$\text{var}(SR_{sum}) = 21.5^2 + 14^2 + \cdots + 14^2 = 4112$$

The standardized statistic is

$$Z = \frac{203}{\sqrt{4112}} = 3.33$$

with an upper-tail *p*-value of .0008. We may apply the paired-comparison permutation test to these 12 scores. The exact *p*-value from StatXact is .0002. Thus, there has been a significant change over time. ∎

6.4.3 Computer Analysis

Resampling Stats code for the multivariate paired-comparison permutation test in Section 6.3 may be modified to obtain *p*-values for multivariate rank tests. Any computer program in which ranking may be done can be used to compute the quantities necessary for SR_{sum} or Z_{sum} and the corresponding normal approximation. We may use StatXact to obtain an exact *p*-value for SR_{sum} or Z_{sum} by applying the paired-comparison permutation test to the sum of the signed ranks or the sum of the *z*-scores on each experimental unit.

6.5
Multiresponse Categorical Data

In some situations individuals may respond to a survey question by checking more than one category. For instance, suppose a survey of home owners asks "Which of the following electronic items do you own?" and the choices are (1) personal computer, (2) VCR, and (3) compact disk player. If the instructions state "Check all that apply," there would be eight possible responses. We may denote responses in the form of trivariate vectors of 0's and 1's where (0, 0, 0) denotes none of the items, (1, 0, 0) denotes personal computer only, (1, 1, 0) denotes personal computer and VCR but not compact disk player, and so on. Data arising in this way are termed *multiresponse* categorical data.

By contrast, in *single-response* categorical data, respondents may choose one and only one category. For instance, in the example on satisfaction with pain relief in Section 5.4.2, the patients could choose one and only one response from the three possibilities: not satisfied, somewhat satisfied, and very satisfied. These possibilities may be denoted by three possible trivariate vectors: (1, 0, 0), (0, 1, 0), and (0, 0, 1).

We are interested in comparing multiresponse data across groups. For instance, several groups of respondents may be asked about electronic components they own. The groups could represent different ages of individuals, such as under 21, 21 to 30, 31 to 40, and over 40. We may wish to know whether or not age is related to ownership of different types of electronic equipment.

The response vector for the ith individual in the hth group is denoted $(X_{ih1}, X_{ih2}, \ldots, X_{ihc})$, where $X_{ihj} = 1$ if the response for the ith individual in the hth group is classified in the jth category, and 0 otherwise, $i = 1, 2, \ldots, n_h$, $h = 1, 2, \ldots, r$, $j = 1, 2, \ldots, c$.

6.5.1 Hypotheses for Multiresponse Data

We now consider three hypotheses in testing multiresponse categorical data. The first is equality of joint probability distributions. Let $f_h(x_1, x_2, \ldots, x_c)$ denote the joint probability distribution of $(X_{ih1}, X_{ih2}, \ldots, X_{ihc})$, $h = 1, 2, \ldots, r$. The hypothesis of equality of joint probability distributions can be stated as

$$H_0: f_1(x_1, x_2, \ldots, x_c) = f_2(x_1, x_2, \ldots, x_c) = \cdots = f_r(x_1, x_2, \ldots, x_c)$$

A second hypothesis is equality of means of the X_{ihj}'s. Let $p_{j|h} = E(X_{ihj})$, which is the probability that an individual from population h checks category j. The hypothesis of equality of means is stated as

$$H_1: p_{j|1} = p_{j|2} = p_{j|r}, \quad j = 1, 2, \ldots, c$$

If we were dealing with single-response categorical data, hypotheses H_0 and H_1 would be equivalent, but this is not the case with multiresponse categorical data. This is illustrated in Table 6.5.1. Here we have two response variables in which the joint probability distributions are not the same but the expected values are the same. H_0 is the stronger hypothesis in the sense that if H_0 is true, then H_1 is true, but not conversely.

TABLE 6.5.1

Unequal Joint Probabilities but Equal Means in Multiresponse Categorical Data

Population Distributions	$f_h(0, 0)$	$f_h(0, 1)$	$f_h(1, 0)$	$f_h(1, 1)$	Expected Values
$h = 1$	.20	.10	.20	.50	(.7, .6)
$h = 2$	.15	.15	.25	.45	(.7, .6)

The third hypothesis of interest is the negation of H_1—that is,

$$H_a: \text{There exist } j, \ h, \text{ and } h' \text{ such that } p_{j|h} \neq p_{j|h'}$$

In this development, we test H_0 against H_a. A statistic is given that is sensitive to H_a and whose null distribution is determined under H_0. The problem of testing H_1 against H_a is not considered.

6.5.2 A Statistical Test

Our test is based on a suggestion by Agresti and Liu (1999). For the jth categorical variable, consider an $r \times 2$ contingency table whose rows are the groups and whose two columns represent the number of respondents who either "checked" or did not check variable j. We compute the chi-square statistic for each of these tables and add them up.

To formally define the statistic, let O_{hj1} and E_{hj1} denote the observed and expected values for row h and the checked-variable-j column, and similarly let O_{hj0} and E_{hj0} denote the observed and expected values for the did-not-check-variable-j column. The statistic is then defined as

$$\phi^2 = \sum_{j=1}^{c} \sum_{h=1}^{r} \left(\frac{\left(O_{hj1} - E_{hj1}\right)^2}{E_{hj1}} + \frac{\left(O_{hj0} - E_{hj0}\right)^2}{E_{hj0}} \right)$$

Bilder (2000) showed that the asymptotic distribution of ϕ^2 under the null hypothesis is that of a linear combination of $c(r-1)$ chi-square random variables, each with 1 degree of freedom. Agresti and Liu (1999) suggested using the bootstrap method to find an approximate null distribution (see Chapter 8). For a comparison of various methods for dealing with this problem, see Bilder, Loughin, and Nettleton (2000).

We use the permutation method to find a reference distribution for testing H_0 against H_a. Randomly permute the multivariate vectors associated with the respondents among the r groups in such a way as to maintain the sample sizes for each of the groups. This is a straightforward extension of the randomization of multivariate vectors between two groups discussed in Section 6.1. Let $\phi^2(\alpha)$ denote the upper $100\alpha\%$ point of the permutation distribution of ϕ^2. If the observed value of ϕ^2 is greater than or equal to $\phi^2(\alpha)$, then conclude that the alternative hypothesis is true at level of significance α. The p-value is the fraction of the permuted values of ϕ^2 that are greater than or equal to the observed value of this quantity.

EXAMPLE 6.5.1   Visitors to garden shows in three cities (1, 2, and 3) were asked where they buy their garden supplies. The possible answers were garden shop (G), discount store (D), and supermarket (S). The respondents were asked to check all the answers that apply. It was of interest to determine whether buying patterns differ among the cities. The responses are summarized in Table 6.5.2. All responded to at least one possibility. The results are arranged in three 3×2 contingency tables shown in Table 6.5.3, which shows both the observed and expected cell counts. To illustrate, consider the entries in the category denoted by city 1 and garden shop G. The observed number who checked G in Table 6.5.2 is $5 + 3 + 2 + 1 = 11$. Since city 1 had 33 respondents, the observed number who did not check G is 22. The remainder of the 3×2 table for variable G is similarly determined. The expected value for row 1 and column G in the 3×2 contingency table is $(33)(50)/117 = 14.1$, and similarly for the other cells.

TABLE 6.5.2
Responses to Garden Show Survey

City	G (1, 0, 0)	D (0, 1, 0)	S (0, 0, 1)	G, D (1, 1, 0)	G, S (1, 0, 1)	D, S (0, 1, 1)	G, D, S (1, 1, 1)	Total
1	5	10	6	3	2	6	1	33
2	13	7	3	8	4	3	2	40
3	5	13	10	2	3	9	2	44
Total	23	30	19	13	9	18	5	117

TABLE 6.5.3
Observed (O) and Expected (E) Cell Counts for Data in Table 6.5.2

City	Table 1 G Checked	Table 1 G Not Checked	Table 2 D Checked	Table 2 D Not Checked	Table 3 S Checked	Table 3 S Not Checked	Sample Size
1	$O = 11$ $E = 14.1$	$O = 22$ $E = 18.9$	$O = 20$ $E = 18.6$	$O = 13$ $E = 14.4$	$O = 15$ $E = 14.4$	$O = 18$ $E = 18.6$	$n_1 = 33$
2	$O = 27$ $E = 17.1$	$O = 13$ $E = 22.9$	$O = 20$ $E = 22.6$	$O = 20$ $E = 17.4$	$O = 12$ $E = 17.4$	$O = 28$ $E = 22.6$	$n_2 = 40$
3	$O = 12$ $E = 18.8$	$O = 32$ $E = 25.2$	$O = 26$ $E = 24.8$	$O = 18$ $E = 19.2$	$O = 24$ $E = 19.2$	$O = 20$ $E = 24.8$	$n_3 = 44$
Chi-square	15.5		1.0		5.2		$n = 117$

The observed value of the statistic is $\phi^2 = 15.5 + 1.0 + 5.2 = 21.7$. The 90th, 95th, 97.5th, and 99th percentiles of the permutation distribution of ϕ^2, based on a random sample of 5000 permutations, are 11.087, 12.995, 15.018, and 18.249, respectively. Thus, the observed statistic is significant with $p < .01$, indicating that the populations differ in where they shop for garden supplies. The p-value is .0032. Tukey-like multiple comparisons may be made to formally test where differences exist. This, procedure is not illustrated here, but the method is analogous to that used in Section 5.4.3. ∎

Remark on Chi-Square Approximations

If the responses from the $r \times 2$ tables are independent of one another and if the sample sizes are large, then ϕ^2 would be the sum of c independent, approximately chi-square statistics each with $r - 1$ degrees of freedom, so its approximate distribution would be that of a chi-square with $c(r - 1)$ degrees of freedom. In Example 6.5.1, the degrees of freedom are 6, and the 90th, 95th, 97.5th, and 99th percentiles from the chi-square distribution with 6 degrees of freedom are 10.6, 12.6, 14.4, and 16.8, respectively. These are roughly the same as the percentiles of the permutation distribution in this case, but the accuracy of the approximation may vary from problem to problem. A conservative approach would be to use the Bonferroni approximation. In Example 6.5.1, we would test each chi-square statistic at the $.05/3 = .0167$ level of significance using 2 degrees of freedom. The critical value is 8.18, and since the chi-square statistic for the first table exceeds this value, the null hypothesis is rejected.

6.5.3   Computer Analysis

The test is not implemented in StatXact, MINITAB, or S-Plus. Figure 6.5.1 shows a Resampling Stats program for doing the computations.

FIGURE 6.5.1

Resampling Stats Program for Obtaining Permutation Distribution of ϕ^2

```
maxsize default 5000
'input data from file
read file "catdat.txt" G D S
'observed statistic is 21.7

'input expected values for checked categories
copy (14.1 18.5 14.4 17.1 22.6 17.4 18.8 24.8 19.2) expect
'compute expected values for not checked categories
subtract (33 33 33 40 40 40 44 44 44) expect nexpect

'index for shuffling 117 vectors
copy 1,117 index

repeat 5000
shuffle index s_index
take s_index 1,33 index_1
take G index_1 G1
take D index_1 D1
take S index_1 S1
take s_index 34,73 index_2
take G index_2 G2
take D index_2 D2
take S index_2 S2
take s_index 74,117 index_3
take G index_3 G3
take D index_3 D3
take S index_3 S3
```

                                    (continued on next page)

FIGURE 6.5.1

Resampling Stats Program for Obtaining Permutation Distribution of ϕ^2 *(continued)*

```
'compute observed shuffled statistics
sum G1 oG1
sum D1 oD1
sum S1 oS1
sum G2 oG2
sum D2 oD2
sum S2 oS2
sum G3 oG3
sum D3 oD3
sum S3 oS3
concat oG1 oD1 oS1 oG2 oD2 oS2 oG3 oD3 oS3 observ

'compute chi-square for checked categories
subtract observ expect diff
square diff diff2
divide diff2 expect chi
sum chi chisq1

'compute chi-square for unchecked categories
subtract (33 33 33 40 40 40 44 44 44) observ nobserv
subtract nobserv nexpect ndiff
square ndiff ndiff2
divide ndiff2 nexpect nchi
sum nchi chisq2

'add chisquare checked and unchecked
add chisq1 chisq2 chisq

'keep track of statistic
score chisq chi_dist
end

'get percentiles and p-value
percentile chi_dist (90 95 97.5 99) pctile
count chi_dist >= 21.7 pval
divide pval 5000 pval

print pctile
print pval

PCTILE = 11.087 12.995 15.018 18.249

PVAL = 0.0032
```

Exercises

1   The data are the heights and weights for two groups of students. Find the permutation distribution of T^2 and the corresponding F statistic. What is the p-value for the observed data?

Group 1	(72, 150), (68, 145), (71, 160)
Group 2	(69, 130), (65, 120), (64, 115)

2   Find the permutation distribution of t_{max} for the data in Exercise 1.

3   Convert the data in Exercise 1 to multivariate ranks, and find the permutation distribution of the statistic Z_{max} defined in Section 6.2.

4   Suppose for the data in Exercise 1 it is expected that group 1 will tend to have greater heights and weights than group 2. Find the permutation distribution of the one-sided W_{sum} statistic of Section 6.2 and obtain the p-value.

5   Lingenfelser (2001) reported data on varieties of soybeans grown under various treatments at different locations. Among the data are heights (cm), seed weights (grams/100 seeds), and yields (kg/ha) for similar varieties and treatments at two locations as shown in the table. Use an appropriate multivariate permutation test to test for differences between locations.

Location	Height	Weight	Yield	Location	Height	Weight	Yield
A	30	12.8	2474	B	28	12.6	652
A	31	13.9	2539	B	29	13.7	390
A	32	12.9	2405	B	31	13.2	2341
A	37	12.9	2432	B	31	12.8	2991
A	32	14.8	3124	B	33	12.9	455
A	36	12.9	2929	B	31	13.0	2276
A	32	13.8	2539	B	29	13.2	910
A	31	13.7	2571	B	30	12.5	1756
A	30	14.9	2600	B	31	13.5	1951
A	33	12.5	2734	B	31	14.1	1430
A	35	13.8	3151	B	34	13.5	1365
A	33	13.5	2499	B	33	13.0	2016
A	34	13.6	2626	B	29	13.3	1821
A	35	12.8	2795	B	28	12.6	455
A	34	13.2	2210	B	36	13.3	715
A	30	13.0	2533	B	31	12.3	2081

6   Refer to the data in Exercise 5. Suppose it is expected that location A will have higher readings on height, weight, and yield due to superior soil conditions. Use the sum of the Wilcoxon statistics in Section 6.2 to test for differences between locations. Compare results of the statistic that has the variance adjusted for ties with the statistic that does not.

7 Suppose we had observed the data configuration in data set 14 in Table 6.3.2. Compare p-values for the tests based on T^2, $t_{\text{max abs}}$, and t_{max}. Comment on circumstances in which one test might be preferred over the others.

8 These data are hypothetical mathematics, English, and social sciences test scores before and after a review program. Find the permutation distributions of the paired T^2, $t_{\text{max abs}}$ and t_{max}.

Student	Before	After
1	(55, 78, 64)	(63, 85, 81)
2	(40, 67, 53)	(55, 63, 69)
3	(80, 77, 78)	(78, 80, 75)
4	(54, 43, 50)	(60, 52, 72)
5	(75, 81, 83)	(76, 86, 91)

9 Refer to the data in Exercise 8.

a Obtain the permutation distributions of the statistics Z_{max} and SR_{sum} defined in Section 6.4, and obtain the p-values for the observed data.

b Compare the exact p-value of SR_{sum} with the approximate p-value obtained from the normal approximation.

10 Two types of electrical devices (A and B) were examined for causes of failure. The devices can fail because of three causes (C1, C2, and C3), and it is possible for there to be more than one cause of failure. The hypothetical data in the table show the numbers of devices that fail under the various causes of failure. Test H_0 versus H_a as defined in Section 6.5.

Cause/Device	C1 only	C2 only	C3 only	C1 & C2	C1 & C3	C2 & C3	C1, C2, & C3
A	5	4	1	5	1	2	0
B	3	5	6	2	5	2	2

Theory and Complements

11 Discuss the problem of comparing more than two treatments with multivariate responses. What statistics would be appropriate for testing differences among the treatments? How would observations be permuted?

7

Analysis of Censored Data

A Look Ahead   Data of the type we consider in this chapter arise naturally in studies of lifetimes of individuals or objects. Such studies occur in medicine when researchers attempt to develop new therapies to prolong the lives of patients. These studies also occur in engineering when researchers try to determine how long mechanical or electrical devices are likely to function before they fail. A complicating feature of lifetime data is called *censoring*. Censoring happens when we are unable to observe the actual lifetime of an individual or object, but instead we simply know that the lifetime extends beyond some measurable point.

      Without censoring, the analysis of lifetime data is essentially no different from the analysis of any other data. The nonparametric techniques developed up to this point apply. With censoring, the nonparametric approach is to assign scores to censored and uncensored observations to reflect in some way the magnitude of the observations. After scores are assigned, we may apply permutation tests to the scores to test for differences among treatments.

      We begin with the one-sample problem of estimating what is called the survival function. We then consider permutation tests and scoring systems that enable us to analyze censored data that arise in two-sample, *k*-sample, and paired-comparison studies.

7.1
Estimating the Survival Function

7.1.1 Censored Data

Suppose a medical researcher is interested in determining the survival time of patients who are undergoing an experimental treatment. If survival time is measured from when a patient enters the study until death, the researcher might not be able to determine this time for all patients. For instance, some patients may withdraw from the study before it is finished, and others may be alive when the study is brought to a close. For such individuals, all that is known is that their survival time is longer than their time in the study. For instance, in a 36-month study, a patient who enters at the 24th month and is alive at the end will have a survival time of at least 12 months, but that is all that can be said. Such data are said to be *censored*.

Censored data can occur in any area where survival time, time-to-death, or time-to-failure is the variable of interest. In an engineering study to determine how long a mechanical or electrical item will run before it fails, items that do not fail during the testing period have censored failure times.

The notion of censoring extends beyond the context of lifetime data. For instance, radioactive counts of a hot substance may exceed the upper limit of an instrument, so all that is known is that the actual count exceeds the maximum measurable limit. Such data are termed *upper-censored*. If a pollutant is known to be present in the environment but the amount falls below the detection limit of the measuring device, the data are termed *lower-censored*. For simplicity, we deal with lifetime data where upper-censoring may occur.

7.1.2 Kaplan–Meier Estimate

Let T denote the survival time of an experimental unit, and let $F(t)$ denote the cdf of T, which is assumed to be continuous. The *survival function* is defined to be $R(t) = 1 - F(t)$, or the probability that an experimental unit survives to time t. In the context of engineering, $R(t)$ is also called the *reliability function*, since it is the probability that an item such as a battery will last at least until time t. If there is no censoring, the estimate of $R(t)$ is

$$\hat{R}(t) = 1 - \hat{F}(t)$$

where $\hat{F}(t)$ is the empirical cdf as defined in Section 1.2.1. In the case of censored data, we use the *Kaplan–Meier estimate* of $R(t)$.

Suppose we have survival measurements on n experimental units. If the ith measurement is uncensored, we denote this value as t_i. If it is censored, we denote it as t_i^+. Let $t_{(1)} \leq t_{(2)} \leq \cdots \leq t_{(n)}$ denote the numerical values, including ties and censored observations, placed in order from smallest to largest, where the censoring notation is suppressed for convenience. To illustrate the ordering when there is censoring, suppose three patients have survival times of 10, 12, and 15 months, and two others leave the study at 12 and 14 months, respectively, so their observations are censored. Then the ordering of survival times is 10, 12, 12+, 14+, and 15.

We assume $t_{(1)}$ is uncensored. Data censored before the first uncensored time do not figure in the estimation of the survival function, so they are ignored. We apply an iterative procedure to find $\hat{R}(t_{(i)})$ for any value of $t_{(i)}$ that is not censored.

For $t = 0$, we set $\hat{R}(0) = 1$. For $t_{(1)}$, we have

$$\hat{R}(t_{(1)}) = \text{fraction of observations} > t_{(1)}$$

If $t_{(i)} < t_{(j)}$ are adjacent uncensored times, we note

$$R\left(t_{(j)}\right) = P\left(T > t_{(j)}\right)$$

$$= P\left(T > t_{(i)}\right)P\left(T > t_{(j)} \middle| T > t_{(i)}\right)$$

$$= R\left(t_{(i)}\right)P\left(T > t_{(j)} \middle| T > t_{(i)}\right)$$

To estimate $P(T > t_{(j)} \mid T > t_{(i)})$, we compute the fraction of observations greater than $t_{(i)}$ that are also greater than $t_{(j)}$. Since $t_{(i)}$ and $t_{(j)}$ are adjacent uncensored failure times, the observations counted in the denominator of this fraction are the times greater than or equal to $t_{(j)}$. We ignore all censored times between $t_{(i)}$ and $t_{(j)}$, since we cannot determine whether they are greater than $t_{(j)}$. Thus, we iteratively compute the estimated survival function using the iterative formula:

$$\hat{R}\left(t_{(j)}\right) = \hat{R}\left(t_{(i)}\right) \frac{\text{number of observations } > t_{(j)}}{\text{number of observations } \geq t_{(j)}}$$

If t is not one of these times and $t_{(i)} \leq t < t_{(j)}$, where $t_{(i)}$ and $t_{(j)}$ are adjacent uncensored times, then $\hat{R}(t) = \hat{R}(t_{(i)})$.

EXAMPLE 7.1.1   Suppose the ordered survival times are 3.1, 4.0+, 5.3, 5.3, 6.1+, 7.5, 7.5+, 8.9, and 9.0. The estimated survival functions at the uncensored times are given in Table 7.1.1.

TABLE **7.1.1**
Estimated Survival Functions

i	$t_{(i)}$	$\hat{R}(t_{(i)})$
1	3.1	8/9 = .889
2	4.0+	
3	5.3	.889(5/7) = .635
4	5.3	.635
5	6.1+	
6	7.5	.635(3/4) = .476
7	7.5+	
8	8.9	.476(1/2) = .238
9	9.0	.238(0/1) = .000

∎

An Alternative Form of the Kaplan–Meier Estimate and Its Variance

For each uncensored observation $t_{(i)}$, let n_i denote the number of observations, both censored and uncensored, greater than or equal to $t_{(i)}$, and let d_i denote the number of uncensored observations that equal $t_{(i)}$. Using the iterative procedure above, we have

$$\hat{R}(t_{(i)}) = \prod_{r:t_{(r)} \leq t_{(i)}} \left(\frac{n_r - d_r}{n_r} \right)$$

where the product is taken over all r for which $t_{(r)}$ is uncensored and $t_{(r)} \leq t_{(i)}$. An approximate variance of the estimated survival function at time $t_{(i)}$ is

$$\text{var}\left[\hat{R}(t_{(i)}) \right] = \left[\hat{R}(t_{(i)}) \right]^2 \sum_{r:t_{(r)} \leq t_{(i)}} \frac{d_r}{n_r(n_r - d_r)}$$

where the sum is taken over all r for which $t_{(r)}$ is uncensored and $t_{(r)} \leq t_{(i)}$. See Lawless (1982) for a discussion of this formula.

EXAMPLE 7.1.2   Here are the computations of the variances for the estimates in Example 7.1.1 at the uncensored times:

$$\text{var}\left[\hat{R}(3.1) \right] = \left(\frac{8}{9} \right)^2 \left(\frac{1}{9 \times 8} \right) = 0.011$$

$$\text{var}\left[\hat{R}(5.3) \right] = \left(\frac{8}{9} \times \frac{5}{7} \right)^2 \left(\frac{1}{9 \times 8} + \frac{2}{7 \times 5} \right) = 0.029$$

$$\text{var}\left[\hat{R}(7.5) \right] = \left(\frac{8}{9} \times \frac{5}{7} \times \frac{3}{4} \right)^2 \left(\frac{1}{9 \times 8} + \frac{2}{7 \times 5} + \frac{1}{4 \times 3} \right) = 0.035$$

$$\text{var}\left[\hat{R}(8.9) \right] = \left(\frac{8}{9} \times \frac{5}{7} \times \frac{3}{4} \times \frac{1}{2} \right)^2 \left(\frac{1}{9 \times 8} + \frac{2}{7 \times 5} + \frac{1}{4 \times 3} + \frac{1}{2 \times 1} \right) = 0.037$$

■

In the case where no observations are censored or tied, the formulas for the estimates and variance follow from the usual binomial formulas; that is,

$$\hat{R}(t_{(i)}) = \frac{n - i}{n}$$

$$\text{var}(\hat{R}_{(i)}) = \frac{i(n - i)}{n^3}$$

7.1.3 Computer Analysis

The Kaplan–Meier estimate is available in MINITAB under the "Reliability/ Survival" and "Nonparametric Dist Analysis-Right Cens..." menu options. In S-Plus it is under the "Survival" menu option. Figure 7.1.1 shows a MINITAB output for the analysis of the data in Example 7.1.1. Included in the output are the Kaplan–Meier estimates, standard errors, and 95% confidence intervals. Note that the standard errors are the square roots of the variances presented in Example 7.1.2. Also included is an estimate of the expected value of the survival time, denoted MTTF for mean time-to-failure, along with its standard error and a 95% confidence interval, which we do not discuss here.

FIGURE **7.1.1**
MINITAB Kaplan–Meier Estimates for Data in Example 7.1.1

```
Nonparametric Estimates

Characteristics of Variable

             Standard       95.0% Normal CI
Mean(MTTF)     Error     Lower      Upper
   7.1429     0.7869    5.6006     8.6851


Median =    7.5000
IQR =       3.6000  Q1 =    5.3000  Q3 =     8.9000


Kaplan-Meier Estimates

            Number    Number    Survival     Standard    95.0% Normal CI
   Time    at Risk   Failed   Probability     Error     Lower    Upper
   3.1000      9        1        0.8889       0.1048    0.6836   1.0000
   5.3000      7        2        0.6349       0.1692    0.3033   0.9666
   7.5000      4        1        0.4762       0.1871    0.1095   0.8429
   8.9000      2        1        0.2381       0.1926    0.0000   0.6156
   9.0000      1        1        0.0000       0.0000    0.0000   0.0000
```

7.2
Permutation Tests for Two-Sample Censored Data

We consider the problem of comparing two survival functions using permutation tests in studies in which censored observations occur. Two issues arise in constructing permutation tests. One has to do with the censoring mechanism, and the other has to do with obtaining a satisfactory statistic for comparing the two sets of survival times.

7.2.1 Censoring Mechanism

A particular difficulty arises when the censoring mechanism depends on the treatments. For instance, suppose two medical treatments are equally effective in prolonging the lives of patients. However, one treatment requires patients to take a medication that causes unpleasant side effects, while the other treatment does not. If the unpleasant side effects cause more patients to drop out of the study early, then the censoring would depend on the treatment. If we were to randomly permute the data as we would in a permutation test, not all permutations would be equally likely to occur even though the medical treatments are equally effective. Smaller censored times would be more likely to occur with the treatment that has the unpleasant side effects. In this situation it is not possible to perform a meaningful permutation test because such a test would require all permutations to be equally likely to occur under the null hypothesis that the distributions of survival times are the same.

A key assumption in this chapter is that the censoring mechanism does not depend on the treatments. This could occur in many ways. Suppose a study of two medical treatments is designed to last for 5 years. Individuals may choose to leave the study simply because they cannot maintain the level of commitment necessary for 5 years, regardless of the treatment. This censoring mechanism does not depend on the treatments. Censoring also occurs when individuals die during the study for reasons unrelated to the treatments—for instance, because of an accident. Here again the censoring mechanism would be independent of the treatment.

If the distributions of survival times of the treatments are not the same, then one treatment may turn out to have considerably more censored observations than another even though the censoring mechanism does not depend on the treatments. For instance, suppose treatment 1 prolongs life more than treatment 2. Also suppose patients tend to leave the study because they cannot maintain a long-term commitment, regardless of the treatment. Then treatment 1 would tend to have more censored observations than treatment 2 even though the censoring mechanism does not depend on the treatment. Thus, we need to examine the experimental setup, not just the data, to determine what causes censoring. The data alone will not provide us with the answer.

Notation and Assumptions

The intuitive notion that the censoring mechanism does not depend on the treatments can be formulated more precisely in terms of statistical assumptions. Suppose there are m observations with treatment 1 and n observations with treatment 2. Let $S_1, S_2, \ldots, S_{m+n}$ denote the survival times for the experimental units in the combined treatment groups assuming no censoring occurs. Let $C_1, C_2, \ldots, C_{m+n}$ denote $m + n$ censoring times. The numerical quantity $t_{(i)}$ that we observe on each experimental unit is the smaller of S_i and C_i. We also observe whether or not the observation is censored.

In *random censoring*, we assume the C_i's have a common cdf $F_C(t)$, and we assume the S_i's and C_i's are mutually independent. The assumption of independence of the S_i's and C_i's and the fact that $F_C(t)$ is the same regardless of the treatment are what we mean when we say the censoring mechanism does not depend on treatments. This is the type of censoring that occurs in medical studies when subjects leave the study at random times for reasons unrelated to the treatment.

Under *fixed-time censoring*, all experimental units begin the study at the same time and the study is terminated at a fixed time $C_i = C$. In engineering studies, this is called *Type I censoring*. It could occur, for instance, in a study of the lifetimes of two brands of batteries, where all the batteries are tested for a period of 4 hours. All batteries that survive the test would have 4 hours as their censored survival time. Here the fact that the study is terminated at a fixed time regardless of the treatment assures us that the censoring mechanism does not depend on the treatments.

Under the null hypothesis of no difference between the treatments, we assume that the S_i's are independent random variables with common cdf $F(t)$. That is, the survival distributions are the same for treatments 1 and 2. With either random censoring or fixed-time censoring, a censored value under the null hypothesis of no difference between the treatments would be as likely to occur under one treatment as the other. Thus, we may assume that all permutations of the data, censored and uncensored, are equally likely to occur under the null hypothesis. This idea provides the basis for doing permutation tests on censored data.

7.2.2 Examples of Tests Based on Medians and Ranks

The question of what is a satisfactory statistic for comparing two treatments when censoring exists is not obvious. For instance, we cannot directly figure the difference between means, since we will not have all the actual survival times with which to compute the means. We now show how in certain circumstances permutation tests that we've already developed may be applied. The two examples illustrate permutation tests based on medians and ranks.

EXAMPLE 7.2.1   The data in Table 7.2.1 from Nelson (1982) are the breakdown times (in seconds) of insulating fluids taken at two voltages. The largest two breakdown times under 30 kV are censored, but not at the same time. So we assume random censoring applies, where in this case the C_i's tend to be large. We can apply the permutation test on medians, since it is possible to compute the median for any permutation of the data. The medians for the two samples are 1880 and 107, with a difference of 1773. An approximate one-sided *p*-value for the difference of medians, based on 10,000 randomly selected permutations of the data, is .0007.

TABLE **7.2.1**
Seconds to Insulating Fluid Breakdown

30 kV	50	134	187	882	1450	1470	2290	2930	4180	15,800	29,200+	86,100+
35 kV	30	33	41	87	93	98	116	258	461	1180	1350	1500

■

The next example illustrates fixed-time censoring. Again in this example it is possible to compute the median for any permutation of the data. It is also possible to apply a rank test in this situation. Here all observations that are censored at time *C* are regarded as tied observations, so that we may apply the Wilcoxon rank-sum test with ties.

EXAMPLE 7.2.2   Suppose two brands of batteries are placed on test for *C* = 4 hours. Either their failure times are observed or the observations are censored at 4 hours. The data in Table 7.2.2 denote the hypothetical lifetimes of such batteries. In ranking the data, we treat the three censored observations as tied. The adjusted rank assigned to these three values is 15. An exact one-sided *p*-value is .0172, which was obtained from StatXact using the Wilcoxon rank-sum test. At the traditional 5% level of significance, the brands differ in the distributions of their lifetimes. If we use the difference between medians as the test statistic, we find that the difference between the medians of brands B and A is 3.8 − 2.8 = 1.0. Based on 10,000 randomly selected permutations of the data, an approximate one-sided *p*-value is .0332.

TABLE **7.2.2**
Lifetimes (in hours) of Two Brands of Batteries with *T* = 4 Hours as a Censoring Time

Brand A	1.5	2.3	2.4	2.6	3.0	3.0	3.4	3.8
Brand B	2.5	2.9	3.3	3.7	3.9	4.0+	4.0+	4.0+

The simple censoring patterns in these two examples allow us to apply previously developed methods. However, other methods need to be developed for more complicated patterns of censoring. We consider this topic in the following sections.

7.2.3 Computer Analysis

Resampling Stats may be used to do the permutation test on medians as shown in Section 2.3.2. Assuming the data are such that the medians may be computed for any permutation, we input the data without regard to which observations are censored. For instance, in Example 7.2.1, the observations 29,200 and 86,100 are included with the other data without denoting that they are censored. For fixed-time censoring problems that use the Wilcoxon rank-sum test with ties, we may use Resampling Stats, StatXact, MINITAB, or S-Plus as discussed in Section 2.4.4.

7.3
Gehan's Generalization of the Mann–Whitney Test

Gehan (1965) introduced a generalization of the Mann–Whitney test to compare two survival functions where censored observations may occur. Let the observations from treatment 1 be denoted $X_1, X_2, \ldots, X_m$, and let the observations from treatment

2 be denoted $Y_1, Y_2, \ldots, Y_n$. For convenience we have suppressed the censoring notation attached to these random variables. If x and y are uncensored observations and $x < y$, or if x is uncensored and y is censored and $x \leq y^+$, we say x is definitely less than y or y^+. Similarly, if $x > y$ or $x^+ \geq y$, we say x is definitely greater than y or y^+. For each pair (X_i, Y_j) such that X_i is definitely less than Y_j, we define $U_{ij} = -1$. Similarly for each pair (X_i, Y_j) such that X_i is definitely greater than Y_j, we define $U_{ij} = 1$. For pairs for which the order cannot be determined (i.e., $x^+ < y$, $y^+ < x$, or both censored) or for pairs that are uncensored and tied, we define $U_{ij} = 0$. The Mann–Whitney statistic for censored data is

$$U_c = \sum_{i=1}^{m} \sum_{j=1}^{n} U_{ij}$$

Note that small values of U_c correspond to the situation in which the X's tend to be smaller than the Y's. The permutation distribution of this statistic may be obtained by permuting the observations between treatments. An alternative form of the statistic discussed next facilitates the process of obtaining the permutation distribution.

7.3.1 An Alternative Form of U_c Based on Gehan Scores

We consider an alternative form of U_c that involves attaching a score to each observation and then summing the scores for one of the groups. The *Gehan score* for an observation is the number of observations in the combined sample that are definitely less than the observation in question minus the number of observations definitely greater than this observation. The data in Table 7.3.1 show seven hypothetical observations along with their Gehan scores.

TABLE **7.3.1**
Combined Observations and Gehan Scores

Data	X_1	X_2	Y_1	X_3	Y_2	X_4	Y_3
Observations	20	25+	29	31	31+	33	34
Definitely Less, Greater	0, 6	1, 0	1, 4	2, 3	3, 0	3, 1	4, 0
Gehan Score	−6	1	−3	−1	3	2	4

It can be shown that the sum of the Gehan scores for treatment 1 (the X's) is equal to U_c. To illustrate, consider the 12 (X, Y) pairs from Table 7.3.1: (20, 29), (20, 31+), (20, 34), (25+, 29), (25+, 31+), (25+, 34), (31, 29), (31, 31+), (31, 34), (33, 29), (33, 31+), and (33, 34). The corresponding values of U_{ij} are −1, −1, −1, 0, 0, 0, 1, −1, −1, 1, 0, and −1, giving $U_c = -4$. The sum of scores for treatment 1 in Table 7.3.1 is $-6 + 1 - 1 + 2 = -4$, which illustrates the equality of these two methods for obtaining a value of the test statistic for censored data.

The *Gehan test* for comparing two treatments with censored observations is the permutation test applied to the Gehan scores. If no observations are censored, the test is equivalent to the Mann–Whitney test and the Wilcoxon rank-sum test.

EXAMPLE 7.3.1 Patients diagnosed with a certain form of cancer were assigned to one of two treatments: the standard treatment (S) and a new treatment (N). The survival time (in days) after diagnosis was obtained for each subject. The data are shown in Table 7.3.2, and the ranked data and their scores are given in Table 7.3.3. The sum of the scores for the standard treatment is −11. A permutation test applied to the scores using StatXact gives an exact lower-tail *p*-value of .2877. Thus, there is no statistically significant difference between the standard treatment and the new treatment.

TABLE **7.3.2**

Survival Time (in days) for Patients Undergoing Treatment

S	94	155	180+	375	741	951+	1133	1198	1261
N	175	382	521	567+	683+	988	1216+	1355+	

TABLE **7.3.3**

Ordered Survival Times and (Gehan Scores) for Data in Table 7.3.2

94	155	175	180+	375	382	521	567+	683+
(−16)	(−14)	(−12)	(3)	(−9)	(−7)	(−5)	(6)	(6)
741	951+	988	1133	1198	1216+	1261	1355+	
(−1)	(7)	(2)	(4)	(6)	(10)	(9)	(11)	

A Derivation of the Equality of U_c and the Sum of Gehan Scores

Let us consider the sum of the scores associated with treatment 1. We have:

score for X_i = number of *Y*'s definitely less than X_i

− number of *Y*'s definitely greater than X_i

+ number of *X*'s definitely less than X_i

− number of *X*'s definitely greater than X_i

Now it can be shown

$$\sum_i \text{number of } X\text{'s definitely less than } X_i$$

$$- \sum_i \text{number of } X\text{'s definitely greater than } X_i = 0$$

Also,

$$\sum_{j=1}^{n} U_{ij} = \text{number of } Y\text{'s definitely less than } X_i$$

$$- \text{ number of } Y\text{'s definitely greater than } X_i$$

Thus,

$$\sum_{i=1}^{n} \text{scores for } X_i = U_c$$

7.3.2 Large-Sample Approximations

The methods in Section 2.10 may be applied to obtain a large-sample approximation for the statistic U_c. Since the sum of the combined scores is 0, we have $E(U_c) = 0$. Let W_k denote the score attached to the kth observation in the combined sample. The population variance of the combined scores is

$$\sigma_c^2 = \frac{1}{m+n} \sum_{k=1}^{m+n} W_k^2$$

It follows that

$$\text{var}(U_c) = \frac{mn}{m+n-1} \sigma_c^2$$

The statistic

$$Z = \frac{U_c}{\sqrt{\text{var}(U_c)}}$$

has an approximate standard normal distribution from which statistical significance may be determined.

EXAMPLE 7.3.2   For the scores in Table 7.3.3, we find

$$\sigma_c^2 = \frac{1240}{17} = 72.94$$

$$\text{var}(U_c) = \frac{(9)(8)}{16} 72.94 = 328.23$$

Thus,

$$Z = \frac{-11}{\sqrt{328.23}} = -0.61$$

The p-value as determined by the standard normal distribution is .27.   ∎

7.3.3 Computer Analysis

Once the Gehan scores have been obtained, any computer program such as StatXact or Resampling Stats that does two-sample or k-sample permutation tests may be used to compare treatments.

7.4
Scoring Systems for Censored Data

We now introduce two scoring systems for censored data. One is a generalization of rank scores as suggested by Prentice (1978). The other is the log-rank scoring system, which is a generalization of the exponential scores discussed in Section 2.7.1.

To illustrate the difficulty in assigning scores to data with censored observations, suppose we have three observations, 10, 11+, and 12, with the middle one being censored. If we were to attempt to rank the observations, then the last two could have ranks 2 and 3, respectively, or ranks 3 and 2, respectively, depending on whether the unobserved, uncensored value associated with 11+ turned out to be less than 12 or greater than 12. Thus, if we wish to assign scores to these observations, where the scores increase with the size of the observations, we would have to have some way to resolve the ambiguity in ranking due to the censoring.

An intuitive way to assign a score to a censored observation is to compute the score one would expect to obtain if the observation were uncensored. For instance, if an observation is censored at t_c, then the score attached to this observation would be the expected value of an uncensored score given that the uncensored time is greater than t_c. A less intuitive, but theoretically sound, way to deal with censoring is to base scores on what is called the *likelihood function* of the observations. We first present the intuitive approach and then outline the likelihood approach.

7.4.1 Prentice–Wilcoxon Scores

To motivate these scores, first consider the case in which the observations have no ties and no censoring. Let $t_{(1)} < t_{(2)} < \cdots < t_{(N)}$ denote the combined observations from two or more treatments. Let $F(t)$ denote the cdf of the survival times assuming that there are no differences among the treatments. A statistical estimate of the cdf at time $t_{(i)}$ is the empirical cdf as discussed in Section 1.2.1—that is,

$$\hat{F}\left(t_{(i)}\right) = \left(\text{fraction of observations} \le t_{(i)}\right) = \frac{i}{N} = \frac{\text{rank of } t_{(i)}}{N}$$

Thus, the ranks, which are the scores for the Wilcoxon rank-sum test and the Kruskal–Wallis test, are proportional to a statistical estimate of the cdf of $F(t_{(i)})$. This idea motivates what will call the *Wilcoxon scores* for censored data.

For an uncensored observation $t_{(i)}$, the Wilcoxon score is $\hat{F}(t_{(i)})$, where $\hat{F}(t_{(i)})$ is a certain nonparametric estimate of $F(t_{(i)})$ to be discussed momentarily. If $t_{(i)}$ is a censored observation, we will compute the expected value of $F(T)$ given that the survival time T is bigger than $t_{(i)}$. It can be shown that this expected value is halfway between $F(t_{(i)})$ and 1—that is,

$$E\left[F(T)\middle|T > t_{(i)}\right] = \frac{F(t_{(i)}) + 1}{2}$$

Thus, the Wilcoxon score for a censored observation is $[\hat{F}(t_{(i)}) + 1]/2$, where again $\hat{F}(t_{(i)})$ is a certain nonparametric estimate of $F(t_{(i)})$.

The *Prentice–Wilcoxon scores* PW are related to the Wilcoxon scores W by the equation

$$PW = 2W - 1$$

The PW scores arise naturally in the likelihood approach, as discussed in Section 7.4.3.

To get at the nonparametric estimate of $F(t)$, Prentice suggested estimating the survival function $1 - F(t) = R(t)$ using something very similar to the Kaplan–Meier estimate. Assuming the data have no ties, the Prentice estimate of $R(t)$ for an uncensored observation is

$$\hat{R}\left(t_{(i)}\right) = \prod_{r:t_{(r)} \leq t_{(i)}} \left(\frac{n_r}{n_r + 1}\right)$$

where n_i is the number of censored and uncensored observations greater than or equal to $t_{(i)}$, and the product is taken over r such that $t_{(r)}$ is an uncensored observation and $t_{(r)} \leq t_{(i)}$. If we replace n_r by $n_r - 1$, we would have the Kaplan–Meier estimate. For a censored observation $t_{(i)}$, let $t_{(j)}$ be the nearest uncensored observation less than $t_{(i)}$. We set $\hat{R}(t_{(i)}) = \hat{R}(t_{(j)})$. Having estimated $R(t_{(i)})$, we may then compute either the Wilcoxon scores or the Prentice–Wilcoxon scores for all the observations.

To summarize, letting $\hat{F}(t_{(i)}) = 1 - \hat{R}(t_{(i)})$ and expressing the scores in terms of $\hat{R}(t_{(i)})$, we find that the Wilcoxon scores are $1 - \hat{R}(t_{(i)})$ for an uncensored observation and $1 - \hat{R}(t_{(i)})/2$ for a censored observation. The Prentice–Wilcoxon scores are $1 - 2\hat{R}(t_{(i)})$ for an uncensored observation and $1 - \hat{R}(t_{(i)})$ for a censored observation. If any observations are tied, we compute the scores as if there were infinitesimal differences among them and then average the scores for the tied observations. Either sets of scores may be used in two-sample or k-sample permutation tests, with the resulting p-value being the same.

EXAMPLE 7.4.1   Table 7.4.1 shows the Prentice estimates of $R(t_{(i)})$ and the PW scores for six observations, two of which are censored. If we were to change the last two observations in Table 7.4.1 so that both were 15, then the score attached to each of these observations would be $(.086 + .542)/2 = .314$.

TABLE **7.4.1**

Prentice–Wilcoxon Scores

Data	5	10+	11	12+	13	15
$\hat{R}(t_{(i)})$	6/7 = .857	.857	.857(4/5) = .686	.686	.686(2/3) = .457	.457(1/2) = .229
Uncensored Observations	1 – 2(.857) = –.714		1 – 2(.686) = –.372		1 – 2(.457) = .086	1 – 2(.229) = .542
Censored Observations		1 – .857 = .143		1 – .686 = .314		

                                                                     ■

7.4.2 Log-Rank Scores

For an uncensored observation $t_{(i)}$, the log-rank score is based on a statistical estimate of the natural logarithm of the survival function—that is, on an estimate of $-\ln[R(t_{(i)})]$. If $t_{(i)}$ is a censored observation, then we compute the expected value of $-\ln[R(T)]$ given that the survival time T is greater than $t_{(i)}$. We use the fact that the random variable $-\ln[R(T)]$ has an exponential distribution with a mean of 1. (See Exercise 14 at the end of the chapter.) It follows that

$$E\left\{-\ln[R(T)]\middle| T > t_{(i)}\right\} = -\ln\left[R(t_{(i)})\right] + 1$$

Thus, the log-rank score for a censored observation is based on a statistical estimate of $-\ln[R(t_{(i)})] + 1$.

    Suppose the data contain no ties. To estimate $-\ln[R(t_{(i)})]$ for an uncensored observation, we replace $R(t_{(i)})$ with the Kaplan–Meier estimate, and then we use the approximation $-\ln(1 - x) \approx x$ for x near 0. Thus,

$$-\ln\left[\hat{R}(t_{(i)})\right] = -\ln\left[\prod_{r:t_{(r)}\leq t_{(i)}}\left(\frac{n_r - 1}{n_r}\right)\right]$$

$$= -\sum_{r:t_{(r)}\leq t_{(i)}}\ln\left(1 - \frac{1}{n_r}\right)$$

$$\approx \sum_{r:t_{(r)}\leq t_{(i)}}\frac{1}{n_r}$$

where n_r is the number of censored and uncensored observations greater than or equal to $t_{(r)}$, and the sum is taken over all r for which $t_{(r)}$ is uncensored and $t_{(r)} \leq t_{(i)}$. For a censored observation $t_{(i)}$, let $t_{(j)}$ be the nearest uncensored observation less than $t_{(i)}$. We set $\hat{R}(t_{(i)}) = \hat{R}(t_{(j)})$. Finally, we subtract 1 from these estimates to obtain the log-rank scores. Subtracting 1 is a matter of convenience, since it will give us a set of scores that sum to 0.

To summarize, the log-rank score for an uncensored observation $t_{(i)}$ is

$$\left(\sum_{r:t_{(r)} \leq t_{(i)}} \frac{1}{n_r} \right) - 1$$

where the sum is taken over all r for which $t_{(r)}$ is uncensored and $t_{(r)} \leq t_{(i)}$. For a censored observation $t_{(i)}$, let $t_{(j)}$ be the nearest uncensored observation less than $t_{(i)}$. The log-rank score is

$$\sum_{r:t_{(r)} \leq t_{(j)}} \frac{1}{n_r}$$

where the sum is taken over all r for which $t_{(r)}$ is uncensored and $t_{(r)} \leq t_{(j)}$. If the data have ties, assign log-rank scores as if there were infinitesimal differences among the tied observations and then average the scores for the tied observations.

EXAMPLE 7.4.2   Log-rank scores for the data in Table 7.4.1 are shown in Table 7.4.2. If the last two observations were 15, then the score attached to these two observations would be $(-0.083 + 0.917)/2 = 0.417$.

TABLE **7.4.2**
Log-Rank Scores

Data	5	10+	11	12+	13	15
$\sum_{r:t_{(r)} \leq t_{(i)}} \dfrac{1}{n_r}$	1/6 = 0.167	0.167	0.167 + 1/4 = 0.417	0.417	0.417 + 1/2 = 0.917	0.917 + 1 = 1.917
Uncensored Observations	0.167 − 1 = −0.833		0.417 − 1 = −0.583		0.917 − 1 = −0.083	1.917 − 1 = 0.917
Censored Observations		0.167		0.417		

7.4.3 The Likelihood Approach

This discussion assumes knowledge of material in a first course in mathematical statistics. The section may be omitted without loss of continuity in the overall development of ideas. Suppose we have two treatments in a completely random design. Let the sets of censored and uncensored observations for treatments 1 and 2 be denoted C_1, U_1, C_2, and U_2, respectively. Suppose the probability density function for treatment 1 is $f(t)$ with cumulative distribution function $F(t)$, and that the probability density function and cumulative distribution function for treatment 2 are $f(t - \theta)$ and $F(t - \theta)$, respectively. That is, if there is a difference, then the distributions of treatment 1 and 2 differ by a location parameter θ.

The likelihood function for θ is given by

$$L(\theta) = \prod_{t_i \in U_1} f(t_i) \prod_{t_i \in C_1} \left[1 - F(t_i)\right] \prod_{t_i \in U_2} f(t_i - \theta) \prod_{t_i \in C_2} \left[1 - F(t_i - \theta)\right]$$

The maximum likelihood estimate of θ is obtained by taking the partial derivative of $\ln[L(\theta)]$ with respect to θ and setting it to 0. The partial derivative is

$$\frac{\partial \ln L(\theta)}{\partial \theta} = \sum_{t_i \in U_2} -\frac{f'(t_i - \theta)}{f(t_i - \theta)} + \sum_{t_i \in C_2} \frac{f(t_i - \theta)}{1 - F(t_i - \theta)}$$

If there is no difference between the two treatments, then the true value of θ is 0, so that

$$\frac{\partial \log L(\theta)}{\partial \theta}\bigg|_{\theta=0} \approx 0$$

Consequently, the statistic

$$T_L = \sum_{t_i \in U_2} -\frac{f'(t_i)}{f(t_i)} + \sum_{t_i \in C_2} \frac{f(t_i)}{1 - F(t_i)}$$

should be around 0 if there is no difference between the two treatments and should depart significantly from 0 if there is a difference. Thus, T_L is a reasonable test statistic for testing whether or not θ is 0.

T_L can be thought of as a sum of scores, where the score for an uncensored observation is

$$-\frac{f'(t_i)}{f(t_i)}$$

and the score for a censored observation is

$$\frac{f(t_i)}{1 - F(t_i)}$$

Although these scores depend on the functional form of $f(t)$, for some $f(t)$'s it is possible to estimate these scores nonparametrically. Doing so leads to the Prentice–Wilcoxon and log-rank scores.

The Prentice–Wilcoxon scores are based on the logistic distribution defined by

$$f(t) = \frac{e^t}{(1 + e^t)^2}, \quad -\infty < t < \infty$$

The cumulative distribution function is

$$F(t) = \frac{e^t}{1 + e^t}$$

Thus, the score for an uncensored observation is

$$\frac{f'(t)}{f(t)} = \frac{1 - e^t}{1 + e^t} = 2F(t) - 1$$

The nonparametric estimate of $2F(t_{(i)}) - 1$ based on the Prentice estimate of $R(t_{(i)}) = 1 - F(t_{(i)})$ gives the Prentice–Wilcoxon score for an uncensored observation. For a censored observation, the score is

$$\frac{f(t)}{1 - F(t)} = \frac{e^t}{1 + e^t} = F(t)$$

The nonparametric estimate of $F(t_{(i)}) = 1 - R(t_{(i)})$ using the Prentice estimate of $R(t_{(i)})$ gives the Prentice–Wilcoxon score for a censored observation.

The log-rank scores are based on the extreme-value distribution defined by

$$f(t) = e^{t - e^t}, \quad -\infty < t < \infty$$

The cumulative distribution function is

$$F(t) = 1 - e^{-e^t}$$

The score for an uncensored observation is

$$\frac{-f'(t)}{f(t)} = e^t - 1$$

Now if T has an extreme-value distribution, then e^T has an exponential distribution with mean 1, which is the same distribution as $-\log[R(T)]$. Replacing $e^{t_i} - 1$ with the statistical estimate of $-\ln[R(t_{(i)})] - 1$ developed in Section 7.4.2, we obtain the log-rank score for an uncensored observation. The score for a censored observation is

$$\frac{f(t)}{1 - F(t)} = e^t$$

Replacing $e^{t_{(i)}}$ by $e^{t_{(j)}}$, where j is the largest index of an uncensored observation such that $t_{(j)} < t_{(i)}$, and replacing $e^{t_{(j)}}$ with the statistical estimate of $-\ln[R(t_{(i)})]$ lead to the log-rank score for a censored observation.

7.4.4 Computer Analysis

StatXact has an option in its "Case Data" menu for computing scores. Both Prentice–Wilcoxon and log-rank scores are available from this menu. StatXact uses the term *Wilcoxon–Gehan scores* for what we have termed Prentice–Wilcoxon scores.

7.5
Tests Using Scoring Systems for Censored Data

7.5.1 Two-Sample and *K*-Sample Tests

For comparing two or more treatments in a completely random design when there is censoring in the data, we assign scores to the observations as in Section 7.4 and then apply a two-sample or *k*-sample permutation test to the scores. To compare two treatments, the test statistic is the sum of the scores for either one of the treatments. If there are more than two treatments, a permutation test may be applied to the scores as discussed in Section 3.2.4, and multiple comparisons may be made as discussed in Section 3.3.2.

EXAMPLE 7.5.1   Table 7.5.1 contains hypothetical data representing the survival times (in weeks) of 20 cancer patients who received three different treatments, and Table 7.5.2 shows the Gehan (G), Prentice–Wilcoxon (PW), and log-rank (LR) scores. The *p*-values for the Gehan, Prentice–Wilcoxon, and log-rank scores are .1587, .1361, and .1044, respectively. These *p*-values were obtained using StatXact with 10,000 randomly selected permutations.

TABLE **7.5.1**
Survival Times of Cancer Patients

Treatment	*Survival Times (in weeks)*
A	1 3 10 13+ 20 25
B	3 8 11 20+ 25 30+ 33
C	9 15 22+ 31 35 40+ 45

■

Selecting Among Scoring Systems

If the censoring is only modest, then the Gehan scores and Prentice–Wilcoxon scores give comparable results. If there is no censoring, they give the same results. However, Prentice and Marek (1979) showed that when the censoring distributions are different for the two treatments, the two scoring systems can result in substantial differences depending on the imbalance in censoring between the two treatments. It should be noted that the permutation approach to finding significance levels would be questionable in this circumstance, since the censoring mechanism would be related to the treatment.

TABLE **7.5.2**
Scores for Data in Table 7.5.1

Treatment	Weeks	G	PW	LR
A	1	−19	−.905	−0.950
A	3	−16	−.762	−0.870
B	3	−16	−.762	−0.870
B	8	−13	−.619	−0.783
C	9	−11	−.524	−0.720
A	10	−9	−.429	−0.654
B	11	−7	−.333	−0.582
A	13+	7	.333	0.418
C	15	−4	−.231	−0.499
A	20	−2	−.128	−0.408
B	20+	9	.436	0.592
C	22+	9	.436	0.592
A	25	3	.060	−0.212
B	25	3	.060	−0.212
B	30+	11	.561	0.860
C	31	7	.269	0.060
B	33	9	.415	0.310
C	35	11	.561	0.643
C	40+	14	.781	1.643
C	45	14	.781	1.643

The choice between Prentice–Wilcoxon scores and log-rank scores depends on the underlying population distributions. For instance, data that come from populations that have Weibull cdf's defined by

$$F_i(t) = 1 - e^{-(\lambda_i t)^\delta}, \quad i = 1, 2$$

would favor use of the log-rank scores. Note that the scale parameter λ_i may vary between treatments, but the shape parameter δ is the same. See Lee, Desu, and Gehan (1975) and Peto (1972). If the observations have a normal distribution or a distribution that is approximately normal, and if the distributions differ only by a location parameter, then the Prentice–Wilcoxon test would be preferred.

EXAMPLE 7.5.2 This example illustrates how conclusions may differ depending on the scoring system. The observations for treatments 1 and 2 were generated from exponential distributions with means of 1 and 3, respectively. These are special cases of the

Weibull distribution with $\lambda_1 = 1$, $\lambda_2 = 1/3$, and $\delta = 1$. Table 7.5.3 contains the observations and the Prentice–Wilcoxon (PW) and log-rank (LR) scores. The test with the Prentice–Wilcoxon scores has a one-sided p-value of .0690, and the test with the log-rank scores has a one-sided p-value of .0258, computed using StatXact. Thus, we would have significant differences between the treatments at the traditional 5% level with the log-rank scores but not with the Prentice–Wilcoxon scores.

TABLE **7.5.3**
Scores for Simulated Exponential Data, Mean of Treatment 1 = 1, Mean of Treatment 2 = 3

Treatment	Observation	Censored Observation = 0	PW	LR
1	0.17	1	−.810	−0.897
1	0.77	1	−.238	−0.505
1	0.67	1	−.524	−0.720
1	0.50	1	−.714	−0.842
1	1.04	1	−.048	−0.331
1	2.40	1	.330	0.155
1	1.96	1	.162	−0.095
1	0.68	1	−.429	−0.654
1	2.27	0	.581	0.905
1	1.03	1	−.143	−0.422
2	1.37	0	.476	0.669
2	0.11	1	−.905	−0.950
2	2.47	0	.665	1.155
2	7.50	0	.665	1.155
2	2.05	0	.581	0.905
2	1.67	1	.057	−0.220
2	2.36	0	.581	0.905
2	5.47	0	.665	1.155
2	0.71	1	−.333	−0.582
2	0.59	1	−.619	−0.783

7.5.2 Paired-Comparison and Blocked Designs

The procedure for paired-comparison and blocked designs in which censoring is present begins by assigning scores to the combined data. These scores are then analyzed using the same permutation tests that would be used for paired data or blocked data.

EXAMPLE 7.5.3   A group of six students each wrote down as many consecutive integers as possible in 10 seconds and as many consecutive letters as possible in 10 seconds. One value was censored because of a broken pencil. The data are listed in Table 7.5.4, and the Prentice–Wilcoxon scores with their paired differences are given in Table 7.5.5. The differences in scores were analyzed using a paired-comparison permutation test. The two-sided *p*-value is .25, indicating no significant difference between the speeds with which the students can write down consecutive letters and numbers.

TABLE **7.5.4**
Number of Consecutive Integers and Letters

Student	1	2	3	4	5	6
Integers	22	18	19	24	17+	31
Letters	20	15	20	22	21	30

TABLE **7.5.5**
Prentice-Wilcoxon Scores and Differences for Data in Table 7.5.4

Student	1	2	3	4	5	6
Integers	0.245	−0.678	−0.511	0.497	0.077	0.832
Letters	−0.259	−0.846	−0.259	0.245	−0.007	0.664
Differences	0.504	0.168	−0.252	0.252	0.084	0.168

∎

7.5.3 Computer Analysis

Once scores have been assigned to the observations, any statistical package such as StatXact or Resampling Stats that does permutation tests may be used to analyze the scores. StatXact has specific menu options for doing two-sample and *k*-sample permutation tests using Prentice–Wilcoxon scores (denoted Wilcoxon–Gehan scores in the menu options) and log-rank scores. With these options, it is not necessary to compute scores before doing the tests because these are computed automatically. However, to use Resampling Stats on censored data, the scores must be computed outside the program and then input in place of the actual data.

Exercises

1   The hypothetical data in the list are the survival days of a group of patients with HIV who were first seen in stage 2 of the disease. Obtain the Kaplan–Meier estimate of the survival function, and obtain the standard deviations of the estimates at the uncensored times.

22 90 256 320+ 428 670+ 910 997 1070 1081 1197 1355+ 1560 1933 2202

2   Consider the two data sets representing observations from treatment 1 and treatment 2, and consider two scenarios that might give rise to the censoring pattern. In one scenario, treatment 1 is worse than treatment 2 in that survival times tend to be shorter and censoring tends to occur early as patients do not wish to stay. In another scenario, treatment 1 is superior to treatment 2, and subjects feel better so quickly that they wish to leave the study early. In both scenarios, treatment 1 tends to have more censoring than treatment 2, as the data show. Discuss the implications of these two scenarios in terms of using a permutation test to determine which treatment is better when the censoring mechanism and the treatment are not independent.

Treatment 1	5+	6+	7+	9	10+
Treatment 2	9	11	12	13	14

3   The data are the times (in minutes) students in two groups took to complete an exam. Some students did not finish at the end of the allotted time of 50 minutes, so their data are censored. Compare the two groups in terms of the distributions of the times to finish the exam.

Group 1	42	37	50+	43	50+	35	40	44	50+	39	38	45	41	50
Group 2	36	46	50+	30	26	25	41	33	37	47	41	28	35	

4   Determine the Gehan, Prentice–Wilcoxon, and log-rank scores for these data.

100	125+	140	150	150+	170	170	180+	200

5   Obtain the permutation distributions of the Gehan, Prentice–Wilcoxon, and log-rank statistics for these data.

Treatment 1	100	125+	150
Treatment 2	140	170	190

6   The data in the table are survival times from a study to compare the effectiveness of a certain treatment for bile duct cancer with a control. See Fleming, O'Fallen, and O'Brien (1980). Compare the two groups using a permutation test applied to the Prentice–Wilcoxon and log-rank scores. In their article, Fleming and colleagues proposed a generalization of the Kolmogorov–Smirnov (K-S) statistic that is sensitive to treatment differences on data in which the survival functions of the treatments may differ substantially at one point in time but not elsewhere. Their procedure was more sensitive to treatment differences on this set of data than either the Prentice–Wilcoxon or the log-rank procedure.

Treatment	30, 67, 79+, 82+, 95, 148, 170, 171, 176, 193, 200, 221, 243, 261, 262, 263, 399, 414, 446, 446+, 464, 777
Control	57, 58, 74, 79, 89, 98, 101, 104, 110, 118, 125, 132, 154, 159, 188, 203, 257, 257, 431, 461, 497, 723, 747, 1313, 2636

7 For the data in Exercise 6, compute the expected value and variance of the sum of the scores for the treatment group using the methodology in Section 2.10. Do this for both the Prentice–Wilcoxon and log-rank scores. Obtain the Z statistics for both sums of scores, and compare p-values with those obtained in Exercise 6.

8 Plot the Kaplan–Meier estimates of the survival functions of the treatment and control groups in Exercise 6. Note that the biggest difference occurs for early deaths. What does comparing the p-values of the Prentice–Wilcoxon and log-rank tests suggest about which test is preferred for detecting differences in survival curves that occur early?

9 The data in the table are from a clinical trial to compare the drug 6-MP with a placebo in terms of remission times of patients with acute leukemia. See Freireich et al. (1963) or Gehan (1965). Apply a nonparametric test to compare 6-MP with the placebo. Is it possible just from the data to tell whether or not the censoring mechanism is independent of the treatment? Is it reasonable to assume that all possible permutations of the data are equally likely to occur under the assumption of no difference between 6-MP and the placebo?

6-MP	6, 6, 6, 7, 10, 13, 22, 23, 6+, 9+, 10+, 17+, 19+, 25+, 32+, 34+, 35+
Placebo	1, 1, 2, 2, 3, 4, 4, 5, 5, 8, 8, 8, 8, 11, 11, 12, 12, 15, 17, 22, 23

10 Subjects were given 60 seconds to complete each of two tasks. Tasks not completed in 60 seconds were given censored times. Test for significant differences between the two tasks using a paired-difference permutation test. Use the Prentice–Wilcoxon scores.

Subject	*1*	*2*	*3*	*4*	*5*	*6*	*7*	*8*	*9*	*10*	*11*
Task 1	44	53	55	47	40	59	60+	43	60+	58	60+
Task 2	38	49	57	41	38	60+	55	39	60+	48	50

Theory and Complements

11 Let U denote the Mann–Whitney statistic defined in Section 2.6, and let U_c denote Gehan's generalization of the Mann–Whitney statistic defined in Section 7.3. If the data contain no censoring and no ties, show that $U_c = mn - 2U$.

12 Derive the expected value and variance of the paired-difference statistic in Exercise 10. Use the fact that the sum of the positive differences of the scores under the hypothesis of no difference between the two tasks may be expressed as

$$\sum_1^n V_i |A_i|$$

where A_i is the difference between the scores of task 1 and task 2 on the ith subject, and the V_i's are independent random variables that can take on the values 0 and 1 with equal probability. Obtain the Z statistic and compare the p-value with that obtained in Exercise 10.

13 Suppose we have random samples from k treatments in which censoring may occur. Discuss how one may carry out multiple comparisons among the treatments. In particular, for the hypothetical survival data in the table, derive a formula for Tukey's HSD using the methodology for general scores discussed in Section 3.3.2. Use log-rank scores in the computation.

Treatment 1	320	310+	350	415	475
Treatment 2	285	325	270+	300	330+
Treatment 3	240	305	250	275	290
Treatment 4	210	195	225	260	180

14 Suppose T has a continuous cdf $F(t)$.

a Show that the random variable $U = F(T)$ has a uniform distribution on the interval $[0, 1]$. Do this by showing that $P[F(T) \le t] = t$, which is the cdf of a uniform $[0, 1]$ random variable.

b Show that $R(T) = 1 - F(T)$ has a uniform distribution on the interval $[0, 1]$.

c Show that if U has a uniform distribution on the interval $[0, 1]$, then $-\ln(U)$ has an exponential distribution. Do this by showing that $P(-\ln(U) \le t) = 1 - e^{-t}$, which is the cdf of an exponential random variable with a mean of 1. Conclude that $-\ln[R(T)]$ has an exponential distribution with a mean of 1.

d For an exponential random variable X with a mean of 1, show that $E(X \mid X > x) = x + 1$. Conclude that

$$E\left\{-\ln[R(T)] \middle| T > t_{(i)}\right\} = -\ln\left[R\left(t_{(i)}\right)\right] + 1$$

8

Nonparametric Bootstrap Methods

A Look Ahead   Suppose we take a sample of size 20, say, from a population. From this sample, we take a random sample of the same size with replacement. That is, from the 20 data values, we select numbers one after the other, putting the number back each time it is selected, until we have another set of 20 numbers. This is called a bootstrap sample. It is intended to mimic a random sample of size 20 from the population. In the use of the bootstrap methodology, we typically select 1000 or more bootstrap samples from the data.

In Section 8.1, we give the rationale for taking bootstrap samples, and we show a simple application of the methodology. In the remaining sections, we develop a number of applications of bootstrap sampling, including its use in analysis of variance and regression.

8.1
The Basic Bootstrap Method

In computing a statistical estimate of a population parameter such as a mean or a standard deviation, we would like to have a measure of how close the estimate is to the population value. For instance, if we were to take a sample of 200 to estimate the mean income of a population, we would like to have an indication of how close the sample mean is to the population mean. In the case of estimating a population mean μ, statistical theory tells us that the variance of the sample mean is

$$\operatorname{var}\left(\overline{X}\right) = \frac{\sigma^2}{n}$$

where σ^2 is the population variance and n is the sample size. Replacing σ^2 by S^2 and using the central limit theorem, we obtain the approximate 95% confidence interval for μ:

$$\overline{X} \pm 2\frac{S}{\sqrt{n}}$$

The quantity $2Sn^{(-1/2)}$ is a measure of how accurately the sample mean estimates the population mean and may be called the *margin of error*.

Our purpose here is to present a simple method for obtaining a margin of error for problems in which analytical solutions are not readily available. The method is called the *bootstrap*, to suggest pulling oneself up by the bootstraps.

8.1.1 Mean Squared Error and Margin of Error

Let θ be a population parameter, and let $\hat{\theta}$ be a statistical estimate of θ based on a random sample of size n. The mean square error (MSE) of the estimate is defined as

$$\mathrm{MSE} = E\left(\hat{\theta} - \theta\right)^2$$

which is the average squared deviation of the estimate from the population quantity that is being estimated. In the case of the sample mean, we have $\mathrm{MSE} = \sigma^2/n$.

The importance of the MSE for our purpose comes from the Chebyshev–Markov inequality, which states that

$$P\left(\left|\hat{\theta} - \theta\right| \le k\sqrt{\mathrm{MSE}}\right) \ge 1 - \frac{1}{k^2}$$

For instance, using $k = 2$, we find that the probability is at least $1 - 1/4 = 3/4$ that the estimate $\hat{\theta}$ is within $\pm 2\sqrt{\mathrm{MSE}}$ of the population value θ. The probability may be much higher than this, as it is in the case of estimating the mean of a population. We refer to the quantity $2\sqrt{\mathrm{MSE}}$ as the *margin of error* of the estimate. It can be used as a rough measure of the accuracy of an estimate when other measures may not be readily available.

Suppose an explicit formula for obtaining the MSE of an estimate is not available. However, imagine that we could take many samples of size n from the population. Let $\hat{\theta}_i$ denote the estimate of θ obtained from the ith repetition of the sampling experiment. The MSE of $\hat{\theta}$ can be estimated by computing

$$\hat{\mathrm{MSE}} = \frac{1}{\mathrm{REP}} \sum_{i=1}^{\mathrm{REP}} \left(\hat{\theta}_i - \theta\right)^2$$

where REP is the number of times the sampling experiment is repeated. Unfortunately, the population distribution is generally not known and thus cannot be sampled, so another approach is needed.

8.1.2 The Bootstrap Estimate of MSE

The bootstrap procedure is based on the idea of resampling the data. When the population distribution is not known, the data may be used as a substitute for the population, and we may resample the data as if we were sampling the population.

To simulate sampling from an infinite population, sampling is done *with replacement*. The resulting sample is called a *bootstrap sample*. The steps for obtaining a bootstrap estimate of MSE are as follows:

1. Compute $\hat{\theta}$ from the original data.
2. Take a number of bootstrap samples of size n from the data. Let REP denote the number of such bootstrap samples. Typically, REP $\geq$ 1000.
3. Compute $\hat{\theta}_{b,i}$, the estimate of θ obtained from the ith bootstrap sample.
4. Obtain the bootstrap MSE as

$$\hat{\mathrm{MSE}} = \frac{1}{\mathrm{REP}} \sum_{i=1}^{\mathrm{REP}} \left(\hat{\theta}_{b,i} - \hat{\theta} \right)^2$$

Note that bootstrap estimates $\hat{\theta}_{b,i}$ are treated as if they were computed from random samples from the actual population, and $\hat{\theta}$ is treated as if it were the true value of θ from the population.

EXAMPLE 8.1.1 Consider the sample mean for the latch failure data in Table 1.2.1. From the original sample of size 20, we find a mean of 38.7. Table 8.1.1 shows three bootstrap samples of size $n = 20$ from the data along with the bootstrap values $\overline{X}_b$ of the sample mean and the square errors $(\overline{X}_b - 38.7)^2$. The bootstrap estimate of MSE is $(29.2 + 62.4 + 49.0)/3 = 46.7$ based on these three bootstrap samples, and the bootstrap estimate of the margin of error is $2(46.7)^{(1/2)} = 13.7$. Applying the bootstrap procedure to 1000 bootstrap samples of size 20 using Resampling Stats, we found an MSE of 28.2 and a bootstrap margin of error of 10.6. The usual estimate of the MSE of the sample mean is $S^2/n = 28.8$ with margin of error of 10.7. For comparison, a second set of 1000 bootstrap samples gave a bootstrap MSE of 29.8 and a margin of error of 10.9.

TABLE 8.1.1

Three Bootstrap Samples Used in Estimating MSE

Original data: 7 11 15 16 20 22 24 25 29 33 34 37 41 42 49 57 66 71 84 90

Mean = 38.7, SD = 24.0

Bootstrap Samples of Size n = 20 (samples selected randomly from the original data with replacement)	*Mean of Bootstrap Sample, $\overline{X}_b$*	$\left(\overline{X}_b - 38.7\right)^2$
22 57 42 16 24 11 20 7 41 90 90 16 66 25 90 66 25 24 84 66	44.1	29.2
90 37 15 84 29 57 57 57 11 49 41 57 84 71 37 20 29 84 7 15	46.6	62.4
49 84 29 41 57 11 49 42 90 34 71 33 41 84 49 66 20 20 29 15	45.7	49.0

EXAMPLE 8.1.2   For the same data as in Example 8.1.1, the sample standard deviation is $S = 24.0$. A bootstrap estimate of the MSE based on 1000 repetitions of the sampling experiment is 13.0, and the bootstrap margin of error is 7.2. The coefficient of variation is defined as

$$CV = 100 \frac{S}{\overline{X}}$$

which is 62.1 for these data. We found a bootstrap estimate of 67.9 for the MSE and an estimated margin of error of 16.5.   ∎

Bootstrap Variance and Bias

The bias, B, of an estimate is the difference between the expected value of an estimate and the quantity being estimated; that is,

$$B = E\left(\hat{\theta}\right) - \theta$$

If var is the variance of the estimate, then

$$MSE = var + B^2$$

Similar to the procedure for the MSE, it is possible to obtain bootstrap estimates of the variance and bias. In place of step 4 earlier, the following quantities are computed:

4′.

$$\hat{E} = \frac{1}{REP} \sum_{i=1}^{REP} \hat{\theta}_{b,i}$$

$$\hat{B} = \hat{E} - \hat{\theta}$$

$$var = \frac{1}{REP} \sum_{i=1}^{REP} \left(\hat{\theta}_{b,i} - \hat{E}\right)^2$$

Number of Bootstrap Samples

The number of bootstrap samples, REP, that should be selected is an issue. Booth and Sarkar (1998) recommend at least 800 bootstrap samples for estimating the variance of $\hat{\theta}$. In the examples in this and subsequent sections, we used at least 1000 repetitions, and in many examples we used 5000 without any significant burden in terms of computing time.

Nonparametric versus Parametric Bootstrap Estimates

The term *nonparametric bootstrap* is used to denote that no assumptions are made about the functional form of the population distribution. Bootstrap samples are drawn from the data. *Parametric bootstrap* methods involve similar procedures in which assumptions are made about the form of the population distribution. Suppose, for instance, it is desired to obtain a bootstrap MSE for the coefficient of variation, and it is known that the population distribution is normal. In the parametric bootstrap, samples would be selected from a normal distribution whose mean and variance are estimated from the data. Otherwise, the procedure would be the same as shown in Example 8.1.2.

Efron (1979) coined the term *bootstrapping* in the context of resampling the data and was the first to investigate the properties of this method. For references, see Efron and Tibshirani (1993) and Davison and Hinkley (1997).

8.1.3 Computer Analysis

Bootstrap sampling in Resampling Stats is easy to do. The "sample" command selects samples of a desired size with replacement from a vector of data. Figure 8.1.1 shows code for obtaining bootstrap estimates of the MSE of the mean, standard deviation, and coefficient of variation for the data in Table 8.1.1. Bootstrap samples may be obtained in S-Plus from the "Resample" menu. An example of an S-Plus printout is given in Section 8.3.3.

FIGURE 8.1.1

Bootstrap Estimates of Mean, SD, and CV for Data in Table 8.1.1 Using Resampling Stats

```
'input data
data (7 11 15 16 20 22 24 25 29 33 34 37 41 42 49 57 66 71 84 90) dat

'get mean, sd, and cv of original data
mean dat m
stdev dat sd
divide sd m cv
multiply cv 100 cv

'get 1000 bootstrap samples
repeat 1000

'take sample with replacement
sample 20 dat bootdat

'get mean, sd, cv of bootstrap sample
mean bootdat bootmean
stdev bootdat bootsd
divide bootsd bootmean bootcv
multiply bootcv 100 bootcv
```

(continued on next page)

FIGURE 8.1.1

Bootstrap Estimates of Mean, SD, and CV for Data in Table 8.1.1 Using Resampling Stats *(continued)*

```
'keep track of bootstrap mean, sd, cv
score bootmean meandist
score bootsd sddist
score bootcv cvdist

end

'get mean squared errors and printout results
sumsqrdev meandist m ssemean
divide ssemean 1000 msemean
sumsqrdev sddist sd ssesd
divide ssesd 1000 msesd
sumsqrdev cvdist cv ssecv
divide ssecv 1000 msecv

print msemean msesd msecv

'print results here


MSEMEAN   =      28.244
MSESD     =      12.999
MSECV     =      67.864
```

8.2
Bootstrap Intervals for Location-Scale Models

Suppose X is a continuous random variable with probability density function

$$g(x) = \frac{1}{\sigma} f\left(\frac{x-\mu}{\sigma}\right)$$

where $f(z)$ is a standard density with mean 0 and standard deviation 1. We may express X as

$$X = \mu + \sigma Z$$

where Z is a random variable with probability density $f(z)$. The mean and standard deviation of X are μ and σ, respectively. Our objective is to obtain bootstrap interval estimates for μ and σ. Let $X_1, X_2, \ldots, X_n$ denote a random sample of the X's. We will take as our estimates of μ and σ the sample mean $\overline{X}$ and the sample standard deviation S.

8.2.1 Interval Estimate of the Mean

In the case in which $f(z)$ is a normal distribution, the t statistic defined by

$$t = \frac{\overline{X} - \mu}{S/\sqrt{n}}$$

has a t-distribution with $n - 1$ degrees of freedom. If $t_{.975}$ is the 97.5th percentile of this t-distribution, then

$$P\left(-t_{.975} < \frac{\overline{X} - \mu}{S/\sqrt{n}} < t_{.975}\right) = .95$$

From this, we can solve the inequality to obtain the familiar 95% confidence interval

$$\overline{X} - t_{.975}\frac{S}{\sqrt{n}} < \mu < \overline{X} + t_{.975}\frac{S}{\sqrt{n}}$$

The important step in this argument from our point of view is to note that the distribution of the t statistic does not depend on either μ or σ. A function of observations and parameters whose probability distribution does not depend on the parameters is called a *pivot quantity*.

In cases in which $f(z)$ is not a normal distribution, the t statistic is still a pivot quantity, which we will refer to as a *t-pivot* quantity. However, we cannot claim that its distribution is a t-distribution with $n - 1$ degrees of freedom. Setting aside this difficulty for the moment, suppose we are able to obtain the distribution of the t-pivot in some manner. Let $t_{.025}$ and $t_{.975}$ be the 2.5th and 97.5th percentiles of this distribution. We can assert

$$P\left(t_{.025} < \frac{\overline{X} - \mu}{S/\sqrt{n}} < t_{.975}\right) = .95$$

By solving the inequality, we have the 95% confidence interval

$$\overline{X} - t_{.975}\frac{S}{\sqrt{n}} < \mu < \overline{X} - t_{.025}\frac{S}{\sqrt{n}}$$

Of course, a similar procedure would yield other levels of confidence.

As an aside, we note that the choices of $t_{.025}$ and $t_{.975}$ are not the only ones that would yield a 95% confidence interval. Any percentiles t_a and t_b such that $b - a = .95$ would work. The choice of $t_{.025}$ and $t_{.975}$ may not produce an interval that is symmetric about the mean as happens with the t-distribution. All of our confidence interval formulas in this chapter require us to choose two percentiles from a sampling distribution. For simplicity of notation, we will illustrate our procedures by constructing 95% confidence intervals using the 2.5th and 97.5th percentiles of the sampling distribution as endpoints. In general, for a $100p\%$ confidence interval ($p > 1/2$), we would select the $[(1 - p)/2]$th and $[(1 + p)/2]$th percentiles.

Steps to Obtain a Bootstrap Confidence Interval for μ

The bootstrap procedure provides a way to approximate the percentiles of distribution of the t-pivot. From these we may construct bootstrap confidence intervals for μ. The steps are as follows:

1. Compute the mean $\overline{X}$ and standard deviation S of the original data.

2. Obtain a bootstrap sample of size n from the data. Compute the mean $\overline{X}_b$ and standard deviation S_b of the bootstrap sample, and compute the bootstrap t-pivot quantity

$$t_b = \frac{\overline{X}_b - \overline{X}}{S_b/\sqrt{n}}$$

3. Repeat step 2 a number of times—say, 1000 or more—to obtain a bootstrap distribution of the t_b's.

4. For a 95% confidence interval, let $t_{b,.025}$ and $t_{b,.975}$ be the 2.5th and 97.5th percentiles of the bootstrap distribution. The bootstrap 95% confidence interval is

$$\overline{X} - t_{b,.975} \frac{S}{\sqrt{n}} < \mu < \overline{X} - t_{b,.025} \frac{S}{\sqrt{n}}$$

Make appropriate modifications for other levels of confidence.

Due to the discrete nature of the bootstrap distribution, there may not be percentiles for the desired level of confidence. We may then either interpolate to find the desired percentiles or simply choose percentiles from the bootstrap distribution as close to the desired ones as possible. We recommend using the traditional 90% or 95% confidence interval and between 1000 and 5000 bootstrap samples. Higher levels of confidence require percentiles in the tails of the distribution of the t-pivot where bootstrap approximations may not be adequate.

EXAMPLE 8.2.1   Using 5000 bootstrap samples, we obtained the 2.5th, 5th, 95th, and 97.5th percentiles of the t-pivot distribution for the latch failure data in Table 1.2.1. These are −2.44, −1.96, 1.55, and 1.85, respectively. The corresponding percentiles for the t-distribution with $n - 1 = 19$ degrees of freedom are −2.09, −1.73, 1.73, and 2.09. The data are somewhat skewed to the right, accounting for the lack of symmetry of the bootstrap t-pivot percentiles. The 95% bootstrap confidence interval is

$$38.7 - 1.85 \frac{24.0}{\sqrt{20}} < \mu < 38.7 + 2.44 \frac{24.0}{\sqrt{20}}$$

or 28.8 to 51.8. This compares to the interval 27.6 to 49.8 using the percentiles from the t-distribution with 19 degrees of freedom. The Resampling Stats code is shown in Figure 8.2.1 (see page 260). ∎

8.2.2 Confidence Intervals for the Variance and Standard Deviation

A pivot quantity for the variance, which we call a χ^2-*pivot*, is defined as

$$\chi^2 = \frac{(n-1)S^2}{\sigma^2}$$

If the observations come from a normal distribution, then this pivot quantity has a chi-square distribution with $n-1$ degrees of freedom. If $\chi^2_{.025}$ and $\chi^2_{.975}$ are the 2.5th and 97.5th percentiles of this chi-square distribution, then

$$P\left(\chi^2_{.025} < \frac{(n-1)S^2}{\sigma^2} < \chi^2_{.975}\right) = .95$$

It follows that a 95% confidence interval for σ^2 is

$$\frac{(n-1)S^2}{\chi^2_{.975}} < \sigma^2 < \frac{(n-1)S^2}{\chi^2_{.025}}$$

If the normality assumption is not plausible, then we may obtain a bootstrap distribution of the χ^2-pivot, and from that obtain approximations for the percentiles required in the computation of the confidence interval.

Steps to Obtain a Bootstrap Confidence Interval for σ^2

1. Compute the sample variance S^2 for the original data.
2. Obtain a bootstrap sample of size n. Compute the sample variance S_b^2 of the bootstrap sample, and compute the bootstrap χ^2-pivot quantity

$$\chi_b^2 = \frac{(n-1)S_b^2}{S^2}$$

3. Repeat step 2 a number of times—say, 1000 or more—to obtain a bootstrap distribution of the χ_b^2's.
4. For a 95% confidence interval, let $\chi^2_{b,.025}$ and $\chi^2_{b,.975}$ be the 2.5th and 97.5th percentiles of the bootstrap distribution. The bootstrap 95% confidence interval is

$$\frac{(n-1)S^2}{\chi^2_{b,.975}} < \sigma^2 < \frac{(n-1)S^2}{\chi^2_{b,.025}}$$

Take the square root for a confidence interval for σ.

EXAMPLE 8.2.2   Bootstrap 2.5th, 5th, 95th, and 97.5th percentiles of the χ^2-pivot for the latch failure data in Table 1.2.1 are 8.10, 9.55, 26.7, and 28.6, respectively, based on 5000 bootstrap samples. The corresponding percentiles of the χ^2 distribution with 19 degrees of freedom are 8.91, 10.1, 30.1, and 32.9. A 95% bootstrap confidence interval for the standard deviation is 19.6 to 36.8. The 95% confidence interval based on the chi-square distribution is 18.2 to 35.0. The Resampling Stats code is shown in Figure 8.2.1. ∎

8.2.3 Coverage Percentages

A bootstrap distribution of a pivot quantity is just an approximation of its true distribution. Thus, the level of confidence asserted by the bootstrap procedure is only an approximation of the actual level of confidence. To investigate how close the stated levels of confidence are to the actual levels of confidence, a simulation study was conducted to estimate the coverage percentages of the bootstrap intervals of the mean and standard deviation.

In this study, 1000 bootstrap 90% confidence intervals for the mean and standard deviation were obtained for simulated data taken from four distributions (exponential, Laplace, uniform, and normal) and two sample sizes ($n = 20, 80$). Each confidence interval was based on 1000 bootstrap samples of the data. The percentages of times the intervals covered the population mean and standard deviation were obtained. If the bootstrap procedure works as desired, these percentages should be around 90%. Estimated coverage percentages are listed in Table 8.2.1.

TABLE 8.2.1

Coverage Percentages for Nominal 90% Nonparametric Bootstrap Confidence Intervals for Mean and Standard Deviation in Location-Scale Models

	n = 20		n = 80	
Error Distribution $f(z)$	*Mean*	*SD*	*Mean*	*SD*
Exponential	90.7	74.1	89.4	81.0
Laplace	88.6	80.7	89.7	84.2
Uniform	92.2	88.7	91.1	89.8
Normal	91.7	84.9	90.4	89.7

*The error distributions are centered and scaled to have mean 0 and standard deviation 1.

The coverage percentages for the mean appear to be reasonable for all the distributions and both sample sizes. However, the coverage percentages for the standard deviation appear to depend on both the error distribution $f(z)$ and the sample size n. For all but the light-tailed uniform distribution, the coverage percentages are low for the standard deviation for small sample sizes. The coverage percentages improve with increasing sample size, but tail weight and skewness affect the results, with results being worse for the heavier-tailed and skewed distributions. The results for the normal distribution are consistent with results found by Schenker (1985).

8.2.4 Derivation of Pivotal Quantities

With the location-scale model, we have

$$X_i = \mu + \sigma Z_i$$

$i = 1, 2, \ldots, n$, where Z_i has the standard distribution $f(z)$ with mean 0 and standard deviation 1. Let $\bar{Z}$ and S_Z denote the sample mean and sample standard deviation of the Z_i's, respectively. Then the probability distribution of the standardized quantity

$$\frac{\bar{X} - \mu}{\sigma/\sqrt{n}} = \bar{Z}\sqrt{n}$$

does not depend on μ or σ. Likewise, the probability distribution of the quantity

$$\frac{(n-1)S^2}{\sigma^2} = \sum_1^n \left(\frac{X_i - \bar{X}}{\sigma}\right)^2 = (n-1)S_Z^2$$

does not depend on μ or σ. Hence what we have called a χ^2-pivot is in fact a pivot quantity for the variance. It follows that the probability distribution of the quantity

$$\frac{\bar{X} - \mu}{S/\sqrt{n}} = \frac{\bar{X} - \mu}{\sigma/\sqrt{n}} \div \frac{S}{\sigma} = \sqrt{n}\,\frac{\bar{Z}}{S_Z}$$

does not depend on μ or σ. Thus, the t-pivot is a pivot quantity for the mean.

If a location-scale model is not explicitly expressed in terms of the mean and standard deviation, it is possible to reparameterize the model so it is. Suppose X has distribution

$$g(x) = \frac{1}{\delta} h\left(\frac{x - \theta}{\delta}\right)$$

so that X may be expressed as

$$X = \theta + \delta \varepsilon$$

where ε is a random variable with density $h(\varepsilon)$. Let μ_ε and σ_ε denote the mean and standard deviation of ε. Then the mean and standard deviation of X are given by $\mu = \theta + \delta\mu_\varepsilon$ and $\sigma = \delta\sigma_\varepsilon$. It follows that X may be expressed as

$$X = \mu + \sigma\left(\frac{\varepsilon - \mu_\varepsilon}{\sigma_\varepsilon}\right) = \mu + \sigma Z$$

where $Z = (\varepsilon - \mu_\varepsilon)/\sigma_\varepsilon$, which is a random variable with mean 0 and standard deviation 1. Therefore, the t-pivot and χ^2-pivot are pivot quantities for the mean and standard deviation for all location-scale models provided the mean and standard deviation exist.

8.2.5 Computer Analysis

Resampling Stats code for the bootstrap *t*-pivot and chi-square pivot is given in Figure 8.2.1. S-Plus allows bootstrap sampling from its "Resample" menu option. The user specifies an expression, such as the *t*-pivot, for which the bootstrap percentiles are desired.

FIGURE 8.2.1

Resampling Stats Code for Bootstrap Percentiles of *t*-Pivot and χ^2-Pivot for Data in Table 1.2.1

```
maxsize default 5000
'input data
data (7 11 15 16 20 22 24 25 29 33 34 37 41 42 49 57 66 71 84 90) dat
'get mean, standard deviation, and variance of original data
mean dat m
stdev dat sd
square sd var

'get 5000 bootstrap samples
repeat 5000
sample 20 dat bootdat

'compute bootstrap t-pivot
mean bootdat bootmean
stdev bootdat bootsd
subtract bootmean m diff
'4.47 = sqrt(20)
divide bootsd 4.47 se
divide diff se tstat

'compute bootstrap chisquare
square bootsd bootvar
divide bootvar var chisq
multiply 19 chisq chisq

'keep track of t and chisquare
score tstat tdist
score chisq chidist
end

'get percentiles of bootstrap distributions
percentile tdist (2.5 5 95 97.5) tpct
percentile chidist (2.5 5 95 97.5) chipct

print tpct chipct


TPCT    =     -2.4403    -1.9574     1.5531     1.8549
CHIPCT  =      8.1042     9.5546    26.682     28.607
```

8.3
BCA and Other Bootstrap Intervals

In this section we present three methods for computing bootstrap confidence intervals. These intervals apply in general—not just to location and scale parameters as discussed in Section 8.2. Let $X_1, X_2, \ldots, X_n$ be a random sample of size n from a population, θ a population parameter to be estimated, $\hat{\theta}$ an estimate of θ from the original data, and $\hat{\theta}_b$ the corresponding estimate based on a bootstrap sample from the data.

8.3.1 Percentile and Residual Methods

The simplest form of a bootstrap confidence interval is based on the *percentile method*, which involves two simple steps:

1. Draw a specified number of bootstrap samples of size n from the data, and for each bootstrap sample compute the estimate $\hat{\theta}_b$ of θ.

2. For a 95% confidence interval for θ, find the 2.5th and 97.5th percentiles $\hat{\theta}_{b,.025}$ and $\hat{\theta}_{b,.975}$ of the bootstrap distribution. The percentile-method bootstrap 95% confidence interval is

$$\hat{\theta}_{b,.025} < \theta \leq \hat{\theta}_{b,.975}$$

Make appropriate modifications for other levels of confidence.

We note that if the bootstrap distribution is reasonably well approximated by a continuous distribution, which is usually the case, then it would make little difference whether the inequalities in the confidence interval are strict or inclusive.

A second method, which we call the *residual method*, is based on obtaining a bootstrap sample of the residuals $\varepsilon = \hat{\theta} - \theta$. If $\varepsilon_{.025}$ and $\varepsilon_{.975}$ are the 2.5th and 97.5th percentiles of the distribution of the ε's, then

$$P\left(\varepsilon_{.025} < \hat{\theta} - \theta \leq \varepsilon_{.975}\right) = .95$$

so that a 95% confidence interval for θ is

$$\hat{\theta} - \varepsilon_{.975} \leq \theta < \hat{\theta} - \varepsilon_{.025}$$

We obtain the bootstrap distribution of the ε's from which the desired percentiles for the confidence interval are obtained. The procedure is as follows.

1. Compute $\hat{\theta}$ from the data.

2. Draw a bootstrap sample of size n from the data. Compute $\hat{\theta}_b$ and the residual $e_b = \hat{\theta}_b - \hat{\theta}$.

3. Repeat step 2 a specified number of times to obtain the bootstrap distribution of the e_b's.

4. For a 95% confidence interval, obtain the 2.5th and 97.5th percentiles, $e_{b,.025}$ and $e_{b,.975}$, of the bootstrap distribution. The confidence interval for θ is

$$\hat{\theta} - e_{b,975} \leq \theta < \hat{\theta} - e_{b,025}$$

Alternatively, the interval may be expressed as

$$\hat{\theta} - \left(\hat{\theta}_{b,.975} - \hat{\theta} \right) \leq \theta < \hat{\theta} - \left(\hat{\theta}_{b,.025} - \hat{\theta} \right)$$

or

$$2\hat{\theta} - \hat{\theta}_{b,.975} \leq \theta < 2\hat{\theta} - \hat{\theta}_{b,.025}$$

Make appropriate modifications for other levels of confidence.

8.3.2 BCA Method

It can happen that the intervals computed by the percentile and residual methods tend not to be centered on the true parameter. It can also happen that the asserted level of confidence is not what is obtained in practice. A third method called the *bias corrected and accelerated method*, or BCA for short, attempts to adjust for these shortcomings. The BCA confidence interval for a parameter θ consists of two percentiles from the bootstrap distribution of $\hat{\theta}_b$, but the endpoints are not chosen in the intuitive way that they are with the percentile and residual methods. For instance, unlike a 95% confidence interval using the percentile method, the endpoints will not necessarily be the 2.5th and 97.5th percentiles, and the fraction of the bootstrap distribution between the two BCA confidence limits will not necessarily be .95. If the bootstrap distribution is symmetric about $\hat{\theta}$, then the BCA method, percentile method, and residual method give the same endpoints. Otherwise, the BCA interval is "corrected" for bias and skewness.

A Sketch of the BCA Method

The idea of the BCA method is to assume that there is a transformation of $\hat{\theta}$ whose distribution is normal and whose mean and standard deviation depend in a particular way on θ. A confidence interval is made on the transformed parameter and then the interval is inverted to obtain an interval for θ. The inversion can be done without knowledge of the explicit form of the transformation using the bootstrap method.

Suppose there is a strictly increasing transformation T such that $T(\hat{\theta})$ has a normal distribution with expected value and standard deviation given by

$$E\left[T\left(\hat{\theta}\right)\right] = T(\theta) - z_0[1 + aT(\theta)]$$

$$\sigma\left[T\left(\hat{\theta}\right)\right] = 1 + aT(\theta)$$

If z_p is the 100pth percentile of the standard normal distribution where $p = 1 - \alpha/2$, then with $1 - \alpha$ probability,

$$-z_p < \frac{T(\hat{\theta}) - T(\theta)}{1 + aT(\theta)} + z_0 < z_p$$

Solving this equation, we obtain a $100(1 - \alpha)\%$ confidence interval for $T(\theta)$ of the form

$$\frac{T(\hat{\theta}) + z_0 - z_p}{1 - a(z_0 - z_p)} < T(\theta) < \frac{T(\hat{\theta}) + z_0 + z_p}{1 - a(z_0 + z_p)}$$

To determine the point on the bootstrap distribution that corresponds to the upper endpoint of the confidence interval, we proceed as follows:

$$P\left(T(\hat{\theta}_b) < \frac{T(\hat{\theta}) + z_0 + z_p}{1 - a(z_0 + z_p)}\right) = P\left(\frac{T(\hat{\theta}_b) - T(\hat{\theta})}{1 + aT(\hat{\theta})} + z_0 < \frac{z_0 + z_p}{1 - a(z_0 + z_p)} + z_0\right)$$

$$= P\left(Z < \frac{z_0 + z_p}{1 - a(z_0 + z_p)} + z_0\right)$$

where Z has a standard normal distribution. The second of the equalities holds, since the bootstrap distribution of

$$\frac{T(\hat{\theta}_b) - T(\hat{\theta})}{1 + aT(\hat{\theta})} + z_0$$

should mimic the distribution of

$$\frac{T(\hat{\theta}) - T(\theta)}{1 + aT(\theta)} + z_0$$

which according to assumption is standard normal. Therefore, the upper limit of the BCA interval for θ has the same cumulative probability as

$$z_U = \frac{z_0 + z_p}{1 - a(z_0 + z_p)} + z_0$$

when referred to the standard normal distribution. Similarly, the lower limit of the BCA interval for θ has the same cumulative probability as

$$z_L = \frac{z_0 - z_p}{1 - a(z_0 - z_p)} + z_0$$

when referred to the standard normal distribution.

For instance, if $z_0 = 0.20$, $z_{.975} = 1.96$, and $a = 0.01$, then $z_L = -0.153$ and $z_U = 2.41$. These are the 6.3th and 99.2th percentiles of the normal distribution. Thus the 6.3th and 99.2th percentiles of the bootstrap distribution would be the BCA 95% confidence limits for θ.

What remains to be determined are the values of z_0 and a. To determine z_0, let p_0 denote the fraction of the observations from the bootstrap distribution for which $\hat{\theta}_b \leq \hat{\theta}$. Then z_0 is the value for which $P(Z \leq z_0) = p_0$, where Z has a standard normal distribution. If the bootstrap distribution has $\hat{\theta}$ as its median, then $z_0 = 0$. Thus, z_0 corrects for the median bias of the bootstrap estimator.

The value of a is based on a measure of skewness of the data. Let $\hat{\theta}_{-i}$ denote the estimate of θ computed from the data $X_1, \ldots, X_{i-1}, X_{i+1}, \ldots, X_n$. That is, it is the estimate of θ using all the data except X_i. Let $\hat{\theta}_{(\ )}$ denote the mean of the $\hat{\theta}_{-i}$'s. Then a is given by

$$a = \frac{\sum_{i=1}^{n}\left(\hat{\theta}_{(\ )} - \hat{\theta}_{-i}\right)^3}{6\left[\sum_{i=1}^{n}\left(\hat{\theta}_{(\ )} - \hat{\theta}_{-i}\right)^2\right]^{3/2}}$$

A simpler version of the BCA interval, called the *bias corrected* or BC interval, does not adjust for skewness. It is obtained by setting $a = 0$ rather than estimating it from the data.

EXAMPLE 8.3.1   The data in Table 8.3.1 are a simulated random sample of size 20 from an exponential distribution with a mean of 10.

TABLE 8.3.1

Simulated Data from an Exponential Distribution with Mean 10

0.77	1.17	2.79	3.13	3.31	3.70	4.13	5.28	5.84	7.38	7.75
8.25	9.00	10.12	11.51	15.82	18.28	21.73	32.74	38.28

The sample mean is 10.55. Bootstrap 95% confidence intervals based on the three methods presented above were obtained using 5000 bootstrap samples. In addition, a bootstrap interval was obtained using the *t*-pivot method of Section 8.2. A parametric confidence interval for the mean of the exponential distribution may be obtained using the fact that the pivot quantity $2n\overline{X}/\theta$ has a chi-square distribution with $2n$ degrees of freedom, where θ is the mean. Placing this pivot between the 2.5th and 97.5th percentiles of the chi-square distribution and solving yield the parametric confidence interval. Results are shown in Table 8.3.2. Computations were done with S-Plus.

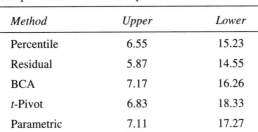

TABLE 8.3.2

Comparison of 95% Bootstrap Confidence Limits

Method	Upper	Lower
Percentile	6.55	15.23
Residual	5.87	14.55
BCA	7.17	16.26
t-Pivot	6.83	18.33
Parametric	7.11	17.27

If it is known that the observations come from an exponential distribution, then the parametric interval is preferred because such intervals have exact 95% coverage of the population mean. In general, of course, we may not know the distribution of the observations, in which case it may be preferable to use a bootstrap interval even though the level of confidence associated with it may be only approximate.   ∎

Manly (1997) reported a simulation study of the various nonparametric methods for constructing a nominal 95% confidence interval for the mean of an exponential distribution with a sample of size 20. The simulated coverage percentages for these methods were: percentile (90.1%), residual (88.8%), BCA (92.4%), and *t*-pivot (95.2%). When the intervals missed the true value, they tended to be on the low side. Our preference is for either the *t*-pivot, assuming the pivot quantity exists, or the BCA interval.

8.3.3 Computer Analysis

BCA intervals are available in S-Plus from its "Resample" menu option. S-Plus output for the analysis of the data in Table 8.3.1 is shown in Figure 8.3.1. The expression for which a bootstrap sample is desired must be specified. For instance, the expression *mean(V1)* provides a bootstrap sample for the mean of the data in column V1. The output includes the bias and standard deviation of the estimate and the endpoints of the 90% and 95% confidence intervals. In addition to the BCA intervals, S-Plus provides an option to obtain empirical percentiles, which are the endpoints of our percentile method. The user may specify the number of bootstrap samples and the seed, or starting point, for the random number generator that produces the bootstrap sample.

Lunneborg (2000) has more details on the computation of the BCA intervals, including code for carrying out the BCA method in Resampling Stats.

FIGURE 8.3.1

S-Plus Percentile and BCA Bootstrap Intervals for the Mean of the Data in Table 8.3.1

```
                    *** Bootstrap Results ***
Call:
bootstrap(data = DS10, statistic = mean(V1), B = 5000,
                seed = 1155, trace = F, assign.frame1 = F,
                save.indices = F)


Number of Replications: 5000


Summary Statistics:
       Observed      Bias   Mean      SE
Param     10.55  -0.06482  10.48   2.215


Empirical Percentiles:
           2.5%      5%        95%      97.5%
Param 6.552388  7.0358  14.40022  15.23417


BCA Percentiles:
           2.5%       5%        95%      97.5%
Param 7.166409  7.632911  15.40842  16.26396
```

8.4
Correlation and Regression

In Section 5.1, we considered permutation tests for the correlation coefficient and the slope of a regression line. Here we consider the bootstrap approach to making inferences about these parameters. As in Section 5.1, we distinguish between two types of sampling: bivariate sampling and fixed-X sampling. We use the notation established in Section 5.1 for our discussion.

8.4.1 Bivariate Bootstrap Sampling

Suppose we have a random sample (X_i, Y_i), $i = 1, 2, \ldots, n$, from a bivariate population. A *bivariate bootstrap sample* is a random sample with replacement of the pairs (X_i, Y_i), $i = 1, 2, \ldots, n$. For instance, if the observations are (1, 1), (2, 5), (4, 7), and (5, 9), then there would be 4^4 equally likely possible bootstrap samples of size 4. One such sample would be ((5, 9), (1, 1), (5, 9), (4, 7)). We may use bivariate bootstrap sampling to obtain bootstrap intervals for bivariate population parameters such as the correlation coefficient ρ and the ratio of the means μ_x/μ_y.

Bootstrap Confidence Interval for ρ

The steps for obtaining a bootstrap interval estimate of the population correlation coefficient ρ are as follows:

1. Draw a specified number of bivariate bootstrap samples of size n from the data.
2. Compute the bootstrap Pearson correlation coefficient r_b for each bootstrap sample.
3. Obtain a confidence interval from the distribution of the r_b's using the BCA method or the percentile method, with the BCA method preferred. The residual method has the undesirable property of possibly producing limits that are beyond the bounds of –1 to 1 for correlation.

EXAMPLE **8.4.1**   Figure 8.4.1 is a plot of the shoe sizes (SS) and heights (HT) of 24 college-age men and a plot of 1000 bootstrap values of r_b. The correlation is $r = .72$. The BCA 95% interval based on these 1000 values is (.39, .87), and the percentile method 95% confidence interval is (.42, .88). The analysis was done with S-Plus.

By comparison, we may obtain limits developed for the bivariate normal distribution. These limits are based on the transformation

$$Z = \frac{1}{2} \ln\left(\frac{1+r}{1-r}\right)$$

which has an approximate normal distribution with mean

$$Z = \frac{1}{2} \ln\left(\frac{1+\rho}{1-\rho}\right)$$

FIGURE 8.4.1
Shoe Size versus Heights of 24 College-Age Men and Bootstrap Sampling Distribution of the Correlation Coefficient

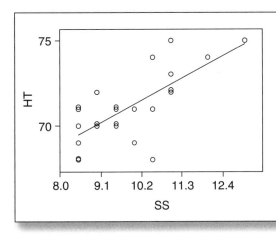
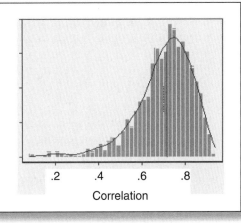

and standard deviation $1/\sqrt{n-3}$. This transformation gives the 95% confidence interval

$$\frac{(1+r)e^{-c} - (1-r)}{(1+r)e^{-c} + (1-r)} < \rho < \frac{(1+r)e^{c} - (1-r)}{(1+r)e^{c} + (1-r)}$$

where $c = 2(1.96)/\sqrt{n-3}$. In this example, the limits are (.45, .87). The BCA interval in particular gives a smaller value for the lower limit and a wider interval than the normal-theory method.   ∎

8.4.2 Fixed-*X* Bootstrap Sampling

We assume *Y* may be expressed in terms of the regression model

$$Y = h(X) + \varepsilon$$

where $h(X)$ is a function of the independent variable—for instance a linear function—and the ε's are independent and identically distributed random variables with mean 0 and standard deviation σ. When the *X*'s are regarded as fixed, the idea is to sample the errors and then add on an estimate of the mean function $h(X)$ to get the bootstrap sample. Here are the steps:

1. Compute an estimate $\hat{h}(x)$ of the mean function $h(x)$.
2. Obtain the observed errors $e_i = Y_i - \hat{h}(X_i)$.
3. Select *n* values of the e_i's at random with replacement. Denote these values as $e_{1,b}, e_{2,b}, \ldots, e_{n,b}$.
4. Compute $Y_i = \hat{h}(X_i) + e_{i,b}$, $i = 1, 2, \ldots, n$. The pairs (X_i, Y_i) comprise a fixed-*X* bootstrap sample of size *n*.

EXAMPLE 8.4.2   Table 8.4.1 shows one fixed-*X* bootstrap sample of size $n = 24$ for the shoe size and height data of 24 college-age men. The mean function $h(x)$ is assumed to be linear, and its least squares estimate is HT = 59.2 + 1.20(SS). The first person in the table has a predicted height P_HT = 59.2 + 1.20(8.5) = 69.4 and the error is *e* = 71 – 69.4 = 1.6. A random selection from all the errors produced a bootstrap error of B_e = –0.6 for the first person in the table, so the bootstrap value of HT is B_HT = 69.4 – 0.6 = 68.8. That is, the first bootstrap pair is (8.5, 68.8), and similarly for the rest of the values. The figures in the table have been rounded for simplicity of the display, so the sum of the e_i's does not exactly equal 0 as would be the case without rounding.   ∎

Use of Fixed-*X* Bootstrap Sampling

We can use fixed-*X* bootstrap sampling in two situations. We can use it if the *X*'s are controlled by the researcher. For instance, if a researcher varied the temperature

TABLE 8.4.1

Bootstrap Sample of Shoe Size (SS) and Height (HT) Data in Figure 8.4.1

P_HT = 59.2 + 1.20(SS) , e = Errors, B_e = Bootstrap Errors,
B_HT = Bootstrap Heights = P_HT + B_e

SS	HT	P_HT	e	B_e	B_HT
8.5	71	69.4	1.6	–0.6	68.8
8.5	68	69.4	–1.4	–0.4	69.0
8.5	71	69.4	1.6	0.6	70.0
8.5	70	69.4	0.6	2.0	71.4
8.5	68	69.4	–1.4	0.2	69.6
8.5	69	69.4	–0.4	–0.6	68.8
9.0	72	70.0	2.0	1.6	71.6
9.0	70	70.0	0.0	0.4	70.4
9.0	70	70.0	0.0	0.4	70.4
9.5	71	70.6	0.4	–0.6	70.0
9.5	70	70.6	–0.6	0.6	71.2
9.5	71	70.6	0.4	–0.6	70.0
9.5	70	70.6	–0.6	–0.4	70.2
10.0	71	71.2	–0.2	0.2	71.4
10.0	69	71.2	–2.2	–0.6	70.6
10.5	68	71.8	–3.8	0.4	72.2
10.5	71	71.8	–0.8	1.6	73.4
10.5	74	71.8	2.2	2.0	73.8
11.0	72	72.4	–0.4	–1.4	71.0
11.0	72	72.4	–0.4	2.6	75.0
11.0	73	72.4	0.6	–0.4	72.0
11.0	75	72.4	2.6	0.4	72.8
12.0	74	73.6	0.4	0.4	74.0
13.0	75	74.8	0.2	2.0	76.8

X to determine its effect on vapor pressure, then we could use fixed-X bootstrap sampling to mimic the way the data are collected.

Fixed-X bootstrap sampling may also be used in some situations in which the X's are not fixed before the data are taken. To illustrate the idea, suppose the mean function $h(x)$ is linear; that is, $h(x) = \beta_0 + \beta_1 x$. Suppose for each fixed set of X's we are able to find a 95% confidence interval for β_1. That is, we are able to find the lower and upper endpoints $L(X_1, X_2, \ldots, X_n)$ and $U(X_1, X_2, \ldots, X_n)$ such that

$$P\left[L\left(X_1, X_2, \ldots, X_n\right) \leq \beta_1 \leq U\left(X_1, X_2, \ldots, X_n\right) \middle| X_1, X_2, \ldots, X_n\right] = .95$$

The interval

$$\left[L(X_1, X_2, \ldots, X_n), U(X_1, X_2, \ldots, X_n)\right]$$

is a conditional 95% confidence interval for β_1 given $X_1, X_2, \ldots, X_n$. However, since the interval has the same probability .95 of capturing the β_1 for any observed set of X's, it is also an unconditional confidence interval, that is,

$$P\left[L(X_1, X_2, \ldots, X_n) \le \beta_1 \le U(X_1, X_2, \ldots, X_n)\right]$$

$$= E\left\{P\left[L(X_1, X_2, \ldots, X_n) \le \beta_1 \le U(X_1, X_2, \ldots, X_n)\middle|X_1, X_2, \ldots, X_n\right]\right\} = .95$$

The various inferential procedures for normal-theory regression analysis are developed by conditioning on the X's. That is, the X's are treated as fixed even if they are not fixed at the time the data are taken. In what follows, we give a bootstrap version of this approach to inference for regression. It requires fixed-X bootstrap sampling.

8.4.3 Bootstrap Inferences for the Slope of a Regression Line

Assume the regression model is

$$Y_i = \beta_0 + \beta_1 X_i + \varepsilon_i, \quad i = 1, 2, \ldots, n$$

where the ε_i's are independent and identically distributed random variables with mean 0 and standard deviation σ. Our interest is in statistical inferences for β_1. A straightforward procedure for making a confidence interval is to apply the BCA method, percentile method, or residual method to the bootstrap distribution of the estimated slope. We may use these methods with either bivariate or fixed-X bootstrap sampling. In the following development, we consider the t-pivot method for making inferences about β_1 based on the conditional inference discussed above and fixed-X sampling.

t-Pivot for the Slope

The least squares estimate of β_1 and the mean square error, MSE, were defined in Section 5.1.2. The estimated standard error of the least squares estimate is

$$SE(\hat{\beta}_1) = \sqrt{\frac{MSE}{(n-1)S_x^2}}$$

where S_X is the standard deviation of the X's. The t-pivot for the slope is

$$t = \frac{\hat{\beta}_1 - \beta_1}{SE(\hat{\beta}_1)}$$

whose distribution does not depend on β_0, β_1, or σ.

When the errors have a normal distribution, the t-pivot has a t-distribution with $n-2$ degrees of freedom. We obtain a 95% confidence interval for β_1 by solving the inequality

$$-t_{.975} < \frac{\hat{\beta}_1 - \beta_1}{\mathrm{SE}\left(\hat{\beta}_1\right)} < t_{.975}$$

where $t_{.975}$ is the 97.5th percentile of the t-distribution with $n-2$ degrees of freedom. Setting $\beta_1 = 0$ in the t-pivot, we have a test statistic for testing H_0: $\beta_1 = 0$ against one-sided or two-sided alternatives.

Bootstrap Distribution of the t-Pivot

Suppose we cannot assume the distribution of the ε_i's is normal, but suppose we could observe their actual values. Apply the least squares procedure to the pairs (X_i, ε_i), $i = 1, 2, \ldots, n$; that is, substitute ε_i for Y_i in the least squares formula for the regression line. Let $\hat{\beta}_{1\varepsilon}$ denote the estimate of the slope obtained this way. It can be shown that the distribution of

$$t_\varepsilon = \frac{\hat{\beta}_{1\varepsilon}}{\mathrm{SE}\left(\hat{\beta}_{1\varepsilon}\right)}$$

is the same as the distribution of the t-pivot for the (X_i, Y_i) data. This suggests a way to obtain a bootstrap approximation for the distribution of the t-pivot without having to explicitly obtain bootstrap values of the Y_i's.

1. Compute the least squares estimates and the residuals $e_i = Y_i - \hat{\beta}_0 - \hat{\beta}_1 X_i$, $i = 1, 2, \ldots, n$.

2. Take a sample of size n of the e_i's with replacement, and form the bootstrap sample $(X_i, e_{i,b})$, $i = 1, 2, \ldots, n$. Compute the least squares estimate of the slope $\hat{\beta}_{1e}$, the estimated standard error $\mathrm{SE}(\hat{\beta}_{1e})$, and the bootstrap t-statistic

$$t_e = \frac{\hat{\beta}_{1e}}{\mathrm{SE}\left(\hat{\beta}_{1e}\right)}$$

3. Repeat step 2 a number of times to obtain the bootstrap distribution of the t_e's. Find the desired percentiles of the distribution of the t_e's and use these to construct a confidence interval or conduct a test of hypothesis for β_1.

For instance, if $t_{e,.025}$ and $t_{e,.975}$ are the 2.5th and 97.5th percentiles of the bootstrap distribution, then a 95% bootstrap confidence interval for β_1 may be obtained by solving for β_1 in the inequality

$$t_{e,.025} < \frac{\hat{\beta}_1 - \beta_1}{\mathrm{SE}\left(\hat{\beta}_1\right)} < t_{e,.975}$$

This gives the 95% bootstrap confidence interval

$$\hat{\beta}_1 - t_{e,.975}\text{SE}\!\left(\hat{\beta}_1\right) < \beta_1 < \hat{\beta}_1 - t_{e,.025}\text{SE}\!\left(\hat{\beta}_1\right)$$

For a two-sided test of hypothesis at the 5% level of significance, reject the null hypothesis if the confidence interval does not contain 0 or, equivalently, if $\hat{\beta}_1 / \text{SE}(\hat{\beta}_1)$ is outside the interval $(t_{e,.025}, t_{e,.975})$.

EXAMPLE 8.4.3   A bootstrap distribution of the t-pivot for the slope of the shoe size and height data in Table 8.4.1 is shown in Figure 8.4.2. It is based on 1000 bootstrap samples treating the X's as fixed. The 2.5th, 5th, 95th, and 97.5th percentiles are -2.18, -1.77, 1.65, and 2.02, respectively. The corresponding percentiles of the t-distribution with 22 degrees of freedom are -2.07, -1.72, 1.72, and 2.07. For the original data, the slope and standard error of the slope are 1.20 and 0.25, respectively. Thus, the bootstrap t-pivot 95% confidence interval for β_1 is $(0.70, 1.75)$. The 95% confidence interval based on the percentiles of the t-distribution is $(0.68, 1.72)$.

FIGURE 8.4.2
Bootstrap Distribution of t-Pivot for Slope of SS and HT Data

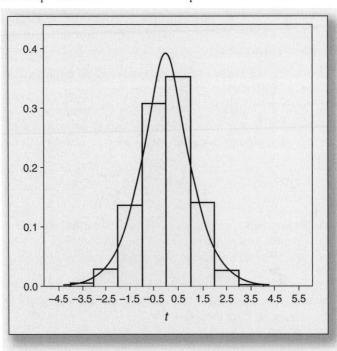

Adjusted Errors

The observed errors e_i, $i = 1, 2, \ldots, n$, which are determined from the data, have variances that depend on the X's. Define the *leverage* of X_i to be

$$h_i = \frac{1}{n} + \frac{\left(X_i - \overline{X}\right)^2}{\sum\limits_{j=1}^{n}\left(X_j - \overline{X}\right)^2}$$

The variance of e_i is given by $\sigma^2(1 - h_i)$. Define r_i as

$$r_i = \frac{e_i}{\sqrt{1 - h_i}}$$

and define the adjusted errors as

$$e_i^* = r_i - \overline{r}$$

Since the adjusted errors have constant variance σ^2 as do the true errors ε_i's, and since they sum to 0 as do the observed errors, it is sometimes recommended to obtain a bootstrap sample of the adjusted errors instead of the observed errors. The adjusted errors should be sampled if a few values of X are extreme relative to the rest of the data or if the sample sizes are small.

Derivation of the Distribution of the *t*-Pivot

Let $\text{SSE}_{(X,Y)}$ and $\text{MSE}_{(X,Y)}$ denote the sum of squared errors and the mean square error for the (X_i, Y_i) data, and let $\text{SSE}_{(X,\varepsilon)}$ and $\text{MSE}_{(X,\varepsilon)}$ denote the sum of squared errors and the mean squared error for the (X_i, ε_i) data. Using the model $Y_i = \beta_0 + \beta_1 X_i + \varepsilon_i$, we have

$$\text{SSE}_{(X,Y)} = \sum_{i=1}^{n}\left[Y_i - \left(\hat{\beta}_0 + \hat{\beta}_1 X_i\right)\right]^2$$

$$= \sum_{i=1}^{n}\left\{\varepsilon_i - \left[\left(\hat{\beta}_0 - \beta_0\right) + \left(\hat{\beta}_1 - \beta_1\right)X_i\right]\right\}^2$$

$$= \text{SSE}_{(X,\varepsilon)}$$

It follows that $\hat{\beta}_{1\varepsilon} = \hat{\beta}_1 - \beta_1$, $\text{MSE}_{(X,Y)} = \text{MSE}_{(X,\varepsilon)}$, $\text{SE}(\hat{\beta}_{1\varepsilon}) = \text{SE}(\hat{\beta}_1)$, and

$$t = \frac{\hat{\beta}_1 - \beta_1}{\text{SE}(\hat{\beta}_1)} = \frac{\hat{\beta}_{1e}}{\text{SE}(\hat{\beta}_{1e})}$$

In a similar way that we defined a *t*-pivot for β_1, we can obtain a *t*-pivot for β_0, but that is not pursued here.

8.4.4 Computer Analysis

The percentile method for making confidence intervals for the correlation and slope of the least squares line using bivariate sampling is easily implemented in Resampling Stats. The code is shown in Figure 8.4.3. The present version of Resampling Stats does not provide standard errors for the regression coefficients, so computing the bootstrap percentiles of the *t*-pivot is somewhat cumbersome. However, the Excel add-in for Resampling Stats does provide the standard errors with its regression option, and this was used in obtaining the bootstrap *t*-pivot intervals in Example 8.4.3.

FIGURE 8.4.3

Resampling Stats Code for Obtaining Percentile Confidence Intervals for the Correlation and Slope Using Bivariate Sampling; Data in Table 8.4.1

```
maxsize default 5000

'input x and y values
copy (8.5 8.5 8.5 8.5 8.5 8.5 9 9 9 9.5 9.5 9.5 9.5 10 10 10.5 10.5 10.5 11 11 11
     11 12 13) x
copy (71 68 71 70 68 69 72 70 70 71 70 71 70 71 69 68 71 74 72 72 73 75 74 75) y
copy (1 2 3 4 5 6 7 8 9 10 11 12 13 14 15 16 17 18 19 20 21 22 23 24) index

'get 5000 bootstrap samples using bivariate sampling

repeat 5000
sample 24 index s_index
take x s_index B_x
take y s_index B_y

'compute correlation
corr B_x B_y r

'compute slope
regress noprint B_y B_x coeff
take coeff 1 slope

'keep track of correlation
score r rdist
score slope slpdist
end

                                                  (continued on next page)
```

FIGURE 8.4.3

Resampling Stats Code for Obtaining Percentile Confidence Intervals for
the Correlation and Slope Using Bivariate Sampling; Data in Table 8.4.1 *(continued)*

```
'get bootstrap percentiles and print out results
percentile rdist (2.5 5 95 97.5) pctile_r
percentile slpdist(2.5 5 95 97.5) pctile_s
print pctile_r
print pctile_s


'percentiles for the correlation and slope respectively
PCTILE_R =     0.40349     0.47266     0.86713     0.88266
PCTILE_S =     0.70536     0.81726     1.5266      1.5949
```

The BCA or percentile method intervals for the correlation and regression coefficients may be obtained in S-Plus from the "Resample" menu option, where the method of sampling is bivariate. The functional expression for the correlation is *cor(V1, V2)* and for the regression coefficients is *coef (lm (V2 ~ V1))*, where the X and Y variables are in columns V1 and V2, respectively. Output for interval estimates of the correlation based on the data in Table 8.4.1 is shown in Figure 8.4.4.

FIGURE 8.4.4

S-Plus Bootstrap Intervals for the Correlation; Data in Table 8.4.1

```
*** Bootstrap Results ***
Call:
bootstrap(data = SDF13, statistic = cor(V1, V2), B = 1000, trace = F,
                assign.frame1 = F, save.indices = F)


Number of Replications: 1000


Summary Statistics:
      Observed      Bias Mean      SE
Param   0.7176 -0.007573 0.71 0.1188


Empirical Percentiles:
          2.5%        5%        95%       97.5%
Param 0.4204215 0.4908637 0.8639083 0.8844412


BCa Percentiles:
          2.5%        5%        95%       97.5%
Param 0.3921959 0.4429506 0.8543887 0.8723936
```

8.5
Two-Sample Inference

We consider the problem of making a confidence interval for the difference between the means of two populations using the bootstrap method. Suppose we have random samples from two populations. Let Y_{ij} denote the jth observation from the ith population, $j = 1, 2, \ldots, n_i$, $i = 1, 2$. Let μ_i denote the mean of the ith population, and let $F_i(y)$ denote the cdf of the population, $i = 1, 2$. Define the errors as $\varepsilon_{ij} = Y_{ij} - \mu_i$.

We consider two cases. In the first case, we assume the distributions of the errors in the two populations are the same. That is, there is a cdf $F(\varepsilon)$ such that the distribution of the ε_{ij}'s is $F(\varepsilon)$, $j = 1, 2, \ldots, n_i$, $i = 1, 2$. With this assumption, it can be shown that the distributions of the two populations differ only by the shift parameter $\Delta = \mu_1 - \mu_2$; that is, $F_1(y) = F_2(y - \Delta)$. In the second case, we allow for the possibility that the distributions of the errors in the two populations are not the same.

In the special case when the errors have a normal distribution, the first case occurs when there is *homogeneity of variances*—that is, when the variances of the populations are the same. The second case occurs when there is *heterogeneity of variances*, or unequal population variances.

8.5.1 *t*-Pivot Method Assuming Equality of Error Distributions

Suppose the distributions of the errors are the same for the two populations. Let

$$t = \frac{\bar{Y}_1 - \bar{Y}_2 - (\mu_1 - \mu_2)}{S_p\sqrt{1/n_1 + 1/n_2}}$$

where $\bar{Y}_1$ and $\bar{Y}_2$ are the sample means and

$$S_p^2 = \sum_{i=1}^{2}\sum_{j=1}^{n_i} \frac{\left(Y_{ij} - \bar{Y}_i\right)^2}{n_1 + n_2 - 2}$$

The distribution of t depends only on the distribution $F(\varepsilon)$ of the ε_{ij}'s. To see this, substitute $\mu_i + \varepsilon_{ij}$ for Y_{ij} and note that the distribution of t is the same as that of

$$t_\varepsilon = \frac{\bar{\varepsilon}_1 - \bar{\varepsilon}_2}{S_{\varepsilon p}\sqrt{\left(1/n_1 + 1/n_2\right)}}$$

where $\bar{\varepsilon}_1$ and $\bar{\varepsilon}_2$ are the means of the errors of the first and second sample, respectively, and

$$S_{\varepsilon p}^2 = \sum_{i=1}^{2} \sum_{j=1}^{n_i} \frac{\left(\varepsilon_{ij} - \overline{\varepsilon}_i\right)^2}{n_1 + n_2 - 2}$$

If the ε_{ij}'s have a normal distribution, then t_ε and t have a t-distribution with $n_1 + n_2 - 2$ degrees of freedom. We may obtain a 95% confidence interval for the difference of the population means by solving the inequality

$$-t_{.975} < \frac{\overline{Y}_1 - \overline{Y}_2 - \left(\mu_1 - \mu_2\right)}{S_p \sqrt{1/n_1 + 1/n_2}} < t_{.975}$$

for $\mu_1 - \mu_2$ where $t_{.975}$ is the 97.5th percentile of the t-distribution with $n_1 + n_2 - 2$ degrees of freedom. Thus, we refer to t as a t-pivot for making inferences for $\mu_1 - \mu_2$.

Bootstrap Interval

If the errors do not have a normal distribution, then we may obtain a bootstrap approximation of the distribution of t_ε and a bootstrap confidence interval for the difference of the means. The steps are as follows:

1. Compute the observed errors $e_{ij} = Y_{ij} - \overline{Y}_i$.
2. Randomly select n_1 errors with replacement from the set of all errors and assign them to the first sample, and similarly select n_2 errors and assign them to the second sample. Denote these as $e_{ij,b}$. Compute t_ε using the $e_{ij,b}$'s instead of the ε_{ij}'s, and denote this bootstrap statistic as t_e.
3. Repeat step 2 a number of times to obtain the bootstrap distribution of the t_e's.
4. For a 95% confidence interval, let $t_{e,.025}$ and $t_{e,.975}$ denote the 2.5th and 97.5th percentiles of the bootstrap distribution. The 95% bootstrap confidence interval is found by solving the inequality

$$-t_{e,.025} < \frac{\overline{Y}_1 - \overline{Y}_2 - \left(\mu_1 - \mu_2\right)}{S_p \sqrt{1/n_1 + 1/n_2}} < t_{e,.975}$$

for $\mu_i - \mu_j$. The solution is given by

$$\overline{Y}_1 - \overline{Y}_2 - t_{e,.975} S_p \sqrt{1/n_1 + 1/n_2} < \mu_1 - \mu_2 < \overline{Y}_1 - \overline{Y}_2 - t_{e,.025} S_p \sqrt{1/n_1 + 1/n_2}$$

EXAMPLE 8.5.1 The data in Table 8.5.1 are the average daily weight gains (in pounds) of cattle under two feeding treatments, and the observed errors. The bootstrap 2.5th, 5th, 95th, and 97.5th percentiles for t_e are -2.16, -1.79, 1.72, and 2.12, respectively. The corresponding percentiles of the t-distribution with 14 degrees of freedom are -2.15, -1.76, 1.76, and 2.15.

The difference of the two means is $1.22 - 1.01 = 0.21$, the variances are 0.024 and 0.017, and the standard error of the difference of the means is

$$\sqrt{\frac{7(0.024) + 7(0.017)}{14}\left(\frac{1}{8} + \frac{1}{8}\right)} = 0.072$$

Thus, the bootstrap lower and upper confidence limits for a 95% confidence interval are $0.21 - 2.12(0.072) = 0.057$ and $0.21 + 2.16(0.072) = 0.366$. In this case, the confidence interval is virtually the same as that given by the t-distribution percentiles.

TABLE 8.5.1

Average Daily Weight Gains (in pounds) of Cattle on Two Treatments

Treatment	Average Daily Gain	Treatment Mean	Observed Error
1	1.40	1.22	0.18
1	1.23	1.22	0.01
1	1.02	1.22	−0.20
1	0.98	1.22	−0.24
1	1.34	1.22	0.12
1	1.36	1.22	0.14
1	1.15	1.22	−0.07
1	1.27	1.22	0.05
2	1.16	1.01	0.15
2	0.99	1.01	−0.02
2	1.04	1.01	0.03
2	1.02	1.01	0.01
2	1.09	1.01	0.08
2	1.12	1.01	0.11
2	0.76	1.01	−0.25
2	0.88	1.01	−0.13

8.5.2 Unequal Distributions of the Errors

If the distributions of the errors are not the same for the two populations, then we change the statistic and the method of bootstrap sampling. The statistic is now

$$z = \frac{\bar{Y}_1 - \bar{Y}_2 - (\mu_1 - \mu_2)}{\sqrt{S_1^2/n_1 + S_2^2/n_2}}$$

where S_1^2 and S_2^2 are the sample variances of the first and second samples, respectively. Its distribution can be expressed in terms of the errors, since it is the same as that of

$$z_\varepsilon = \frac{\bar{\varepsilon}_1 - \bar{\varepsilon}_2}{\sqrt{S_{\varepsilon 1}^2/n_1 + S_{\varepsilon 2}^2/n_2}}$$

where

$$S_{\varepsilon i}^2 = \sum_{j=1}^{n_1} \frac{\left(\varepsilon_{ij} - \bar{\varepsilon}_i\right)^2}{n_i - 1}, \quad i = 1, \ 2$$

The bootstrap errors are selected with replacement from within each of the samples, rather than being selected from the combined errors. For instance, if sample 1 has errors −1, 3, −2, and sample 2 has errors −5, 5, 0, then a bootstrap selection of the three values for the first sample could contain only −1, 3, and −2, whereas if the errors were combined, the bootstrap selection could include any of the errors −1, 3, −2, −5, 5, or 0. Thus, steps 2, 3, and 4 of the procedure are replaced by the following steps:

2′. Randomly select n_1 errors with replacement from the first sample; similarly select n_2 errors from the second sample. Compute z_ε using the errors obtained in step 2′ and denote this bootstrap statistic as z_e.

3′. Repeat step 2′ a desired number of times to obtain the bootstrap distribution of z_e.

4′. For a 95% confidence interval, find the $z_{e,.025}$ and $z_{e,.975}$ percentiles and solve the inequality

$$z_{e,.025} < \frac{\bar{Y}_1 - \bar{Y}_2 - \left(\mu_1 - \mu_2\right)}{\sqrt{S_1^2/n_1 + S_2^2/n_2}} < z_{e,.975}$$

for $\mu_1 - \mu_2$.

Other Methods

Another way to handle the unequal variance problem, and one that is commonly used, is called the *Satterthwaite approximation*. Instead of a bootstrap approximation for the distribution of z_ε, its distribution is approximated by a *t*-distribution where the degrees of freedom are given by

$$df = \frac{\left(S_1^2/n_1 + S_2^2/n_2\right)^2}{\dfrac{\left(S_1^2/n_1\right)^2}{n_1 - 1} + \dfrac{\left(S_2^2/n_2\right)^2}{n_2 - 1}}$$

from which the desired percentiles are obtained.

We may also use the percentile, residual, or BCA method to obtain confidence intervals for the difference of means. For these methods, bootstrap sampling is done directly from the Y_{ij}'s. That is, we randomly select n_1 of the Y_{1j}'s with replacement from the first sample, similarly select n_2 of the Y_{2j}'s from the second sample. The endpoints of the confidence intervals are obtained from the appropriate percentiles of the bootstrap distribution of $\overline{Y}_1 - \overline{Y}_2$ as outlined in Section 8.3. For instance, for a 95% confidence interval using the percentile method, we would select the 2.5th and 97.5th percentiles of the bootstrap distribution of $\overline{Y}_1 - \overline{Y}_2$ as the endpoints of the interval.

EXAMPLE 8.5.2   Table 8.5.2 contains two series of measurements used to determine of the gravitational constant g. See Cressie (1982). The bootstrap 2.5th and 97.5th percentiles of z_e are -3.26 and 2.04, respectively, which were obtained from 5000 bootstrap samples produced by S-Plus. The difference between the means of series H and A is $80.38 - 66.38 = 14$, and the standard error is $\sqrt{370.55/8 + 11.26/13} = 6.87$. Thus, the endpoints of the 95% confidence interval are $14 - 2.04(6.87) = -0.01$, and $14 + 3.26(6.87) = 36.4$.

Table 8.5.3 compares this interval with confidence intervals based on the Satterthwaite approximation and the percentile method. The Satterthwaite approximation has 7.3 degrees of freedom. The percentile method used the same 5000 bootstrap samples as above. The percentile method had a positive lower endpoint for its confidence interval. Since the series represent measurements of the same constant, we might expect the confidence interval to include 0. However, the issue is complicated by the fact that the measurements were taken at different times under presumably different conditions.

TABLE 8.5.2
Two Series for Determining Gravitational Constant g

Series														Mean	Variance
A	76	82	83	54	35	46	87	68						66.38	370.65
H	84	86	85	82	77	76	77	80	83	81	78	78	78	80.38	11.26

Errors													
A	9.63	15.6	16.6	−12	−31	−20	20.6	1.63					
H	3.62	5.62	4.62	1.62	−3.4	−4.4	−3.4	−0.4	2.62	0.62	−2.4	−2.4	−2.4

TABLE 8.5.3
Comparison of Confidence Intervals for Differences of Series Means

Method	Lower 95%	Upper 95%
z_ε	−0.01	36.4
Satterthwaite	−2.1	30.1
Percentile	2.1	27.1

8.5.3 Computer Analysis

Resampling Stats may be used to obtain percentiles of the distributions of the statistics t_e, z_e, and $\overline{Y}_1 - \overline{Y}_2$. Figure 8.5.1 shows code for obtaining the bootstrap percentiles of t_e for the data in Table 8.5.1. Figure 8.5.2 shows code for obtaining the percentiles of z_e and $\overline{Y}_1 - \overline{Y}_2$ for the data in Table 8.5.2, which requires sampling with replacement separately from the two series. The resulting percentiles, computed using 15,000 bootstrap samples, are comparable to those obtained by S-Plus for this data set.

FIGURE 8.5.1

Resampling Stats Code for Obtaining Percentiles of a *t*-Pivot for Data in Table 8.5.1

```
maxsize default 5000

'input observed errors
copy (.18 .01 -.20 -.24 .12 .14 -.07 .05 .15 -.02 .03 .01 .08 .11 -.25 -.13) e

repeat 5000
sample 8 e e1
sample 8 e e2

'get means and variance for two groups
mean e1 m1
mean e2 m2
variance e1 v1
variance e2 v2

'compute pooled variance
concat v1 v2 v
multiply (7 7) v wt
sum wt wtss
divide wtss 14 vpool

'compute t stat
subtract m1 m2 diff
'1/n1 + 1/n2 = .25 in this case
multiply vpool .25 verror
sqrt verror standerr
divide diff standerr t

score t tdist

end

percentile tdist (2.5 5 95 97.5) pctile
print pctile

PCTILE   =    -2.1603    -1.7899    1.7245    2.1244
```

FIGURE 8.5.2

Resampling Stats Code for Analysis of Data in Table 8.5.2;
Sampling Done Separately from Two Data Series

```
maxsize default 15000
'input the data series A and H
copy (76 82 83 54 35 46 87 68) A
copy (84 86 85 82 77 76 77 80 83 81 78 78 78) H

'generate 15000 bootstrap samples
repeat 15000
sample 8 A b_A
sample 13 H b_H

'get bootstrap dist for percentile method
mean b_A mA
mean b_H mH
subtract mH mA diff
score diff diffdist

'create errors for the two samples
'means of the samples are 66.38 and 80.38
subtract b_A (66.38) eA
subtract b_H (80.38) eH

'get bootstrap dist of ze
mean eA meA
mean eH meH
subtract meH meA diffz
variance eA vA
variance eH vH
divide vA 8 vnA
divide vH 13 vnH
add vnA vnH v
sqrt v se
divide diffz se z
score z zdist
end

'printout percentiles
percentile diffdist (2.5 97.5) pctile
print pctile
percentile zdist (2.5 97.5) zpctile
print zpctile

PCTILE   =     1.9615      27.212
ZPCTILE  =    -3.2358      2.0801
```

In its "Resample" menu option, S-Plus allows sampling from the combined data set or separately from two or more sets of data. The statistic for which the bootstrap method is to be applied may be specified as an S-Plus expression. For instance, if the data in Table 8.5.2 are entered as variable V1, with the first 8 being series A and the next 13 being series H, then the difference between the means may be written as the expression *mean(V1[1:8]) – mean(V1[9:21])*.

8.6
Bootstrap Sampling from Several Populations

Suppose we have random samples from each of k populations. Extending the notation of Section 8.5, we let Y_{ij} denote the jth observation from the ith population, $j = 1, 2, \ldots, n_i$, $i = 1, 2, \ldots, k$, and let μ_i denote the mean of the ith population. The errors are $\varepsilon_{ij} = Y_{ij} - \mu_i$. We will consider the null hypothesis $H_0: \mu_1 = \mu_2 = \cdots = \mu_k = 0$, which is tested against the alternative hypothesis that not all means are equal.

8.6.1 Equal Error Distributions

We assume the errors have a common cdf $F(\varepsilon)$. The F statistic for testing equality of means is

$$F = \frac{\sum_{i=1}^{k} n_i \left(\overline{Y}_i - \overline{Y} \right)^2 \big/ (k-1)}{\sum_{i=1}^{k} \sum_{j=1}^{n_i} \left(Y_{ij} - \overline{Y}_i \right)^2 \bigg/ \left(\sum_{i=1}^{k} n_i - k \right)}$$

where

$$\overline{Y}_i = \frac{1}{n_i} \sum_{j=1}^{n_i} Y_{ij}, \quad \overline{Y} = \frac{\sum_{i=1}^{k} n_i \overline{Y}_i}{\sum_{i=1}^{k} n_i}$$

If $\mu_i = \mu$, $i = 1, 2, \ldots, k$, and if we substitute $\mu + \varepsilon_{ij}$ for Y_{ij} in the equations above, then we see that the null distribution of F is the same as that of

$$F_\varepsilon = \frac{\sum_{i=1}^{k} n_i \left(\overline{\varepsilon}_i - \overline{\varepsilon} \right)^2 \big/ (k-1)}{\sum_{i=1}^{k} \sum_{j=1}^{n_i} \left(\varepsilon_{ij} - \overline{\varepsilon}_i \right)^2 \big/ (n-k)}$$

where

$$\bar{\varepsilon}_i = \frac{1}{n_i} \sum_{j=1}^{k} \varepsilon_{ij}, \quad \bar{\varepsilon} = \frac{\sum_{i=1}^{k} n_i \bar{\varepsilon}_i}{\sum_{i=1}^{k} n_i}$$

When the ε_{ij}'s have a normal distribution, F_ε has an *F*-distribution with $k - 1$ degrees of freedom for the numerator and $n - k$ degrees of freedom for the denominator. If the normality assumption is not tenable, then we may obtain the bootstrap distribution of F_ε and use this distribution to carry out the desired test of hypothesis. The steps are as follows:

1. Compute the observed errors $e_{ij} = Y_{ij} - \bar{Y}_i$.
2. Randomly select n_i errors with replacement from the set of all errors and assign them to the *i*th sample, $i = 1, 2, \ldots, k$. Denote these values as $e_{ij,b}$.
3. Compute F_ε using the $e_{ij,b}$'s instead of the ε_{ij}'s, and denote this bootstrap statistic as F_e. Repeat this process a predetermined number of times to obtain the bootstrap distribution of F_e.
4. If F_{obs} denotes the observed *F* statistic for the original data, then the bootstrap *p*-value for testing equality of means is the fraction of the F_e's obtained in step 3 that are greater than or equal to F_{obs}.

EXAMPLE 8.6.1 The data in Table 8.6.1 are the percent oil contents of samples of flax seed grown in a greenhouse under three different conditions (C1, C2, and C3). The means and errors are also listed. The *F* statistic for testing equality of means is 4.81. The bootstrap *p*-value based on 5000 bootstrap samples using Resampling Stats is .015.

TABLE 8.6.1
Percent Oil Contents of Flax Seed and Errors

	Data			*Errors*	
C1	C2	C3	C1	C2	C3
37.8	37.0	35.6	−1.3	0.0	1.0
37.1	38.7	32.4	−2.0	1.7	−2.2
45.3	39.1	32.7	6.2	2.1	−1.9
36.9	38.3	33.2	−2.2	1.3	−1.4
38.8	36.9	33.1	−0.3	−0.1	−1.5
38.6	38.6	39.0	−0.5	1.6	4.4
39.3	30.7	36.0	0.2	−6.3	1.4
Means					
39.1	37.0	34.6			

8.6.2 Unequal Error Distributions

In Example 8.6.1, we selected errors with replacement from the 21 pooled observed error terms, since the distributions of the errors were assumed to be identical. If this assumption cannot be made, then we do not pool the errors but instead select randomly with replacement from the errors within each sample.

One may consider more than one statistic for testing equality of means when the variances are unequal. In the case of equal sample sizes, Box (1954) suggested computing the usual F statistic but adjusting the degrees of freedom. Welch (1951) suggested a modified form of the F statistic. Many practitioners apply the F statistic as if the observations have a normal distribution with equal variances, even when these assumptions may be in question. The justification for this is the robustness of the F statistic. Its distribution is not greatly affected by moderate departures from either the equal variance assumption or the normality assumption.

The bootstrap approach that we propose uses the usual F statistic, but the random selection of errors is done within samples; that is, step 2 above is replaced with the following:

2'. Randomly select n_i errors with replacement from the errors within the ith sample, $i = 1, 2, \ldots, k$. Denote these values as $e_{ij,b}$.

The other steps remain the same.

EXAMPLE 8.6.2   The data in Table 8.6.2 are eight series of measurements of the gravitational constant g expressed as deviations from $980,000 \times 10^{-3}$ cm/sec^2. See Cressie (1982). The standard deviations vary from 19.25 for series A to 3.36 for series H. The 90th, 95th, 97.5th, and 99th percentiles of the bootstrap distribution of F_e are 2.39, 3.03, 3.45, and 4.15, respectively, where the sampling of the errors was done within each sample. The observed F from the data is 3.57, so there is a significant difference among the means of the series at the 2.5% level ($p = .023$). Since the series were the result of measurements of the same physical constant, the difference among means is apparently due to differences in experimental procedures. If we ignore the unequal variance problem and use the F-distribution with 7 degrees of freedom for the numerator and 73 degrees of freedom for the denominator, the statistic would be significant at less than the 0.5% level ($p = .0024$).   ∎

8.6.3 Regression with Unequal Error Variances

Suppose the populations are defined by the levels of a quantitative variable, and suppose we can express the means as

$$\mu_i = \beta_0 + \beta_1 X_i, \quad i = 1, 2, \ldots, k$$

Note that for each X_i there are n_i observations of the Y's, so the setup is just that of sampling from k populations. The only thing we are doing differently here is

TABLE 8.6.2

Eight Series for Determining Gravitational Constant g

Series														Means
A	76	82	83	54	35	46	87	68						66.38
B	87	95	98	100	109	109	100	81	75	68	67			89.91
C	105	83	76	75	51	76	93	75	62					77.33
D	95	90	76	76	87	79	77	71						81.38
E	76	76	78	79	72	68	75	78						75.25
F	78	78	78	86	87	81	73	67	75	82	83			78.91
G	82	79	81	79	77	79	79	78	79	82	76	73	64	77.54
H	84	86	85	82	77	76	77	80	83	81	78	78	78	80.38

Errors														
A	9.63	15.6	16.6	–12	–31	–20	20.6	1.63						
B	–2.9	5.09	8.09	10.1	19.1	19.1	10.1	–8.9	–15	–22	–23			
C	27.7	5.67	–1.3	–2.3	–26	–1.3	15.7	–2.3	–15					
D	13.6	8.63	–5.4	–5.4	5.63	–2.4	–4.4	–10						
E	0.75	0.75	2.75	3.75	–3.3	–7.3	–0.3	2.75						
F	–0.9	–0.9	–0.9	7.09	8.09	2.09	–5.9	–12	–3.9	3.09	4.09			
G	4.46	1.46	3.46	1.46	–0.5	1.46	1.46	0.46	1.46	4.46	–1.5	–4.5	–14	
H	3.62	5.62	4.62	1.62	–3.4	–4.4	–3.4	–0.4	2.62	0.62	–2.4	–2.4	–2.4	

imposing a regression structure on the mean. If the error variances are unequal, we may use either ordinary or weighted least squares to estimate the coefficients. In using weighted least squares, we will assume that the weights w_i depend on the population from which the sample is selected. Thus, weighted least squares estimates of β_0 and β_1 are the values that minimize

$$\sum_{i=1}^{k}\sum_{j=1}^{n_i} w_i\left(Y_{ij} - \beta_0 - \beta_1 X_i\right)^2 = \sum_{i=1}^{k} w_i n_i\left(\overline{Y}_i - \beta_0 - \beta_1 X_i\right)^2 + \sum_{i=1}^{k} w_i(n_i - 1)S_i^2$$

where S_i^2 is the sample variance of the Y's in the ith sample. Since only the first term on the right-hand side of the equation contains β_0 and β_1, the weighted least squares estimates may be based on fitting a line to the pairs $(X_i, \overline{Y}_i)$, $i = 1, 2, \ldots,$ k; that is, we may fit the line to the sample means.

The estimated slope of the line is

$$\hat{\beta}_1 = \frac{\displaystyle\sum_{i=1}^{k} w_i n_i\left(X_i - \overline{X}_w\right)\overline{Y}_i}{\displaystyle\sum_{i=1}^{k} w_i n_i\left(X_i - \overline{X}_w\right)^2}$$

where

$$\overline{X}_w = \frac{\sum_{i=1}^{k} w_i n_i X_i}{\sum_{i=1}^{k} w_i n_i}, \quad \overline{Y}_w = \frac{\sum_{i=1}^{k} w_i n_i Y_i}{\sum_{i=1}^{k} w_i n_i}$$

The estimated intercept is

$$\hat{\beta}_0 = \overline{Y}_w - \hat{\beta}_1 \overline{X}_w$$

If $w_i = 1$, then we have the ordinary least squares estimates. If $w_i = 1/\sigma_i^2$ where σ_i^2 is the variance of the ith population, then the estimates are unbiased and have the smallest variances among unbiased estimates that are linear combinations of the Y's.

A difficulty in using the optimal weights is that the σ_i^2 are usually not known. We may estimate them with S_i^2, the "pure error" estimates. However in doing so, we can no longer be assured that the weighted least squares estimates have smaller variances than the ordinary least squares estimates. The bootstrap method provides a way to compare the variances of two estimates. We apply bootstrap sampling to the data, where random selection is done with replacement within each sample, and we obtain the bootstrap variances of both estimates. The size of the variances will provide a guideline as to which estimate to use.

EXAMPLE 8.6.3 The data in Table 8.6.3 are the logarithms of the dry weights of strawberry plants that have been treated with various levels of a herbicide. A plot of the data would show a linearly decreasing trend, but the assumption of equal variances appears to be questionable. The weighted least squares estimate of the slope is –0.37 using the sample variances S_i^2 as the weights. The ordinary least squares estimate is –0.31. We obtained 1000 bootstrap samples of the data and computed the weighted least squares and ordinary least squares estimates of the slope for each sample. In obtaining the weighted least squares estimates, we recomputed the variances for each sample in order to mimic the process of using the sample variances as weights. For weighted least squares, the average of the bootstrap slopes is –0.37 and the variance is 0.00620. The average and variance of the bootstrap slopes for ordinary least squares are –0.31 and 0.00060, respectively. Since the averages of both the weighted least squares and the ordinary least squares bootstrap estimates are close to their respective slopes from the original data, it suggests the estimates are unbiased. However, the variances are an order of magnitude different. Here ordinary least squares would appear to be the preferred estimate. Thus, when weights have to be estimated from the data, the bootstrap procedure suggests that weighted least squares may not have an advantage over ordinary least squares, as they would if the weights were based on known variances. This is especially true with small sample sizes.

TABLE 8.6.3
Herbicide versus Log Dry Weights of Strawberry Plants

Herbicide Levels			
0	2.25	4.5	9
−0.43	−2.04	−2.12	−3.22
−0.53	−1.9	−2.21	−2.53
−0.80	−1.39	−2.81	−4.61
−0.51	−1.47	−2.41	−3.00
−0.46	−2.41	−2.81	−3.00
−0.8	−1.27	−2.53	−3.91
−0.76	−2.04	−1.97	−4.61
−0.54	−1.35	−3.22	−3.51
−0.65	−1.56	−2.66	−3.22
−0.62	−1.56	−2.12	−3.51
Means and Standard Deviations			
−0.61	−1.70	−2.49	−3.51
0.14	0.38	0.39	0.69

8.6.4 Computer Analysis

It is not particularly convenient, although it is possible, to do the analyses in this section with Resampling Stats due to the limited selection of predefined functions. However, with the Resampling Stats add-in to Excel, computations are easily managed. This is how the examples in this section were done. It is also possible to use the "Resample" option in S-Plus to carry out these procedures.

8.7
Bootstrap Sampling for Multiple Regression

A multiple regression model relating a response variable Y to a set of predictor variables $X_1, X_2, \ldots, X_k$ is a model of the form

$$Y_i = \beta_0 + \beta_1 X_{1i} + \beta_2 X_{2i} + \cdots + \beta_k X_{ki} + \varepsilon_i, \quad i = 1, 2, \ldots, n$$

where Y_i is the response on the ith unit, $(X_{1i}, X_{2i}, \ldots, X_{ki})$ is the set of predictor variables measured on the ith unit, and ε_i is the ith unobserved error. We assume the errors are independent and identically distributed with mean 0 and standard deviation σ. Such a model might describe the relationship between a daughter's height Y_i, and the heights (X_{1i}, X_{2i}) of her mother and father.

Various hypotheses about the β's are of interest. We focus on two of them. The hypothesis $H_{0(M)}$: $\beta_1 = \beta_2 = \cdots = \beta_k = 0$ versus $H_{a(M)}$: not all coefficients are 0 tests whether or not at least one coefficient in the model is significant. Failure to reject $H_{0(M)}$ is an indication that the model is not useful for predicting Y. The test $H_{0(j)}$: $\beta_j = 0$ versus $H_{a(j)}$: $\beta_j \neq 0$ tests whether or not the jth coefficient is significant given that the other variables are included in the model. Most multiple regression computer programs provide tests of these hypotheses.

We use the same statistics that are used when the errors have a normal distribution. We will discuss the computational forms of these statistics in Section 8.7.3, but now we simply assume that a computer program is available for doing the computation. The statistic for testing $H_{0(M)}$ versus $H_{a(M)}$ is an F statistic that we denote F_M. The test statistic for testing $H_{0(j)}$ versus $H_{a(j)}$ is also an F statistic, which we denote F_j. Under the assumption that the errors have a normal distribution, F_M has an F-distribution with degrees of freedom k for the numerator and $n - k - 1$ for the denominator, and F_j has an F-distribution with degrees of freedom 1 for the numerator and $n - k - 1$ for the denominator. Moreover, F_j may be expressed in terms of a t statistic t_j with $n - k - 1$ degrees of freedom using the relationship $F_j = t_j^2$.

We first describe how to carry out bootstrap tests of these hypotheses in an intuitive way. Then we follow with a more precise discussion of the methodology, including how to handle more general tests involving the β's. Sampling is fixed-X to mimic sampling from the conditional distributions of the Y_i's given $(X_{1i}, X_{2i}, \ldots, X_{ki})$, $i = 1, 2, \ldots, n$. The rationale for doing this is the same as discussed in Section 8.4.2.

8.7.1 The Bootstrap Procedure for Testing $H_{0(M)}$ and $H_{0(j)}$

If the ε's don't have a normal distribution, we may use the bootstrap method to obtain an approximate null distribution for either F_M or F_j. Suppose we could observe the ε_i's. Let $F_{M\varepsilon}$ and $F_{j\varepsilon}$ denote the statistics F_M and F_j applied to the data $(X_{1i}, X_{2i}, \ldots, X_{ki})$ and ε_i, $i = 1, 2, \ldots, n$. That is, wherever Y_i is used in the regression computations, use ε_i instead. It can be shown that the distributions of $F_{M\varepsilon}$ and $F_{j\varepsilon}$ are the same as those of F_M and F_j under their respective null hypothesis $H_{0(M)}$ and $H_{0(j)}$.

Using this result, we can obtain bootstrap approximations of the distributions of these statistics under their null hypotheses and use these distributions to compute p-values. The procedure is as follows:

1. Fit the regression model to the data and obtain the observed errors e_i, where $e_i = Y_i - (\hat{\beta}_0 + \hat{\beta}_1 X_{1i} + \hat{\beta}_2 X_{2i} + \cdots + \hat{\beta}_k X_{ki})$ and $\hat{\beta}_j$ is the least squares estimate of β_j, $j = 1, 2, \ldots, n$.

2. Take a random sample of size n of the e_i's with replacement, and let $e_{i,b}$, $i = 1, 2, \ldots, n$, denote these values. Form the fixed-X bootstrap sample $(X_{1i}, X_{2i}, \ldots,$

X_{ki}) and $e_{i,b}$, $i = 1, 2, \ldots, n$. Apply the computational formulas for the statistics F_M and F_j to these data, and denote the results F_{Me} and F_{je}, respectively.

3. Repeat step 2 a sufficient number of times to generate the bootstrap distributions of the F_{Me}'s and F_{je}'s.

4. Let $F_{M,obs}$ and $F_{j,obs}$ denote the observed values of F_M and F_j, respectively, from the original data. The bootstrap p-value for $F_{M,obs}$ is the fraction of the F_{Me}'s that are greater than or equal to $F_{M,obs}$. The bootstrap p-value for $F_{j,obs}$ is the fraction the F_{je}'s that are greater than or equal to $F_{j,obs}$.

EXAMPLE 8.7.1   Table 8.7.1 lists the collar sizes (CS), shoe sizes (SS), and weights (WT) of 24 college-age men. It also gives the predicted values and errors, where the equation is WT = $-156.572 + 15.0738$(CS) $+ 8.7928$(SS). The observed F statistic for H_0: $\beta_1 = \beta_2 = 0$ is $F_M = 22.96$, the observed F statistic for $H_{0(1)}$: $\beta_1 = 0$ is $F_1 = 9.61$, and the observed F-statistic for $H_{0(2)}$: $\beta_2 = 0$ is $F_2 = 15.95$. All statistics are significant at the 1% level. Table 8.7.2 lists bootstrap 90th, 95th, 97.5th, and 99th percentiles and the corresponding percentiles of the F-distributions. The two sets of percentiles agree well for the 90th and 95th percentiles, but generally not so well for the higher percentiles. Computations were based on 5000 bootstrap samples using the Resampling Stats add-in for Excel.   ∎

Practitioners often use regression methodology that is based on the assumption of normally distributed errors even if the errors are not normally distributed. Justification for this is the so-called robustness of the normal-theory methods. That is, the presumed levels of significance, based on the use of the F-distribution as a reference distribution, often agree quite well with the actual levels. The bootstrap methodology gives a way to verify robustness for a particular problem. If the bootstrap percentiles agree with those based on normal theory, then the use of normal-theory methodology would be supported. However, if there is substantial disagreement between percentiles, then one would have to look carefully at the data to see whether the violation of normality assumptions might be severe enough to cause concern. If so, the bootstrap p-values would present an alternative to normal-theory p-values in carrying out tests of hypotheses.

8.7.2 A Confidence Interval for β_j

A method for obtaining bootstrap confidence intervals for a coefficient β_j is based on the t-pivot

$$ t = \frac{\hat{\beta}_j - \beta_j}{\mathrm{SE}(\hat{\beta}_j)} $$

where $\mathrm{SE}(\hat{\beta}_j)$ is the standard error of $\hat{\beta}_j$. The standard error is provided by most regression programs. The steps for obtaining a bootstrap approximation of the

TABLE 8.7.1

Collar Sizes, Shoe Sizes, Weights, Predicted Values,
and Errors for 24 College-Age Men

CS	SS	WT	Predicted	Error
14.5	9.5	140	145.53	−5.53
15.5	9.5	155	160.60	−5.60
15.5	10.5	153	169.40	−16.40
15	10.5	150	161.86	−11.86
16.5	11	180	188.87	−8.87
16.5	8.5	160	166.88	−6.88
15.5	8.5	155	151.81	3.19
14.5	9.5	145	145.53	−0.53
15	10	163	157.46	5.54
15	9	150	148.67	1.33
15	8.5	140	144.27	−4.27
15.5	9.5	170	160.60	9.40
15.5	11	180	173.79	6.21
15.5	11	175	173.79	1.21
15.5	10.5	155	169.40	−14.40
15.5	8.5	150	151.81	−1.81
15	8.5	180	144.27	35.73
15.5	10	160	165.00	−5.00
15	9	145	148.67	−3.67
16	12	190	190.12	−0.12
16.5	13	228	206.45	21.55
15	8.5	150	144.27	5.73
15	11	165	166.26	−1.26
15	9	145	148.67	−3.67

TABLE 8.7.2

Bootstrap Percentiles for Data in Table 8.7.1 and Percentiles of the *F*-Distributions

Hypothesis	Distribution	90th	95th	97.5th	99th
$H_{0(1)}$: $\beta_1 = 0$	Bootstrap: *F*-dist,	2.95	4.33	6.19	8.53
	df (1, 21):	2.96	4.32	5.83	8.02
$H_{0(2)}$: $\beta_2 = 0$	Bootstrap: *F*-dist,	2.84	4.16	5.52	7.69
	df (1, 21):	2.96	4.32	5.83	8.02
$H_{0(M)}$: $\beta_1 = \beta_2 = 0$	Bootstrap: *F*-dist,	2.55	3.50	4.71	6.16
	df (2, 21):	2.57	3.47	4.42	5.78

distribution of t are the same as those for F_j, the only difference being the statistic. If we let $t_{e,.025}$ and $t_{e,.975}$ denote the 2.5th and 97.5th percentiles of the bootstrap distribution, then a 95% confidence interval may be found by solving for β_j in the inequality

$$ t_{e,.025} < \frac{\hat{\beta}_j - \beta_j}{SE(\hat{\beta}_j)} < t_{e,.975} $$

EXAMPLE 8.7.2   Bootstrap 95% confidence intervals were obtained for the coefficients of CS and SS in Example 8.7.1. The bootstrap t-pivot 2.5th and 97.5th percentiles for β_1 are -1.93 and 2.25. The standard error of the estimate of β_1 from the original data is 4.86. Thus, the 95% bootstrap lower and upper confidence limits are $15.07 - 2.25(4.86) = 4.14$ and $15.07 + 1.93(4.86) = 24.45$. The corresponding confidence limits based on the t-distribution with 21 degrees of freedom are 4.96 and 25.18. The 2.5th and 97.5th bootstrap percentiles of the t-pivot for β_2 are -2.06 and 2.04. The standard error of the estimate of β_2 is 2.20. Thus the bootstrap 95% confidence interval for β_2 is $(4.30, 13.29)$. The corresponding confidence interval based on the t-distribution is $(4.21, 13.37)$.   ∎

8.7.3 Theoretical Development

It is convenient for our purposes to express the multiple regression model in matrix form. Let vector Y, matrix X, vector β, and vector ε be defined as

$$ Y = \begin{pmatrix} Y_1 \\ Y \\ \dots \\ Y_n \end{pmatrix}, \quad X = \begin{pmatrix} 1 & X_{11} & X_{21} & \dots & X_{k1} \\ 1 & X_{12} & X_{22} & \dots & X_{k2} \\ & & \dots & & \\ 1 & X_{1n} & X_{2n} & \dots & X_{kn} \end{pmatrix}, \quad \beta = \begin{pmatrix} \beta_0 \\ \beta_1 \\ \dots \\ \beta_k \end{pmatrix}, \quad \varepsilon = \begin{pmatrix} \varepsilon_1 \\ \varepsilon_2 \\ \dots \\ \varepsilon_n \end{pmatrix} $$

The regression model can be expressed as

$$ Y = X\beta + \varepsilon $$

The least squares estimate of β can be expressed as

$$ \hat{\beta} = (X^T X)^{-1} X^T Y $$

where the superscript T denotes the transpose and the superscript -1 denotes matrix inverse. It is, of course, assumed that the inverse exists. We refer to the data as the pair (X, Y).

We are concerned with the null hypotheses that may be expressed in the form $H_0\colon C\beta = 0$, where C is a $q \times (k + 1)$ matrix of full rank q. For instance, for the hypothesis $H_0\colon \beta_1 = \beta_2 = \cdots = \beta_k = 0$, the matrix C is the $k \times (k + 1)$ matrix given by

$$C = \begin{pmatrix} 0 & 1 & 0 & \cdots & 0 \\ 0 & 0 & 1 & \cdots & 0 \\ & & \cdots & & \\ 0 & 0 & 0 & \cdots & 1 \end{pmatrix}$$

The matrix C for the null hypothesis H_0: $\beta_1 - \beta_2 = 0$ is the $1 \times (k + 1)$ matrix given by

$$C = (0 \quad 1 \quad -1 \quad 0 \quad \cdots \quad 0)$$

The alternative hypothesis is, of course, H_a: $C\beta \neq 0$.

A test statistic for testing H_0: $C\beta = 0$ is given by

$$F = \frac{\left(C\hat{\beta}\right)^T \left[C(X^T X)^{-1} C^T\right]^{-1} C\hat{\beta}}{q\text{MSE}}$$

where

$$\text{MSE} = \frac{\left(Y - X\hat{\beta}\right)^T \left(Y - X\hat{\beta}\right)}{n - k - 1}$$

If the ε's have a normal distribution, then under the null hypothesis H_0: $C\beta = 0$, F has an F-distribution with q degrees of freedom for the numerator and $n - k - 1$ degrees of freedom for the denominator. Various tests of this form are provided by most multiple regression software.

Distribution of F Under H_0

To set up the bootstrap approach, we need to consider a property of the null distribution of F. Let F_ε denote the F statistic obtained by using (X, ε) instead of (X, Y). That is, wherever Y is used in the computation of F, ε is used instead. We show that the distribution of F_ε is the same as the distribution of F under the null hypothesis H_0: $C\beta = 0$.

Let $\hat{\beta}_\varepsilon$ denote the least squares estimates of β applied to (X, ε). Then

$$\hat{\beta} - \beta = \left(X^T X\right)^{-1} X^T Y - \beta$$

$$= \left(X^T X\right)^{-1} X^T \left(X\beta + \varepsilon\right) - \beta$$

$$= \left(X^T X\right)^{-1} X^T \varepsilon$$

$$= \hat{\beta}_\varepsilon$$

Let MSE_ε denote the mean square error for (X, ε). We have

$$(n-k-1)\text{MSE} = \left(Y - X\hat{\boldsymbol{\beta}}\right)^T \left(Y - X\hat{\boldsymbol{\beta}}\right)$$

$$= \left(X\boldsymbol{\beta} + \boldsymbol{\varepsilon} - X\hat{\boldsymbol{\beta}}\right)^T \left(X\boldsymbol{\beta} + \boldsymbol{\varepsilon} - X\hat{\boldsymbol{\beta}}\right)$$

$$= \left(\boldsymbol{\varepsilon} - X\hat{\boldsymbol{\beta}}_\varepsilon\right)^T \left(\boldsymbol{\varepsilon} - X\hat{\boldsymbol{\beta}}_\varepsilon\right)$$

$$= (n-k-1)\text{MSE}_\varepsilon$$

Thus, $\text{MSE} = \text{MSE}_\varepsilon$.

Using the results above, and under the null hypothesis H_0: $\boldsymbol{C\beta} = \mathbf{0}$, the F statistic can be expressed as

$$F = \frac{\left(C\hat{\boldsymbol{\beta}}\right)^T \left[C(X^T X)^{-1} C^T\right]^{-1} C\hat{\boldsymbol{\beta}}}{q\text{MSE}}$$

$$= \frac{\left[C\left(\hat{\boldsymbol{\beta}} - \boldsymbol{\beta}\right)\right]^T \left[C(X^T X)^{-1} C^T\right]^{-1} C\left(\hat{\boldsymbol{\beta}} - \boldsymbol{\beta}\right)}{q\text{MSE}}$$

$$= \frac{\left(C\hat{\boldsymbol{\beta}}_\varepsilon\right)^T \left[C(X^T X)^{-1} C^T\right]^{-1} C\hat{\boldsymbol{\beta}}_\varepsilon}{q\text{MSE}_\varepsilon}$$

$$= F_\varepsilon$$

It follows that the distribution of F under H_0: $\boldsymbol{C\beta} = \mathbf{0}$ is the same as the distribution of F_ε. This result is the basis for determining a bootstrap approximation of the distribution of F under H_0.

Steps for the Bootstrap Test of H_0: $\boldsymbol{C\beta} = \mathbf{0}$

1. Compute the vector $\hat{\boldsymbol{\beta}}$ of least squares estimates of $\boldsymbol{\beta}$, and obtain the vector of observed errors $e = Y - X\hat{\boldsymbol{\beta}}$.

2. Take a random sample of size n of the e_i's with replacement. Let e_b denote the $n \times 1$ vector of these values. Form the bootstrap sample (X, e_b) by matching the rows of the matrix X with the n randomly selected values of e_b. For each bootstrap sample (X, e_b), obtain the vector of least squares estimates $\hat{\boldsymbol{\beta}}_e$, the mean square error MSE_e, and the statistic F_e defined by

$$F_e = \frac{\left(C\hat{\boldsymbol{\beta}}_e\right)^T \left[C(X^T X)^{-1} C^T\right]^{-1} C\hat{\boldsymbol{\beta}}_e}{q\text{MSE}_e}$$

3. Repeat step 2 a sufficient number of times to determine the bootstrap distribution of F_e.

4. Let F_{obs} denote the observed value of F from the data. The p-value is the fraction of the F_e's greater than or equal to F_{obs}.

In the case in which C is a row vector, that is, when $q = 1$, the quantity

$$t = \frac{C\hat{\beta} - C\beta}{\sqrt{(\text{MSE})C(X^T X)^{-1} C^T}}$$

is a t-pivot quantity that has a t-distribution with $n - k - 1$ degrees of freedom when the errors have a normal distribution. This statistic is used to make confidence intervals for the individual coefficients β_j or to test hypotheses of the form H_0: $\beta_j = 0$ against one-sided or two-sided alternatives. If the assumption of normality of the error distributions is in question, then we may use the bootstrap procedure to obtain critical values of the statistic under the null hypothesis. We compute

$$t_e = \frac{C\hat{\beta}_e}{\sqrt{(\text{MSE}_e)C(X^T X)^{-1} C^T}}$$

instead of F_e in step 4 above and use the percentiles of the distribution of the t_e's for confidence intervals and tests of hypotheses.

Adjusted Errors

As in the case of simple linear regression, it may be advisable in some circumstances to use adjusted errors. Define the matrix H, commonly called the "hat" matrix, as

$$H = X(X^T X)^{-1} X^T$$

Let h_i denote the ith diagonal element of H, and let r_i be defined as

$$r_i = \frac{e_i}{\sqrt{1 - h_i}}$$

Define the adjusted errors as

$$e_i^* = r_i - \bar{r}$$

The adjusted errors may be used instead of the observed errors to obtain a bootstrap distribution for F_e.

8.7.4 Other Methods for Regression Analysis

Permutation Tests

Instead of sampling the observed errors with replacement to form bootstrap samples, we may permute the observed errors randomly among the rows of X.

Permuting the observed errors amounts to sampling them without replacement. If we carry out the steps above using the permuted errors, then we have an approximate permutation test for testing the hypothesis H_0: $C\beta = 0$. The test is approximate in the sense that the permutation distribution approximates the one that would be obtained from permuting the actual, unobservable errors. This method is discussed by ter Braak (1992) and Manly (1997).

Multivariate Bootstrap Sampling

Multivariate bootstrap sampling is a direct extension of bivariate sampling described in Section 8.4. That is, we select vectors $(X_1, X_2, \ldots, X_k, Y)$ randomly with replacement from the set of observed data vectors. We may use the percentile method or the BCA method in conjunction with multivariate bootstrap sampling to make confidence intervals for the regression coefficients. The procedure is as follows:

1. Take a bootstrap multivariate sample and compute estimates of the coefficients, denoted $\hat{\beta}_{j,b}$, $j = 0, 1, \ldots, k$. The method of estimation may be least squares or it may be other methods as discussed in Section 10.3. Also compute any functions of the coefficients, such as linear combinations, for which confidence intervals are desired.

2. Form the bootstrap distributions of the $\hat{\beta}_{j,b}$'s and other functions of interest by taking repeated bootstrap samples of the data.

3. Apply the BCA method, percentile method, or residual method to the bootstrap distributions.

EXAMPLE **8.7.3**   The BCA method and percentile method were applied to the data in Table 8.7.1. The 95% BCA confidence interval for β_1 is (8.09, 25.99), and for β_2 it is (3.92, 13.08). The corresponding percentile method confidence intervals are (7.80, 25.56) and (2.02, 12.22). S-Plus output is shown in Figure 8.7.1.   ■

We should note that there are no restrictions on the distributions of the errors for the application of the BCA and percentile methods with multivariate sampling. In particular, there is no requirement that the errors have constant variance.

8.7.5 Computer Analysis

The stand-alone version of Resampling Stats does not provide the commands or output that would permit all the methodology of this section to be applied. However, the Resampling Stats add-in to Excel does. A regression function is included in this add-in that allows the user to access the usual regression output, including coefficients, standard errors, the statistic F_M and the t-statistics for the individual coefficients, from which the F_j's may be determined by $F_j = t_j^2$. Examples in this section were done using this add-in.

There is controversy in the statistics community over using Excel for data analysis. Several statistical algorithms implemented in Excel do not pass certain benchmark tests for accuracy. For instance, the regression routine fails to recognize the situation in which one variable is a linear function of another. In this case, the regression solution is indeterminate, but Excel will produce an output just the same with no warning. See McCullough and Wilson (1999, 2002) for a comprehensive discussion of the issues.

When we compute least squares estimates, the matrix $X^T X$ must be inverted. If the matrix is singular, as occurs when one variable is a linear function of another, or if the matrix is ill-conditioned, being in some sense nearly singular, then the Excel output is questionable. Under fixed-X sampling, this matrix stays the same for any resampling of the errors. Thus, if there is no problem with singularity or ill-conditioning in analyzing the original data, there is no problem of this type in bootstrap sampling or in permuting the errors. As a check, one may analyze the original data using both Excel and another statistical package that is generally regarded as having better algorithms. If there are discrepancies in the answers, Excel should not be used for the bootstrap procedure. The Excel regression routine applied to the original data in Table 8.7.1 gave the same results as MINITAB and S-Plus.

Multivariate bootstrap sampling presents another level of difficulty. The matrix X changes with the bootstrap sample. It may happen that a bootstrap sample yields a result in which $X^T X$ is singular or ill-conditioned even if the original matrix is not. This can be a problem especially when the original sample is small. Potential computational problems are inherent in the bootstrap methodology in this situation and are not limited to difficulties with any particular statistical package.

The BCA and percentile methods for regression with multivariate sampling are available in S-Plus. The output used for Example 8.7.3 is shown in Figure 8.7.1. The percentile method can be carried out in Resampling Stats with multivariate sampling. The code is similar to that in Figure 8.4.3.

FIGURE 8.7.1

S-Plus Percentile Method and BCA Confidence Intervals for the Regression Coefficients of CS and SS for Data in Table 8.7.1

```
           *** Bootstrap Results ***
Call:
bootstrap(data = DS11, statistic = coef(lm(WT ~ CS + SS)), B = 5000,
trace = F, assign.frame1 = F, save.indices = F)


Number of Replications: 5000

                                       (continued on next page)
```

FIGURE 8.7.1

S-Plus Percentile Method and BCA Confidence Intervals for the Regression Coefficients
of CS and SS for Data in Table 8.7.1 *(continued)*

```
Summary Statistics:
            Observed    Bias      Mean       SE
(Intercept) -156.572  2.9611  -153.611   73.005
         CS   15.074  0.2609    15.335    4.359
         SS    8.793 -0.7264     8.066    2.641


Empirical Percentiles:
                  2.5%          5%          95%         97.5%
(Intercept) -297.976605 -272.774786  -34.87292    -6.037103
         CS    7.804306    9.138459   23.33926    25.556805
         SS    2.017903    3.241345   11.75147    12.220366


BCa Percentiles:
                  2.5%          5%          95%         97.5%
(Intercept) -321.763778 -289.145633  -48.20242   -28.95374
         CS    8.086032    9.441646   23.87199    25.98675
         SS    3.921916    4.818282   12.49853    13.07813
```

Exercises

1   The data are the eosinophil counts taken from blood samples of 40 healthy rabbits. Obtain
bootstrap estimates of the MSE, standard error, and margin of error of the sample mean, the
sample standard deviation, and the coefficient of variation.

55	140	91	122	111	185	203	101
76	145	95	101	196	45	299	226
65	70	196	72	121	171	151	113
112	67	276	125	100	81	122	71
158	78	162	128	96	79	67	119

2   Use the *t*-pivot and BCA methods to make 95% confidence intervals for the mean of the
data in Exercise 1.

3   Use bivariate bootstrap sampling and the BCA method applied to the heterophil and lymphocyte data in Table 5.1.2 to find a 95% confidence interval for the correlation coefficient.

4   For the data in Table 5.1.2, use fixed-*X* sampling in conjunction with the bootstrap *t* statistic to make 95% confidence intervals for the slope of the least squares line relating
heterophils (*x*) to lymphocytes (*y*).

5 For the shrew abundance data in Exercise 5 of Chapter 5, use an appropriate bootstrap method to make a 90% confidence interval for the slope of the regression line.

6 For the data in Table 8.4.1, compute the adjusted errors e_i^* defined in Section 8.4 and compare with the unadjusted errors. Compare confidence intervals for the slope using adjusted and unadjusted errors in the bootstrap sampling.

7 The data in the table are random samples from two normal populations with equal variances. The first population has a mean of 15 and the second a mean of 18. The population variance is 11.2. Use the data to obtain bootstrap t-pivot percentiles for a 95% confidence interval for the difference of the population means. Compute the confidence interval, and compare it with the interval we get using the fact that the t-pivot has a t-distribution with 18 degrees of freedom.

Sample 1	13.4	11.8	18.5	13.1	7.1	17.2	19.8	15.8	13.6	14.2
Sample 2	16.5	17.3	16.3	16.3	16.7	17.1	17.9	16.6	23.7	19.5

8 Refer to the sibling data in Exercise 7 of Chapter 2.

    a Make a 90% confidence interval for the difference between the means of the populations using the percentile and BCA methods.

    b Make a 90% bootstrap confidence interval using the z statistic method of Section 8.5.2. Compare the results with those of part a.

9 Refer to the automobile crash data in Exercise 2 of Chapter 3. Obtain the bootstrap distributions of the F statistic assuming both equal variances and unequal variances in bootstrap sampling. Compare the 90th, 95th, 97.5th, and 99th percentiles with those of the F-distribution.

10 In addition to the bootstrap F statistic for comparing several populations, we may analyze the data using the permutation F-test and the Kruskal–Wallis test. Compare these tests by applying them to the data in Table 8.6.1.

11 The data in the table are the heights (in inches) of 34 daughters (DH) along with the heights of their mothers (MH) and fathers (FH).

DH	MH	FH	DH	MH	FH	DH	MH	FH	DH	MH	FH
70	67	72	70	74	73	63	65	68	66	70	72
69	64	74	66	60	72	64	62	68	64	63	72
65	62	71	66	66	67	65	61	71	68	65	68
64	64	71	60	60	65	66	65	76	62	60	71
66	69	68	68	65	65	65	64	72	62	66	66
65	70	71	65	65	71	64	63	64	67	65	72
64	65	71	65	66	70	62	61	66	69	69	71
66	66	70	67	67	69	66	62	68	64	63	68
60	63	66	64	63	74						

a Use fixed-X bootstrap sampling to test the following three null hypothesis H_0: $\beta_1 = 0$, H_0: $\beta_2 = 0$, and H_0: $\beta_1 = \beta_2 = 0$ in the model

$$DH = \beta_0 + \beta_1(MH) + \beta_2(FH) + \varepsilon$$

b Make 95% confidence intervals for β_1 and β_2 using the t-pivot method with fixed-X sampling and using the BCA and percentile methods with multivariate sampling.

12 Use multivariate bootstrap sampling and either the percentile method or the BCA method to make a 95% confidence interval for the square of the multiple correlation coefficient (R^2) for the data and model in Exercise 11.

Theory and Complements

13 Discuss the bootstrap approach to doing multiple comparisons when comparing the means of several populations. Use a bootstrap version of Tukey's HSD to compare the means of the data in Table 8.6.1.

9

Multifactor Experiments

A Look Ahead   In this chapter, we introduce nonparametric methods for the analysis of multifactor experiments. Bootstrap, permutation, and rank-based methods are considered. In Sections 9.1 and 9.2, we develop nonparametric tests for the same types of hypotheses that are tested using the normal-theory analysis of variance. In particular, we test for so-called main effects and interactions. In Section 9.3, we consider different types of alternatives called lattice-ordered alternatives. These alternatives take advantage of prior knowledge about the potential effects of the treatments.

9.1
Analysis of Variance Models

In planned experiments researchers often manipulate several factors to observe their effects on response variables. A horticultural scientist, for instance, might grow shrubs in three types of soil with two types of fertilizer to see how the six combinations of these factors affect root biomass. For simplicity of notation, we will deal with just two factors, although the ideas extend directly to experiments with three or more factors.

We denote the two factors A and B. We assume there are "a" levels of factor A and "b" levels of factor B. We have a *factorial treatment structure* when all combinations of the levels of factors A and B are included in the experiment. Let Y_{ijk} denote the kth observation for the ith level of factor A and the jth level of factor B, $k = 1, 2, \ldots, n_{ij}$. Let $E(Y_{ijk}) = \mu_{ij}$. We assume the additive error model applies; that is,

$$Y_{ijk} = \mu_{ij} + \varepsilon_{ijk}$$

where the errors are independent and identically distributed with mean 0 and standard deviation σ.

The analysis appropriate for this model depends on the structure we are willing to impose on the μ_{ij}'s. If the levels of the factors are quantitative—for instance, amounts of nitrogen and amounts of phosphorus applied to shrubs—then we may wish to assume that $\mu_{ij} = \beta_0 + \beta_1 X_{1i} + \beta_2 X_{2j}$ where X_{1i} is the value of factor A at level

i and X_{2j} is the value of factor B at level j. In this case, an appropriate analysis would be the fixed-X bootstrap multiple regression as discussed in Section 8.7.

Another approach is to express μ_{ij} as a sum of three effects: a main effect for factor A, a main effect for factor B, and the interaction effect between factors A and B. We first define the population means for level i of factor A, level j of factor B, and the overall mean as

$$\bar{\mu}_{i.} = \frac{1}{b}\sum_{j=1}^{b}\mu_{ij}, \quad \bar{\mu}_{.j} = \frac{1}{a}\sum_{i=1}^{a}\mu_{ij}, \quad \mu = \frac{1}{ab}\sum_{i=1}^{a}\sum_{j=1}^{b}\mu_{ij}$$

The main effects α_i and β_j and the interaction effect γ_{ij} are defined as

$$\alpha_i = \bar{\mu}_{i.} - \mu, \quad \beta_j = \bar{\mu}_{.j} - \mu, \quad \gamma_{ij} = \mu_{ij} - \bar{\mu}_{i.} - \bar{\mu}_{.j} + \mu$$

Thus we may express μ_{ij} as

$$\mu_{ij} = \mu + \alpha_i + \beta_j + \gamma_{ij}$$

We term this model the *fixed-effects model* for an $a \times b$ factorial treatment structure.

The fixed-effects model can be expressed as a fixed-X regression model where the X's are based on *indicator variables*. An indicator variable is a function $I(u, v)$ such that $I(u, v) = 1$ if $u = v$, and $I(u, v) = 0$ otherwise. We may express the fixed-effects model as

$$Y_{ijk} = \mu + \sum_{r=1}^{a}\alpha_r I(r,i) + \sum_{s=1}^{b}\beta_s I(s,j) + \sum_{r=1}^{a}\sum_{s=1}^{b}\gamma_{rs}I(r,i)I(s,j) + \varepsilon_{ijk}$$

Because of our definitions of the main effects and interactions, we see that the parameters satisfy the *sum-to-zero restrictions*: $\Sigma_i\alpha_i = 0$, $\Sigma_j\beta_j = 0$, $\Sigma_i\gamma_{ij} = 0$, and $\Sigma_j\gamma_{ij} = 0$. The model has more parameters than needed. This redundancy can be eliminated by solving for α_a, β_b, γ_{aj}, and γ_{ib} in terms of the other parameters. When we substitute into the equation for Y_{ijk}, the fixed-effects model becomes

$$Y_{ijk} = \mu + \sum_{r=1}^{a-1}\alpha_r X_a(r,i) + \sum_{s=1}^{b-1}\beta_s X_b(s,j) + \sum_{r=1}^{a-1}\sum_{s=1}^{b-1}\gamma_{rs}X_a(r,i)X_b(s,j) + \varepsilon_{ijk}$$

where $X_a(u, v) = I(u, v) - I(u, a)$ and $X_b(u, v) = I(u, v) - I(u, b)$. To illustrate, Table 9.1.1 shows all possible values of the X's in the fixed-effects model for an experiment that has three levels of factor A and two levels of factor B.

9.1.1 The Bootstrap Approach to the Analysis of the Fixed-Effects Model

The null hypotheses of interest for the fixed-effects model are

$$H_0: \alpha_1 = \cdots = \alpha_{k-1} = 0$$

$$H_0: \beta_1 = \cdots = \beta_{k-1} = 0$$

$$H_0: \gamma_{1,1} = \cdots = \gamma_{a-1,b-1} = 0$$

TABLE 9.1.1

All Possible Values of the X's in the Fixed-Effects Model
for a 3×2 Factorial Treatment Structure

Levels of Factor A	Levels of Factor B	$X_a(i, 1)$	$X_a(i, 2)$	$X_b(j, 1)$
$i = 1$	$j = 1$	1	0	1
$i = 2$	$j = 1$	0	1	1
$i = 3$	$j = 1$	−1	−1	1
$i = 1$	$j = 2$	1	0	−1
$i = 2$	$j = 2$	0	1	−1
$i = 3$	$j = 2$	−1	−1	−1

The first two hypotheses are main effect hypotheses, and the third is the interaction hypothesis. We use the normal-theory F statistic for each of these hypotheses. Since the fixed-effects model can be expressed as a fixed-X regression model, the methodology developed in Section 8.7.3 applies here. In the special case in which all treatments have an equal number of observations, there are closed-form expressions for the sums of squares that make up the F statistics. See Ott and Longnecker (2001).

Let us assume we have some method of computing the F statistic for the hypothesis of interest. To obtain a bootstrap approximation of the null distribution of this statistic, we randomly select errors with replacement from the observed errors and then substitute these values into the formula for the F statistic. We repeat this process a sufficiently large number of times. The distribution of these values is an approximation to the null distribution of the F statistic, which we may use to compute approximate p-values.

We note that in the case of the model with main effects and interactions, the observed errors are $e_{ijk} = Y_{ijk} - \bar{Y}_{ij.}$, where $\bar{Y}_{ij.}$ is the mean of the observations at level i of factor A and level j of factor B. If the interaction terms are not included, then the parameters must be estimated and the errors obtained as deviations from the fitted model. The multiple regression and analysis of variance models are special cases of the general linear model. See Graybill (1976). The bootstrap methodology outlined here and in Section 8.7 applies to the general linear model.

Rather that randomly sampling the errors with replacement, we may permute the observed errors among the treatments. This gives us an approximate permutation test in the sense that the resulting permutation distributions of the F statistics approximate the distributions we would obtain if we could permute the actual, unobserved errors.

EXAMPLE 9.1.1   The data in Table 9.1.2 are the logarithms of bacteria counts of meat samples that were treated with one of three types of preservatives (P1, P2, and P3). Independent samples were prepared for each of three times (week 1, week 2, and week 4), with the objective being to look at how well the preservatives prevent bacterial growth

304 Chapter 9: Multifactor Experiments

over time. Three meat samples were used for each of the nine preservative–time combinations. Another way the experiment might have been run would have been to use the same meat sample for each of the three times. This would have been a repeated-measures design that should not be analyzed by the methods we use here.

TABLE 9.1.2

Effect of Preservative and Time on Bacteria Counts

		Log Bacteria Count			Observed Errors		
Preservative	Time	Sample 1	Sample 2	Sample 3	Sample 1	Sample 2	Sample 3
P1	Week 1	1.38	2.54	2.14	−0.64	0.52	0.12
P1	Week 2	3.83	3.85	4.60	−0.27	−0.24	0.51
P1	Week 4	5.45	6.57	6.33	−0.67	0.45	0.22
P2	Week 1	2.32	3.77	2.59	−0.57	0.88	−0.30
P2	Week 2	4.95	3.86	3.62	0.81	−0.28	−0.52
P2	Week 4	5.46	6.26	6.52	−0.62	0.18	0.44
P3	Week 1	4.57	3.69	4.41	0.35	−0.53	0.19
P3	Week 2	4.09	4.87	5.24	−0.64	0.13	0.51
P3	Week 4	6.06	5.39	6.20	0.18	−0.49	0.32

The F statistics are 5.18 for preservative, 59.41 for time, and 3.48 for interaction. If we use the standard normal-theory analysis of variance, the p-values for these effects are .017 for preservative, less than .001 for time, and .028 for interaction. Based on a bootstrap sample of 1000, the bootstrap p-values are .017 for preservative, less than .001 for time, and .031 for interaction. The interaction can be explained by the fact that preservatives differ at week 1 but come progressively closer with time. See Figure 9.1.1 for an interaction plot. Table 9.1.3 compares the percentiles of the bootstrap and permutation distributions with those of the F-distribution. In this rather well-behaved data set, all methods give comparable results.

TABLE 9.1.3

Comparison of Percentiles; A = Preservative, B = Time, AB = Interaction

	90%			95%		
	A	B	AB	A	B	AB
Bootstrap F	2.60	2.53	2.33	3.56	3.69	2.93
Permutation F	2.75	2.65	2.38	3.98	3.59	2.94
F-distribution	2.62	2.62	2.29	3.55	3.55	2.93

FIGURE 9.1.1
Plots of Mean Log Bacteria Counts for Data in Table 9.1.2

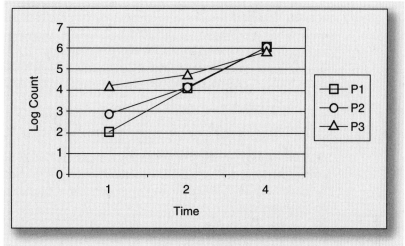

9.1.2 A Permutation Test for the Overall Effect of a Factor

Suppose in a multifactor survey a researcher wishes to investigate the effects of various factors such as race, gender, education level, and job status on a response such as income. It would not be unusual for some of the possible combinations of levels of these factors to have no observations at all. With missing combinations like this, investigating main effects and interactions could be problematic. However, it would make sense to ask a simpler question such as: Does education level, adjusted for the other factors, affect income? We now consider a permutation test for judging whether factor A has an effect on the response, given that factor B is present. The method can easily be extended to multiple factors.

Suppose $F_{ij}(x)$ is the cdf of the observations taken at level i of factor A and level j of factor B. For each level j of factor B, let $a(j) \leq a$ denote the number of levels of factor A for which there are observations, and let $i_1, i_2, \ldots, i_{a(j)}$ denote the indices of these levels. The null hypothesis that factor A does not have an effect in the presence of factor B is

$$H_0: F_{i_1 j}(x) = F_{i_2 j}(x) = \cdots = F_{a(j) j}(x), \ \ j = 1, 2, \ldots, b$$

If all possible combinations of levels of factor A and factor B have at least one observation, then this test is equivalent to simultaneously testing whether the A main effect and the AB interaction are 0.

The test statistic for this hypothesis can be computed by doing the normal-theory one-way analysis of variance computations twice. For the first computation, give labels 1, 2, . . . to all the treatment combinations, perform the one-way analysis of variance to test for equality of means of the treatments, and get the sum of squared errors and the degrees of freedom of the errors. We denote these values as SSE_{full} and df_{full} for the *full model*. For the second computation, group together the observations of factor A for each level of factor B, labeling the groups $1, 2, \ldots, b$. If a group does not have any observations, it will not figure into the computations. Perform the one-way analysis of variance to test for equality of means of these groups, and get the sum of squared errors and the degrees of freedom of the errors. We denote these values as $\text{SSE}_{\text{reduced}}$ and $\text{df}_{\text{reduced}}$ for the *reduced model*. The F statistic is defined as

$$F_{A|B} = \frac{\left(\text{SSE}_{\text{reduced}} - \text{SSE}_{\text{full}}\right) / \left(\text{df}_{\text{reduced}} - \text{df}_{\text{full}}\right)}{\text{SSE}_{\text{full}} / \text{df}_{\text{full}}}$$

If the usual normality assumptions are satisfied, then this statistic has an F-distribution with degrees of freedom $\text{df}_{\text{reduced}} - \text{df}_{\text{full}}$ for the numerator and df_{full} for the denominator. If the normality assumption is not reasonable, we may apply a permutation test.

Steps for Doing the Permutation Test

1. Permute the observations for the jth level of factor B among the levels $i_1, i_2, \ldots, i_{a(j)}$ of factor A in such a way that the sample sizes n_{ij}, $i = i_1, i_2, \ldots, i_{a(j)}$, are maintained. Do this for all levels of factor B to obtain a permutation of the data.

2. Take a suitably large random sample of the permutations. For each permutation, compute $F_{A|B}$. Form the permutation distribution of this statistic.

3. The p-value is the fraction of the $F_{A|B}$'s computed in step 2 that are greater than or equal to the observed statistic $F_{A|B,\text{obs}}$.

EXAMPLE **9.1.2**   To illustrate the technique, suppose we have the data in Table 9.1.2 except that the data for P3 and week 4 are missing. We wish to see whether the preservatives are different as adjusted for time effects. The data with the full and reduced labels are shown in Table 9.1.4. They are arranged to show the grouping by week.

The observations within each week are the ones that are permuted. For instance, within week 1, the values are 1.38, 2.54, 2.14, 2.32, 3.77, 2.59, 4.57, 3.69, and 4.41, which are permuted among the levels of preservative in such a way that three observations are assigned to each level as with the original data. The sums of squared errors for the full and reduced models are 5.73 and 13.87 with 16 and 21 degrees of freedom, respectively. The observed value of $F_{A|B}$ is 4.55. Based on 5000 randomly selected permutations, we found a p-value of .027. The normal-theory p-value is .006.   ∎

An Alternative Form of the Test Statistic

In permuting the observations within weeks, the sum of squared errors for the reduced model does not change. Thus, a permutation test may be based on just the sum of squared errors for the full model. Since a small value of SSE_{full} corresponds

TABLE 9.1.4

Data for Testing the Effect of Preservative in Presence of the Effect of Time

Preservative	Time	Log Bacteria Count			Labels	
		Sample 1	Sample 2	Sample 3	Full	Reduced
P1	Week 1	1.38	2.54	2.14	1	1
P2	Week 1	2.32	3.77	2.59	2	1
P3	Week 1	4.57	3.69	4.41	3	1
P1	Week 2	3.83	3.85	4.60	4	2
P2	Week 2	4.95	3.86	3.62	5	2
P3	Week 2	4.09	4.87	5.24	6	2
P1	Week 4	5.45	6.57	6.33	7	3
P2	Week 4	5.46	6.26	6.52	8	3
P3	Week 4	Missing	Missing	Missing		

to a large value of $F_{A|B}$, we would use SSE_{full} with a lower-tail test. Let S_{ij}^2 denote the sample variance for the ijth treatment combination. The sum of squared errors for the full model is

$$\text{SSE}_{\text{full}} = \sum_{ij} \left(n_{ij} - 1 \right) S_{ij}^2$$

If $n_{ij} \le 1$, $i = i_1, i_2, \ldots, i_{a(j)}$, for a given level j of factor B, then permuting observations within this level of B does not affect the value of SSE_{full}. At least one level of factor A for a given level of factor B must have more than one observation in order for the observations at this level of B to contribute to the test statistic.

9.1.3 Computer Analysis

In the case in which there are equal numbers of observations for all combinations of levels of factors A and B, Resampling Stats has sufficient functionality to carry out the bootstrap analysis of variance with the fixed-effects model. For an unequal numbers of observations, we may express the fixed-effects model as a regression model using the sum-to-zero restrictions and then apply the methods outlined in Section 8.7.3. The Resampling Stats add-in to Excel may be used to carry out the computations. We should note that the sum-to-zero restrictions are not the only ones that may be used to express the fixed-effects analysis of variance model as a regression model, but they are the ones that we recommend. For a discussion of the issues, see Blair and Higgins (1978).

Figure 9.1.2 shows Resampling Stats code for the permutation test of the overall effect of preservative adjusted for week for the data in Table 9.1.4. The alternative statistic is used in the computation.

FIGURE 9.1.2

Resampling Stats Code for Test of Overall Effect of Preservative Adjusted for Week for
Data in Table 9.1.4

```
  maxsize default 5000

'enter data by week
copy (1.38 2.54 2.14 2.32 3.77 2.59 4.57 3.69 4.41) w1
copy (3.83 3.85 4.60 4.95 3.86 3.62 4.09 4.87 5.24) w2
copy (5.45 6.57 6.33 5.46 6.26 6.52) w3

'get permutation distribution
repeat 5000
shuffle w1 sw1
shuffle w2 sw2
shuffle w3 sw3

'obtain shuffled data for the 8 treatments
take sw1 1,3 p1w1
take sw1 4,6 p2w1
take sw1 7,9 p3w1
take sw2 1,3 p1w2
take sw2 4,6 p2w2
take sw2 7,9 p3w2
take sw3 1,3 p1w3
take sw3 4,6 p2w3

'get variances for the 8 treatments
variance p1w1 v11
variance p2w1 v21
variance p3w1 v31
variance p1w2 v12
variance p2w2 v22
variance p3w2 v32
variance p1w3 v13
variance p2w3 v23

'get sum of squared errors for full model
concat v11 v21 v31 v12 v22 v32 v13 v23 v
multiply (2 2 2 2 2 2 2 2) v ssqr
sum ssqr ssefull

'keep track of ssefull
score ssefull ssedist
end

'get p-value ssefull is 5.708 for original data
count ssedist <= 5.708 pval
divide pval 5000 pval
print pval

PVAL    =      0.0272
```

9.2
Aligned-Rank Transform

The rank transform methodology was popularized by Iman and Conover (1981) as a way to construct rank tests for a variety of experimental designs. The idea is to convert observations to ranks and then apply the normal-theory analysis of variance or regression methods to the ranks. This would allow a researcher to perform a rank test simply by plugging ranks of the data into a standard analysis of variance or regression program and using the p-values from the program to draw conclusions. In some cases, this appealingly simple methodology gives about the same result as the corresponding nonparametric rank test. For instance, a two-sample t-test applied to ranks will yield essentially the same conclusion as the Wilcoxon rank-sum test, provided the sample sizes are large enough. However, there are difficulties in testing for main effects and interactions in multifactor experiments.

Consider the data in Table 9.2.1 which have two levels of factor A and three levels of factor B. The original data and the ranks are identical except for the data in the A2–B3 combination. Here the data have been rigged so there is a strong interaction. However, when the data are converted to ranks, the observations in the A2–B3 combination have ranks 16, 17, and 18, and the test for interaction applied to ranks is not significant. The reason is that ranking of the data is a nonlinear transformation, and interaction in the original data is not necessarily preserved under such a transformation.

TABLE 9.2.1

Comparison of Tests for Interaction on Original Data and Ranks

	$B1$	$B2$	$B3$
$A1$	1, 2, 3	4, 5, 6	7, 8, 9
$A2$	10, 11, 12	13, 14, 15	20, 21, 22
Original Data	$F_A = 480.5, p < .0001; F_B = 98.00, p < .0001; F_{AB} = 8.00, p = .0062$		
Ranks	$F_A = 364.5, p < .0001; F_B = 54.00, p < .0001; F_{AB} = 0.00, p = 1.000$		

Whether the researcher wishes to apply analysis of variance to ranks is a matter of judgment. However, the researcher needs to be aware that the nature of the hypotheses being tested can be fundamentally changed when analysis of variance is applied to ranks. For a theoretical discussion of the rank transform method, see Akritas (1990) and Thompson (1991). For simulation studies that show the effects of the rank transform method, see Blair, Higgins, and Sawilowsky (1987) and Sawilosky, Blair, Higgins (1989).

9.2.1 Aligning the Data in a Completely Random Design

A fix-up for the problem with the rank transform is to *align* the data before ranking. Using the notation of Section 9.1, we have

$$Y_{ijk} = \mu + \alpha_i + \beta_j + \gamma_{ij} + \varepsilon_{ijk}$$

With the sum-to-zero restrictions, we may estimate the parameters from the data. In the case where the sample sizes are equal to n for all the cells, the estimates are

$$\hat{\alpha}_i = \hat{\mu}_{i.} - \hat{\mu}, \quad \hat{\beta}j = \hat{\mu}_{.j} - \hat{\mu}, \quad \hat{\gamma}_{ij} = \hat{\mu}_{ij} - \hat{\mu}_{i.} - \hat{\mu}_{.j} + \hat{\mu}$$

where

$$\hat{\mu}_{i.} = \frac{1}{nb} \sum_{j=1}^{b} \sum_{k=1}^{n} Y_{ijk}, \quad \hat{\mu}_{.j} = \frac{1}{na} \sum_{i=1}^{a} \sum_{k=1}^{n} Y_{ijk}, \quad \hat{\mu} = \frac{1}{nab} \sum_{i=1}^{a} \sum_{j=1}^{b} \sum_{k=1}^{n} Y_{ijk}$$

To align the Y_{ijk}'s for testing for interaction, we compute

$$AB_{ijk} = Y_{ijk} - \hat{\mu} - \hat{\alpha}_i - \hat{\beta}_j$$

The aligned-rank transform test for interaction is carried out by ranking the AB_{ijk}'s and then applying the usual analysis of variance to these ranks. An approximate p-value for the test may be read from the output of any software package that does two-way analysis of variance. See Section 9.2.3 for references that justify the use of this methodology. When we test for interaction, the tests for main effects should not be significant because the main effects have been subtracted out.

To test for the factor A main effect, the data are aligned as

$$A_{ijk} = Y_{ijk} - \hat{\mu} - \hat{\beta}_j - \hat{\gamma}_{ij}$$

and ranked. Similarly, to test for the factor B main effect, the data are aligned as

$$B_{ijk} = Y_{ijk} - \hat{\mu} - \hat{\alpha}_i - \hat{\gamma}_{ij}$$

and ranked. Thus, three separate aligned-rank tests would have to be performed to test for main effects and interaction. In each case, an approximate p-value for the test can be taken directly from the output of an analysis of variance program.

EXAMPLE 9.2.1   The data in Table 9.2.2 have been aligned and ranked to test for interaction. For instance, the data value of 3 for the $(A1, B1)$ combination has the aligned value of $3 - 5.00 - 6.50 + 10.17 = 1.67$. As expected when aligning for interaction, the F statistics for the A and B main effects would be judged not significant at the 5% level with $p = .6918$ for A and $p = 1.000$ for B. The F statistic for interaction would be judged significant at the 5% level with $p = .0052$. These values may be obtained from any standard analysis of variance program.

TABLE 9.2.2
Data Aligned for Testing for Interaction

		B1	B2	B3	Row Means
Data		1, 2, 3	4, 5, 6	7, 8, 9	5.00
Aligned Data	A1	−0.33, 0.67, 1.67	−0.33, 0.67, 1.67	−2.33, −1.33, −0.33	
Aligned Ranks		8, 13.5, 16.5	8, 13.5, 16.5	1, 4, 8	
Data		10, 11, 12	13, 14, 15	20, 21, 22	15.333
Aligned Data	A2	−1.67, −0.67, 0.33	−1.67, −0.67, 0.33	0.33, 1.33, 2.33	
Aligned Ranks		2.5, 5.5, 11	2.5, 5.5, 11	11, 15, 18	
Column Means		6.50	9.50	14.50	10.167
F statistics		$F_A = 0.1650$ $p = .6918$	$F_B = 0.0000$ $p = 1.0000$	$F_{AB} = 8.4175$ $p = 0.0052$	

∎

9.2.2 Aligning the Data in a Split-Plot Design

In split-plot designs, there are two sizes of experimental units, called the whole plots and the subplots. For instance, in an agricultural experiment, a watering treatment might be applied to large plots, and a set of nitrogen treatments might be applied to subdivisions of each large plot. A mathematical model for the split-plot design is

$$Y_{ijk} = \mu + \alpha_i + \delta_{ik} + \beta_j + \gamma_{ij} + \varepsilon_{ijk}$$

The δ_{ik}'s are called the whole-plot error terms and the ε_{ijk}'s are called the subplot error terms. The δ_{ik}'s are assumed to be independent and identically distributed random variables with mean 0 and standard deviation σ_δ, and they are assumed to be independent of the ε_{ijk}'s, which are independent and identically distributed with standard deviation σ_ε. Factor A is the whole-plot factor, and factor B is the subplot factor. We assume that the design is balanced—that is, that there are n observations for each of the levels of factors A and B.

To test for the factor A effect, we first average over the subplots; that is, we compute

$$\bar{Y}_{i.k} = \frac{1}{b} \sum_{j=1}^{b} Y_{ijk}$$

We then apply the aligned-rank transform to the $\bar{Y}_{i.k}$'s. The analysis is comparable to the Kruskal–Wallis test applied to these averages. To test for the factor B effect and AB interaction, we first remove the whole-plot error term; that is, we compute

$Y_{ijk} - \overline{Y}_{i.k}$. Then we align the data, removing the effect of B in testing for interaction and removing the effect of interaction in testing for B. The aligned data for testing for interaction are

$$AB_{ijk}(\text{split plot}) = Y_{ijk} - \overline{Y}_{i.k} - \hat{\beta}_j$$

and for the B effect,

$$B_{ijk}(\text{split plot}) = Y_{ijk} - \overline{Y}_{i.k} - \hat{\gamma}_{ij}$$

After alignment, the data are ranked and the usual split-plot analysis of variance is applied to the aligned ranks.

EXAMPLE 9.2.2   The data in Table 9.2.3 are the dry weights of wheat plants grown in peat pots in a greenhouse. The pots were placed in trays (the whole plots), and different levels of water were put in the bottoms of the trays to be absorbed by the pots. Within each tray were four pots. Four levels of fertilizer were randomly assigned to the four pots per tray (the subplots).

TABLE 9.2.3
Dry Weights from a Split-Plot Experiment

Moisture	Tray	Fertilizer 1	Fertilizer 2	Fertilizer 3	Fertilizer 4	Tray Means
	1	3.3	4.3	4.5	5.8	4.475
1	2	4.0	4.1	6.5	7.3	5.475
	3	1.9	3.8	4.4	5.1	3.800
	4	5.0	7.9	10.7	13.5	9.275
2	5	5.9	8.5	10.3	13.9	9.650
	6	6.9	7.0	10.9	15.2	10.000
	7	6.5	10.7	12.2	15.7	11.275
3	8	8.2	8.9	13.4	14.9	11.350
	9	5.2	8.6	11.1	15.6	10.125
	10	6.8	9.0	10.3	12.5	9.650
4	11	6.4	6.0	10.7	12.5	8.900
	12	4.0	3.8	9.4	10.2	6.850
Fertilizer Means		5.342	6.883	9.533	11.850	8.402

To see how the data are aligned, consider the observation in tray 1, moisture 1, and fertilizer 1. The aligned value for testing for interaction is 3.3 − 4.475 − 5.342 + 8.402 = 1.885, and this has rank 47 among the aligned observations. The aligned observations and ranks are shown in Table 9.2.4.

TABLE 9.2.4
Aligned Data and Ranks for Testing for Interaction in a Split-Plot Design

Moisture	Tray	Fertilizer 1		Fertilizer 2		Fertilizer 3		Fertilizer 4	
	1	1.885	47	1.344	42	−1.106	10	−2.123	2
1	2	1.585	45	0.144	26.5	−0.106	20	−1.623	5
	3	1.160	41	1.519	44	−0.531	14	−2.148	1
	4	−1.215	9	0.144	26.5	0.294	31	0.777	35
2	5	−0.690	12	0.369	32	−0.481	15.5	0.802	36
	6	−0.040	23	−1.481	7	−0.231	17	1.752	46
	7	−1.715	4	0.944	39	−0.206	18	0.977	40
3	8	−0.090	22	−0.931	11	0.919	38	0.102	25
	9	−1.865	3	−0.006	24	−0.156	19	2.027	48
	10	0.210	29.5	0.869	37	−0.481	15.5	−0.598	13
4	11	0.560	33	−1.381	8	0.669	34	0.152	28
	12	0.210	29.5	−1.531	6	1.419	43	−0.098	21

The analysis of variance is shown in Table 9.2.5. In the usual split-plot analysis, tray within moisture, denoted Tray(Moisture), is the error term to test for moisture effect, and the residual, denoted Error, is used as the error term to test for the fertilizer main effect and moisture-by-fertilizer interaction. In this case, the only meaningful test is the one for interaction, since the other effects have been subtracted out. The strong interaction is due to the fact that increasing levels of fertilizer have a greater effect on dry weights when the moisture level is 2, 3, or 4 than when the moisture level is 1. The data for this experiment came from Milliken and Johnson (1984).

TABLE 9.2.5
Analysis of Variance for Aligned-Rank Test for Interaction

Source	df	SS	MS	F	p-value
Moisture	3	4.125	1.38	Not applicable	
Tray(Moisture)	8	26.875	3.36	Not applicable	
Fertilizer	3	41.167	13.72	Not applicable	
Moisture*Fertilizer	9	5965.71	662.86	5.01	0.0007
Error	24	3172.63	132.19		

■

9.2.3 Properties of the Aligned-Rank Procedure

The aligned-rank transform tests are not distribution-free in the sense that the distributions of the test statistics in general depend on the population distribution from which the observations are selected. However, simulation studies by Higgins, Blair, and Tashtoush (1990), Salter and Fawcett (1993), and Higgins and Tashtoush (1994) show that the tests maintain their Type I error rates reasonably well. Puri and Sen (1985) consider the theory of aligned tests in the context of the general linear model. If we let F_A, F_B, and F_{AB} denote the aligned-rank F statistics for the A main effect, B main effect, and AB interaction, Mansoui (1999) shows that the statistic $(a-1)F_A$, $(b-1)F_B$, and $(a-1)(b-1)F_{AB}$ have asymptotic chi-square distributions with degrees of freedom $a-1$, $b-1$, and $(a-1)(b-1)$ respectively, as $n \to \infty$. He also reports simulations that show the tests maintain reasonable Type I error rates. The tests enjoy power advantages over the standard analysis of variance in situations where other rank tests might also be expected to do well—namely, when the distributions of the observations have heavier tails or when outliers may be present.

9.2.4 Computer Analysis

The alignment and ranking can be carried out by any program that allows transformations of the data and ranking. Once the data are aligned, then any standard analysis of variance program may be applied to the data. The computations can be done in MINITAB, for instance.

9.3
Testing for Lattice-Ordered Alternatives

In Section 3.4 we introduced the notion of ordered alternatives in the context of a single-factor experiment. Here we extend the idea to two factors. For example, if different levels of nitrogen and phosphorus are applied to experimental plots, it may be reasonable to assume that yields will increase with increasing amounts of either. Our purpose is to develop a nonparametric statistical test that takes advantage of this type of ordering among treatments.

We use the assumptions and notation of Section 9.1. For pairs of integers (i, j) and (s, t), we say $(i, j) \le (s, t)$ if $i \le s$ and $j \le t$. The means in the model $Y_{ijk} = \mu_{ij} + \varepsilon_{ijk}$ are *lattice-ordered* and increasing if $\mu_{ij} \le \mu_{st}$ whenever $(i, j) \le (s, t)$. The means are lattice-ordered and decreasing if $\mu_{ij} \ge \mu_{st}$ whenever $(i, j) \le (s, t)$. Table 9.3.1 shows treatment combinations for a 3×4 layout. The means of the shaded treatments are those that can be compared to μ_{23} under lattice-ordering. For instance, with increasing means, $\mu_{12} \le \mu_{23}$ and $\mu_{23} \le \mu_{34}$. However, the inequalities between μ_{23} and the means μ_{14}, μ_{31}, and μ_{32} are not specified under the lattice ordering.

TABLE 9.3.1

Means of Shaded Treatment Combinations Can Be Compared to μ_{23} Under Lattice Ordering

	$B = 1$	$B = 2$	$B = 3$	$B = 4$
$A = 1$	μ_{11}	μ_{12}	μ_{13}	μ_{14}
$A = 2$	μ_{21}	μ_{22}	μ_{23}	μ_{24}
$A = 3$	μ_{31}	μ_{32}	μ_{33}	μ_{34}

The hypotheses of interest are

$$H_0: \text{All means are equal}$$

$$H_a: \mu_{ij} \leq \mu_{st}, \quad (i,j) \leq (s,t)$$

with strict inequality holding for at least one pair of means. That is, the alternative hypothesis is lattice-ordered and increasing means. A similar procedure applies to decreasing means. We may formulate the hypotheses in terms of the distribution functions of the observations. If $F_{ij}(x)$ is the distribution function of the observations under the ijth treatment combination, then the hypotheses are

$$H_0: F_{11}(x) = \cdots = F_{ab}(x)$$

$$H_a: F_{ij}(x) \geq F_{st}(x), \quad (i,j) \leq (s,t)$$

where strict inequality holds for at least one pair (i, j) and (s, t), and one x.

9.3.1 A Test Statistic for Lattice-Ordered Alternatives

The development of a test statistic follows along the lines of the Jonckheere–Terpstra statistic in Section 3.4.1. Let $T_{ij,st}$ denote any statistic that tests $H_0: F_{ij}(x) = F_{st}(x)$ versus $H_a: F_{ij}(x) \geq F_{st}(x)$. The general form of the test statistic for lattice-ordered alternatives is

$$T_L = \sum_{(i,j) \, < \, (s,t)} \sum T_{ij,st}$$

where the double summation is taken over all distinct pairs of indices (i, j) and (s, t) such that $(i, j) < (s, t)$. We use the one-sided Wilcoxon rank-sum statistic for comparing two groups. The permutation distribution of the test statistic is formed by permuting the observations among the treatment combinations while maintaining the sample sizes and then computing T_L each time. The number of test statistics $T_{ij,st}$ that must be computed for each permutation of the data is

$$\binom{a+1}{2}\binom{b+1}{2} - ab$$

Asymptotic normality of T_L holds in two situations. One is where the levels of A and B are fixed and the number of observations per treatment combination is allowed to increase. The other is where the number of observations per treatment combination is fixed and the levels of A and B are allowed to increase. As shown in Figure 9.3.1, the normal distribution is a reasonable approximation to the permutation distribution even for relatively small numbers of observations.

In the special case in which there are no ties and each treatment combination has n observations, the expected value of T_L is given by

$$E(T_L) = \frac{n(2n+1)}{2}\left[\binom{a+1}{2}\binom{b+1}{2} - ab\right]$$

and the variance is given by

$$\text{var}(T_L) = \frac{n^2}{12}\left[\binom{a+1}{2}\binom{b+1}{2} - ab\right]$$

$$+ \frac{n^3}{54}\left[\binom{a+1}{2}\binom{b+1}{2}\left[(2a+1)(2b+1) - (a+2)(b+2)\right]\right]$$

EXAMPLE 9.3.1   A permutation distribution of T_L was obtained for the bacteria count data in Table 9.1.2. Here we assume that responses tend to increase as time increases and as we go from preservative P1 to P2 to P3. A plot of the distribution is shown in Figure 9.3.1. The value of T_L for the original data is 374 with a permutation p-value less than .001. Computations were based on 1000 randomly selected permutations using the Resampling Stats add-in to Excel. The expected value and variance of T_L for the data in Table 9.1.2 are 283.5 and 452.25, respectively. The z statistic is $(374 - 283.5)/21.27 = 4.25$ with $p < .0001$.   ∎

9.3.2 Lattice Correlation

We may rescale the statistic T_L so that its value ranges from -1 to 1. We refer to this as *lattice correlation*. This statistic is analogous to Kendall's tau. It measures the degree to which the observations conform to a lattice ordering, either increasing or decreasing. In the case of an equal number of observations per treatment combination, the rescaled statistic is

$$r_L = \frac{2[T_L - E(T_L)]}{n^2\left[\binom{a+1}{2}\binom{b+1}{2} - ab\right]}$$

FIGURE 9.3.1

Permutation Distribution of T_L

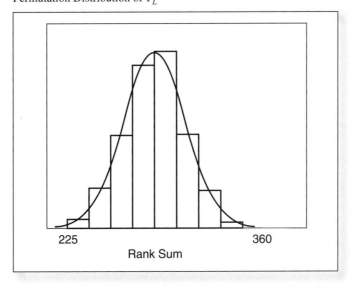

Rank Sum

The denominator of r_L is the number of pairs of the form (Y_{ijk}, Y_{stu}), where $(i, j) < (s, t)$.

This statistic may also be expressed in terms of the Mann–Whitney statistic. Let $\Psi(Y_{ijk}, Y_{stu}) = 1$ if $Y_{ijk} \le Y_{stu}$ and 0 otherwise. Define $U_{ij,st}$ and U_L as

$$U_{ij,st} = \sum_{k=1}^{n} \sum_{u=1}^{n} \Psi\left(Y_{ijk}, Y_{stu}\right)$$

$$U_L = \sum_{(i,j) < (s,t)} \sum U_{ij,st}$$

Then

$$r_L = \frac{2U_L}{n^2\left[\left(\dfrac{a+1}{2}\right)\left(\dfrac{b+1}{2}\right) - ab\right]} - 1$$

Suppose a response Y_{ijk} can be represented as

$$Y_{ijk} = f\left(s_i, t_j\right) + \varepsilon_{ijk}$$

where (s_i, t_j) are points on a rectangular grid. The extent to which Y tends to increase or decrease with increasing s and t depends on both the deterministic component $f(s, t)$ and the random error. The lattice correlation r_L is a measure of the monotonicity of the observations. It can help give us a picture of the surface created by the observations (s_i, t_j, Y_{ijk}). If the function $f(s, t)$ increases smoothly as s and t increase and if the random error is small, then the correlation r_L will be near 1. If the surface is choppy due to either a large random error or an irregular function $f(s, t)$, the lattice correlation will be near 0.

EXAMPLE 9.3.2   Figure 9.3.2 is a plot of $Y = 1 + 0.30s + 0.55t + \varepsilon$, where ε has a uniform distribution on the interval 0 to 2, $s = 1, 2, \ldots, 5$, and $t = 1, 2, \ldots, 5$. We see a tendency for the surface to increase with s and t consistent with the deterministic part of the model, but fluctuate due to the random error component. The lattice correlation is $r_L = .75$.

FIGURE 9.3.3
Surface with $r_L = .75$

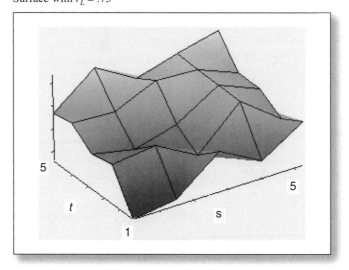

9.3.3 References for Lattice-Ordered Alternatives

Agar and Brent (1978) developed a test statistic for testing for partially ordered alternatives, of which lattice ordering is a special case. They derived the standard deviation of their Kendall-type statistic and included an adjustment for ties. They applied their method to a psychological study. They also established the asymptotic normality of the statistic under the assumption that the number of observations per treatment increases.

Strand (2000) considered lattice-ordered alternatives for multifactor experiments. He derived an explicit formula for the standard deviation in the case of equal sample sizes for the treatment combinations, and he presented two biological applications. Higgins and Bain (1999) established the asymptotic normality of r_L as the levels of factors A and B increase in the case of singly replicated experiments. They also developed a Spearman-type correlation for lattice-ordered alternatives. The notion of lattice correlation also extends to more than two dimensions.

9.3.4 Computer Analysis

Computations for the permutation test may be carried out in Resampling Stats, but coding is lengthy except for small data sets due to the number of Wilcoxon statistics that must be computed. When the data have no ties and contain an equal number of observations for all the treatment combinations, then the normal approximation is quite good even for small sample sizes, and we recommend it.

Exercises

1 The data in the table are the nitrate uptakes (in parts per million) of barberry plants grown individually in containers using one of three growing media and one of two types of fertilizer. There were three replications for each of the media–fertilizer combinations. Measurements were made at 40 and 70 days. The data are courtesy of Dr. Houchang Khatamian, Department of Horticulture, Forestry, and Recreation Resources, Kansas State University. Carry out a bootstrap analysis of variance on the day 40 and day 70 data to test for main effects and interaction.

Media	Fertilizer	Day 40	Day 70
1	1	240	47
1	1	320	99
1	1	227	57
1	2	710	416
1	2	800	216
1	2	1152	360
2	1	502	23
2	1	134	7
2	1	240	43
2	2	192	352
2	2	214	363
2	2	384	64
3	1	147	103
3	1	6	27
3	1	332	63
3	2	294	69
3	2	70	280
3	2	74	235

2   For the data in Exercise 1, conduct an approximate permutation test to test for main effects and interaction. That is, permute the errors among the treatments rather than bootstrap sampling from them. Do the test for both the day 40 and day 70 data.

3   For the data in Exercise 1, carry out a permutation test to test for fertilizer effect adjusted for media effect as discussed in Section 9.1. Do so for both the day 40 and day 70 data.

4   Analyze the day 40 data in Exercise 1 using the aligned-rank transform method of Section 9.2.

5   A researcher selected six plots of ground for experimentation. Three randomly selected plots were burned in the spring, and the other three were unburned. Each plot was divided into three subplots. One subplot was clipped to simulate heavy grazing (H), another was clipped to simulate medium grazing (M), and the third light grazing (L). Use the aligned-rank method for split-plot data to test for main effects and interaction. The data represent forage quality.

Plot	Burn	Graze	Quality
1	No	L	135
1	No	M	125
1	No	H	105
2	No	L	130
2	No	M	115
2	No	H	110
3	No	L	120
3	No	M	120
3	No	H	100
4	Yes	L	155
4	Yes	M	140
4	Yes	H	140
5	Yes	L	180
5	Yes	M	170
5	Yes	H	160
6	Yes	L	160
6	Yes	M	140
6	Yes	H	135

6   Data are arranged in the 2×2 table shown here. Obtain the permutation distribution of T_L, the test statistic for lattice-ordered alternatives.

	B1	B2
A1	1	2
A2	3	4

7 The data in the table from Nandi and Nelson (1992) are the percentages of low-birth-weight babies cross-classified according to mother's age and mother's weight for nonsmokers. The percentages should increase as mother's age increases and increase as mother's weight decreases. Use the test for lattice-ordered means to see whether this is a plausible assumption.

Weight/Age	High	Medium	Low
20–24	3.8	4.2	7.8
25–29	3.6	6.0	6.1
30–34	6.7	4.1	9.2
35–39	4.4	3.9	6.8

8 Find the lattice correlation for the data in Exercise 7.

9 Simulated blood alcohol readings are given on two individuals in each of four weight categories who were given from one to four drinks. It is expected that the blood alcohol level will increase with decreasing weight and increase with increasing number of drinks. Use the asymptotic test for lattice-ordered alternatives to verify these assumptions for the data in the table. Find the lattice correlation.

Weight/Number of Drinks	200	180	160	140
1	.021, .027	.025, .036	.050, .031	.032, .049
2	.046, .035	.048, .058	.041, .067	.057, .069
3	.068, .053	.063, .059	.077, .101	.086, .103
4	.098, .072	.084, .066	.090, .099	.111, .085

Theory and Complements

10 Extend the notion of lattice ordering to three factors. How would the three-dimensional analogue of the statistic T_L be defined? Explain how to obtain the permutation distribution of this statistic.

10

Smoothing Methods and Robust Model Fitting

A Look Ahead   The ideas in Sections 10.1 and 10.2 have a different flavor from those of the preceding chapters. Here we are interested in curve fitting when the functional form of the curve is not specified. Section 10.1 deals with estimating probability density functions, and Section 10.2 presents nonparametric regression when the form of the regression function is not specified. In Section 10.3, we return to the more familiar realm of fitting linear regression models to data, but we abandon the least squares criterion in favor of criteria that are less sensitive to outlying observations.

10.1
Estimating the Probability Density Function

Let us assume we have a random sample $X_1, X_2, \ldots, X_n$ from a population that has a continuous probability density function $f(x)$. We will consider the problem of estimating $f(x)$.

The simplest such estimate is a histogram in which the axes have been scaled to represent relative frequency. Partition the range of the data into subintervals $a_1 < a_2 < \cdots < a_k$. The density estimate for a value of x in the interval $a_i < x \leq a_{i+1}$ is

$$\hat{f}(x) = \frac{\text{number of observations} \in (a_i, a_{i+1}]}{n(a_{i+1} - a_i)}$$

Typically, the intervals are of equal length.

If the intervals are too narrow, the histogram will have a choppy appearance unrepresentative of the population density, and if the intervals are too wide, the histogram may obscure interesting features of the distribution. Scott (1992) suggested the interval width $d = 3.5S/n^{1/3}$, where S is the sample standard deviation, and discussed other possibilities. Generally, if the number of intervals in the histogram is in the range 5 to 10, the histogram will give a reasonable representation of the distribution of the data. More intervals may be needed for very large data sets.

10.1.1   Kernel Method

The lack of smoothness of the histogram has led to other nonparametric procedures for estimating densities. We present the *kernel method* of density estimation, which involves taking a certain weighted average of data points near x to estimate $f(x)$. Let $w(z)$ be a symmetric probability density function centered at 0 [think of $w(z)$ as the standard normal probability density function]. The kernel estimate of $f(x)$ is

$$\hat{f}(x) = \frac{1}{n\Delta} \sum_{i=1}^{n} w\left(\frac{x - X_i}{\Delta}\right)$$

The function $w(z)$ is called the *kernel* and the quantity Δ is called the *bandwidth*. The bandwidth plays the same role as the interval length in the histogram. If the bandwidth is too narrow, the density estimate will have a choppy appearance, and if the bandwidth is too wide, the estimate will be overly smooth. In the remaining discussion, we will assume that $w(z)$ is the standard normal probability density function.

Hardle (1991) suggests several rules of thumb for choosing Δ. The one that serves our purposes is

$$\Delta = \frac{1.06}{n^{1/5}} S$$

where S is the sample standard deviation. Other possibilities suggested by Hardle involve replacing S with estimates of the population standard deviation that are less sensitive to outlying observations, such as the scaled interquartile range $(X_{(.75n)} - X_{(.25n)}) / 1.34$.

In practice, it is a good idea to try values of Δ on either side of the choice given by the rule of thumb above. Smaller values of Δ may reveal important details in the distribution, such as multiple modes, while larger values of Δ may give a pleasing, smooth appearance to the density. One should not be timid about varying Δ. The kernel density estimate is relatively insensitive to small variations in Δ. Decreasing or increasing Δ by a factor of 2 is not unreasonable in examining the shape of the estimated density function.

EXAMPLE 10.1.1   The data in Table 10.1.1 are 107 durations (in minutes) of the eruptions of Old Faithful geyser. Figure 10.1.1 shows two histograms and density estimates of these data. The ones on the left use the suggested values of histogram width and bandwidth, and the ones on the right are half the suggested values. The computations were done in S-Plus. The kernels used by S-Plus are scaled so that the upper and lower quartiles are ±0.25. As a result, the normal kernel used by S-Plus has a standard deviation of 0.37 instead of 1.

TABLE 10.1.1

Durations (in minutes) of Eruptions of Old Faithful Geyser

1.7	1.9	3.4	3.8	4.1	4.5
1.7	1.9	3.5	3.9	4.1	4.5
1.7	2.0	3.5	3.9	4.1	4.5
1.7	2.0	3.5	3.9	4.1	4.6
1.7	2.0	3.5	3.9	4.2	4.6
1.8	2.0	3.6	4.0	4.2	4.6
1.8	2.0	3.6	4.0	4.3	4.6
1.8	2.3	3.7	4.0	4.3	4.6
1.8	2.3	3.7	4.0	4.3	4.6
1.8	2.3	3.7	4.0	4.3	4.6
1.8	2.3	3.7	4.0	4.3	4.6
1.8	2.5	3.7	4.0	4.3	4.6
1.8	2.9	3.7	4.0	4.4	4.7
1.8	2.9	3.7	4.1	4.4	4.7
1.9	3.1	3.8	4.1	4.4	4.7
1.9	3.2	3.8	4.1	4.4	4.8
1.9	3.3	3.8	4.1	4.4	4.9
1.9	3.4	3.8	4.1	4.5	

FIGURE 10.1.1

Histogram and Kernel Density Estimates of Old Faithful Data; Histogram Interval
and Bandwidth on Left Are Suggested Values; Those on Right Are Half Those on Left

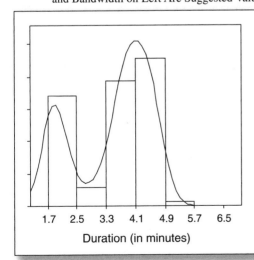

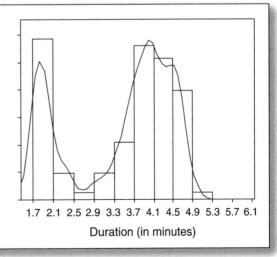

The choice of the kernel is not as important as the choice of the bandwidth. Smooth kernels will give density estimates that are similar to the ones given by the normal distribution kernel. It is also possible to vary the bandwidths so that they are wider where observations are sparse. Early theoretical work on kernel estimates was carried out by Parzen (1962). Also see Bean and Tsokos (1980).

10.1.2 Computer Analysis

S-Plus is especially useful for curve fitting and plotting, with many choices available from its menu.

10.2
Nonparametric Curve Smoothing

Suppose we wish to investigate the relationship between variables X and Y and we have taken n pairs of data (X_i, Y_i), $i = 1, 2, \ldots, n$. In some cases, a functional form of a model, such as linear or quadratic, may not be apparent. Then we may use nonparametric regression. We assume the responses can be represented as $Y = \phi(x) + \varepsilon$, where $E(\varepsilon) = 0$. The objective is to obtain an estimate of $\phi(x)$ without specifying its functional form. These methods involve smoothing the data.

Several types of smooth nonparametric estimates of $\phi(x)$ involve giving greater weight to pairs of observations (X_i, Y_i) for which X_i is near x and lesser weight to other pairs. Regression *splines* piece together lower-order polynomials, typically cubic polynomials, in a smooth way. The *loess method* (for *local regression*) uses weighted linear regression applied to pairs (X, Y) for which the X's are near x. The *kernel method* is analogous to the method used for nonparametric density estimation. We now consider the loess and kernel methods.

10.2.1 Loess Method

To estimate $y = \phi(x)$ at $x = x_0$, we note that $\phi(x)$ can often be adequately approximated by a linear function $l(x) = \beta_0 + \beta_1(x - x_0)$ when x is near x_0. To fit $l(x)$ locally, first determine the k values of the X_i's that are nearest to x_0, where k/n is a specified fraction of the total number of points. The fraction, called the *span*, determines the smoothness of the approximation. The larger the span, the smoother the curve. Let $N_k(x_0)$ denote this set of k points. Let $W(u)$, $0 \le u \le 1$, be a nonnegative weighting function with mode at $u = 0$. One such function is $W(u) = (1 - u^3)^3$. The loess approximation uses weighted least squares to find the $l(x)$ that minimizes

$$\sum_{X_i \in N_k(x_0)} \left[Y_i - l(X_i) \right]^2 W\left(\frac{|x_0 - X_i|}{\Delta_{x_0}} \right)$$

where

$$\Delta_{x_0} = \max_{X_i \in N_k(x)} |X_i - x|$$

The loess estimate of $y_0 = \phi(x_0)$ is $\hat{y}_0 = \hat{l}(x_o)$. If there is considerable curvature, then a quadratic function $q(x) = \beta_0 + \beta_1(x - x_0) + \beta_2(x - x_0)^2$ may be used instead of $l(x)$, but the procedure is otherwise the same.

The loess method can be extended to functions of more than one variable. For instance, if we wish to estimate $z = \phi(x, y)$ at $x = x_0$ and $y = y_0$, we approximate $\phi(x, y)$ by $l(x, y) = \beta_0 + \beta_1(x - x_0) + \beta_2(y - y_0)$ and use weighted least squares to obtain $\hat{z} = \hat{l}(x_0, y_0)$.

10.2.2 Kernel Method

Let the joint probability density function of (X, Y) be denoted $f(x, y)$ and the marginal density function of X be denoted $f_X(x)$. Then the conditional expectation of Y given $X = x$ is

$$\phi(x) = \int y \frac{f(x, y)}{f_X(x)} dy$$

An estimate of $\phi(x)$ is obtained by estimating $f(x, y)$ and $f_X(x)$ by the kernel method, and then performing the integration to get an approximation of the conditional expectation. The joint density function $f(x, y)$ is estimated using a product of univariate kernels; that is,

$$\hat{f}(x, y) = \frac{1}{n\Delta_1\Delta_2} \sum_{i=1}^{n} w\left(\frac{x - X_i}{\Delta_1}\right) w\left(\frac{y - Y_i}{\Delta_2}\right)$$

where $w(z)$ is a symmetric probability density function centered at 0. If the same bandwidth Δ_1 is used to estimate $f_X(x)$, then after integration with respect to y, the kernel estimate turns out to be

$$\hat{\phi}(x) = \frac{\sum_{i=1}^{n} Y_i w\left(\frac{x - X_i}{\Delta_1}\right)}{\sum_{i=1}^{n} w\left(\frac{x - X_i}{\Delta_1}\right)}$$

EXAMPLE 10.2.1   Table 10.2.1 lists the horsepower (HP) and miles per gallon (MPG) of 82 automobiles reported by Heavenrich, Murrell, and Hellman (1991). Figure 10.2.1 shows the loess and kernel estimates of the regression function. The span for the loess estimate is 0.5 with linear regression for the local approximation. The kernel estimate is based on the standard normal weight function, and the bandwidth is 25 using the formula suggested in Section 10.1. Computations and plotting were done in S-Plus.

TABLE 10.2.1

Horsepower and Miles Per Gallon of 82 Automobiles

HP	MPG	HP	MPG	HP	MPG	HP	MPG
49	65.4	74	40.7	95	32.2	140	25.3
55	56.0	95	40.0	102	32.2	140	23.9
55	55.9	81	39.3	95	32.2	150	23.6
70	49.0	95	38.8	93	31.5	165	23.6
53	46.5	92	38.4	100	31.5	165	23.6
70	46.2	92	38.4	100	31.4	165	23.6
55	45.4	92	38.4	98	31.4	165	23.6
62	59.2	90	29.5	130	31.2	245	23.5
62	53.3	52	46.9	115	33.7	280	23.4
80	43.4	103	36.3	115	32.6	162	23.4
73	41.1	84	36.1	115	31.3	162	23.1
92	40.9	84	36.1	115	31.3	140	22.9
92	40.9	102	35.4	180	30.4	140	22.9
73	40.4	102	35.3	160	28.9	175	19.5
66	39.6	81	35.1	130	28.0	322	18.1
73	39.3	90	35.1	96	28.0	238	17.2
78	38.9	90	35.0	115	28.0	263	17.0
92	38.8	102	33.2	100	28.0	295	16.7
78	38.2	102	32.9	100	28.0	236	13.2
90	42.2	130	32.3	145	27.7		
92	40.9	95	32.2	120	25.6		

FIGURE 10.2.1

Loess and Kernel Smoothers of HP versus MPG Data

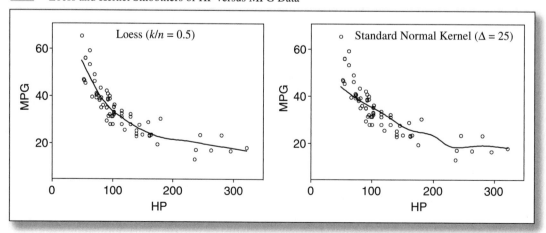

See Hart (1997) for a discussion of smoothing techniques. The kernel estimate of the regression function was developed by Nadaraya (1964) and Watson (1964). See Cleveland and Devlin (1988) for a discussion of the loess method and its applications.

10.2.3 Computer Analysis

S-Plus has a variety of two-dimensional and three-dimensional curve fitting and plotting options accessible from its graphing menu, including the loess and kernel methods.

10.3
Robust and Rank-Based Regression

Suppose we have the multiple regression model

$$Y_i = \beta_0 + \beta_1 X_{1i} + \beta_2 X_{2i} + \cdots + \beta_k X_{ki} + \varepsilon_i, \quad i = 1, 2, \ldots, n$$

as discussed in Section 8.7. Let $\hat{Y}_i = \hat{\beta}_0 + \hat{\beta}_1 X_{1i} + \hat{\beta}_2 X_{2i} + \cdots + \hat{\beta}_k X_{ki}$ denote the predicted value of Y_i. In least squares analysis we measure the distance between Y_i and $\hat{Y}_i$ by $(Y_i - \hat{Y}_i)^2$, and we estimate the β_i's by choosing those values that minimize $\Sigma_i (Y_i - \hat{Y}_i)^2$. Though optimal when the errors are independent and identically distributed normal random variables, least squares estimates can have some undesirable properties when outliers are present or when the error distributions have heavy tails. In particular, a few observations that may appear not to follow the pattern of the others can sometimes have a large influence on the values of the estimated coefficients. In robust regression we use measures that place less weight on extreme observations than the squared distance.

10.3.1 *M*-Estimation

Perhaps the most obvious modification of the method of least squares is to use absolute deviation to measure distance—that is, to choose β_i's to minimize $\Sigma_i |Y_i - \hat{Y}_i|$. A more general procedure called *M*-estimation is carried out by choosing the β_i's to minimize functions of the form

$$\sum_{i=1}^n \rho\left(\frac{Y_i - \hat{Y}_i}{\hat{\sigma}_i}\right)$$

where $\rho(x)$ is a symmetric function with a unique minimum at $x = 0$, and $\hat{\sigma}_i$ is an estimate of the standard deviation of the ε_i's. One such function among several implemented in S-Plus is the Tukey bisquare function

$$\rho(x) = \left(\frac{x}{c}\right)^6 - 3\left(\frac{x}{c}\right)^4 + 3\left(\frac{x}{c}\right)^2, \quad |x| \le c$$

$$= 1, \quad |x| > c$$

S-Plus also implements what is termed an optimal weight function proposed by Yohai and Zamar (1998). Early work on *M*-estimators was carried out by Huber (1964) and developed extensively since then. See Rousseeuw and Leroy (1987) for a discussion of this and other methods of robust regression.

EXAMPLE 10.3.1   Figure 10.3.1 is an edited S-Plus printout of the analysis of the collar size, shoe size, and weight data in Table 8.7.1 using *M*-estimation with the optimal weight function of Yohai and Zamar (1998). Ordinary least squares analysis is shown for comparison. Figure 10.3.2 shows a plot of the least squares estimates (*x* axis) versus the *M*-estimates (*y* axis) of weights, along with a plot of the line $y = x$. We see that the least squares estimates are higher for the higher weights.

FIGURE 10.3.1

M-Estimates and Least Squares Estimates of Coefficients of Collar Size and Shoe Size for Predicting Weights in Table 8.7.1

```
                *** Robust MM Linear Regression ***
Final M-estimates.

Residuals:
    Min     1Q Median    3Q    Max
 -12.91 -3.162 0.1801 5.794 36.67

Coefficients:
             Value Std. Error   t value  Pr(>|t|)
(Intercept) -123.2133    50.2618   -2.4514    0.0231
        cs   13.2502     3.3834    3.9162    0.0008
        ss    7.9756     1.6895    4.7208    0.0001

Residual scale estimate: 7.941 on 21 degrees of freedom
Proportion of variation in response explained by model: 0.504

                *** Linear Model ***
Residuals:
   Min     1Q Median    3Q    Max
 -16.4 -5.548 -1.533 3.776 35.73

Coefficients:
             Value Std. Error   t value  Pr(>|t|)
(Intercept) -156.5722    67.8862   -2.3064    0.0314
        cs   15.0738     4.8622    3.1002    0.0054
        ss    8.7928     2.2020    3.9932    0.0007

Residual standard error: 11.6 on 21 degrees of freedom
Multiple R-Squared: 0.6862
F-statistic: 22.96 on 2 and 21 degrees of freedom, the p-value is 5.183e-006
```

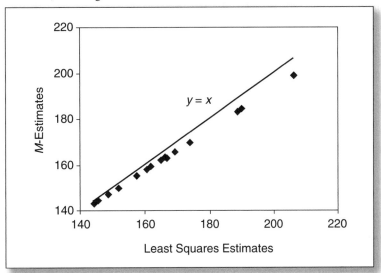

Least Squares versus *M*-Estimates of Weights for Collar Size,
Shoe Size, and Weight Data in Table 8.7.1

10.3.2 Rank-Based Regression

Following along the lines of Jaeckel (1972), Hettmansperger and McKean (1998)
extensively developed rank-based robust methods that extend in a natural way the
various rank-based statistics that are used for comparing two groups.

We first consider a special case. Let $(u_1, u_2, \ldots, u_k)$ be a $1 \times k$ vector. We define
a measure of the dispersion of u_i's as

$$D(\boldsymbol{u}) = \sum_{i=1}^{k} \sum_{j=1}^{k} |u_i - u_k|$$

If we assume for the moment that $u_1 < u_2 < \cdots < u_k$, we see

$$D(\boldsymbol{u}) = 2 \sum_{i=1}^{k} \sum_{j=i+1}^{k} \left(u_j - u_i \right)$$

$$= 2 \sum_{i=1}^{k} \left[-k + (2i - 1) \right] u_i$$

$$= 4 \sum_{i=1}^{k} \left(i - \frac{k+1}{2} \right) u_i$$

In general,

$$D(u) = 4 \sum_{i=1}^{k} \left(R(i) - \frac{k+1}{2} \right) u_i$$

where $R(i)$ is the rank of u_i. Thus, this measure of dispersion is a sum of the observations weighted by rank scores.

Let $e_i = Y_i - \hat{Y}_i$ denote the observed errors, and let $e = (e_1, e_2, \ldots, e_k)$. The Jaeckel–Hettmansperger–McKean (JHM) estimates of the coefficients $\beta_1, \beta_2, \ldots, \beta_k$ are those values that minimize $D(e)$. Note that the intercept cannot be estimated this way because the difference $e_i - e_j$ does not depend on β_0. It may be estimated by the median of the terms of the form $Y - \hat{\beta}_1 X_1 + \hat{\beta}_2 X_2 + \cdots + \hat{\beta}_k X_k$, or if the errors can be assumed to have a symmetric distribution, it may be estimated by the Hodges–Lehmann estimate, which is the median of pairs of the form $(e_i + e_j)/2$ (see Exercise 14 of Chapter 4).

To obtain the asymptotic standard errors of the estimates, we first define a quantity τ by

$$\tau = \left[\sqrt{12} \int_{-\infty}^{\infty} f^2(x) dx \right]^{-1}$$

where $f(x)$ is the probability density function of the ε's. This quantity may be estimated from the data. In the case of symmetric error distributions, in which the Hodges–Lehmann estimate is used to estimate β_0, the estimates $\hat{\beta}_i$ from rank-based regression have, under mild conditions, an asymptotically multivariate normal distribution with mean vector $(\beta_0, \beta_1, \ldots, \beta_k)^T$ and variance-covariance matrix given by

$$\frac{\tau^2}{\sigma^2} \text{cov}\left(\hat{\beta}_{LS} \right)$$

where $\text{cov}(\hat{\beta}_{LS})$ is the covariance of the least squares estimates of the coefficients and σ^2 is the variance of the ε's. In particular, the asymptotic standard errors of the rank-based estimates differ from the standard errors of the least squares estimates by a factor of τ/σ. If the median of the residuals is used to estimate β_0, then the standard error of $\hat{\beta}_0$ must be modified, but the variance-covariance matrix of the $\hat{\beta}_i$, $i = 1, 2, \ldots, k$, remains the same.

EXAMPLE 10.3.2   Table 10.3.1 lists the JHM estimates for the analysis of the collar size, shoe size, and weight data in Table 8.7.1. The median of the residuals is used as the estimate of the intercept. These values were obtained from a web program developed by McKean (2002). Figure 10.3.3 shows a plot of the least squares estimates versus the rank-based estimates of weights and a plot of the line $y = x$. The rank-based estimates in this case are close to the least squares estimates. Rank-based estimates are also available in MINITAB. Figure 10.3.4 shows the syntax and output of the MINITAB command RREG for rank-based regression. The default option uses the Hodges–Lehmann estimate of the intercept.

TABLE 10.3.1

JHM Estimates of Regression Coefficients for Collar Size, Shoe Size, and Weight Data in Table 8.7.1

df error = 21	Estimate	Standard Error	t-Value	p-Value
Intercept	−135.94	55.04	−2.47	.0222
CS	13.36	3.94	3.39	.0028
SS	9.16	1.79	5.13	<.0001

FIGURE 10.3.3

Least Squares versus JHM Estimates of Weights for Collar Size, Shoe Size, and Weight Data in Table 8.7.1

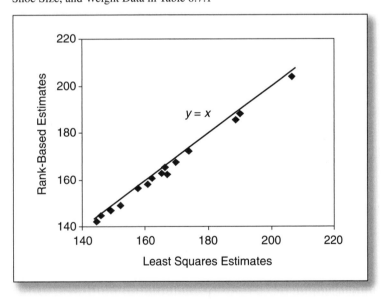

Jaeckel (1972) and Hettmansperger and McKean (1998) consider general measures of dispersion of the form $\Sigma_{i=1}^{k} a(R(u_i))u_i$, where $R(u_i)$ is the rank of u_i and $a(1) \leq a(2) \leq \cdots \leq a(k)$ is a nondecreasing set of weights such that $\Sigma_{i=1}^{k} a(i) = 0$. This allows, for instance, the use of van der Waerden scores weights, scaled to sum to zero, for doing rank-based regression. Their methodology applies to all the usual regression and design models that are typically considered under the heading of the

FIGURE 10.3.4

MINITAB Syntax and Output for Rank-Based Regression for Collar Size, Shoe Size, and Weight Data in Table 8.7.1

```
MTB > RREG C3 2 C1 C2
(Note C3 is weight, C1 is collar size, and C2 is shoe size)
Rank Regression: C3 versus C1, C2

The regression equation is
C3 = - 138 + 13.6 C1 + 9.10 C2

                    Coefficient                 SE Coef
Predictor        Rank    Least-sq         Rank    Least-sq
Constant       -138.32    -156.57        56.66      67.89
C1               13.594     15.074        4.058       4.862
C2                9.102      8.793        1.838       2.202

Hodges-Lehmann estimate of tau = 9.679    Least-squares S = 11.60

Unusual observations
Observation      C1        C3       Pseudo       Fit      SE Fit   Residual
          6     16.5    160.00      157.05    163.34       6.41     -3.34 X

X denotes an observation whose X value gives it large influence.
```

general linear model and to censored data models. They have developed large-sample theory for various statistics of interest and report simulation studies that show good small-sample approximations of *p*-values over a range of conditions.

10.3.3   Computer Analysis

S-Plus implements *M*-estimation, which may be accessed from its graph menu. Rank-based regression is available at the web site developed by McKean (2002). In addition to regression analysis, rank-based analysis of variance and analysis of covariance methods are included in the procedures available at this site. Rank-based regression is also available in MINITAB using the RREG command, as shown in Figure 10.3.4.

Exercises

1   These data are the average March temperatures (Fahrenheit) for Kansas City from 1961 to 1990. Estimate the probability density function using the histogram and the kernel density estimate.

43.8 40.1 49.2 41.8 34.0 49.1 47.8 48.1 37.6 42.0 43.7 47.1 47.7 46.9 36.5
45.0 48.0 37.6 42.2 38.7 45.2 42.5 43.1 36.0 47.4 48.5 47.1 43.2 43.8 45.7

2   The data are the heights (in inches) of 35 college-age women. Estimate the probability density function using the kernel density estimate.

70 69 65 64 66 65 64 66 66 60 70 66 60 68 65 65 67
64 63 64 65 66 65 64 62 66 66 64 68 62 62 67 69 64

3   The data in the table are the miles per gallon (MPG) and top speed (SP) of 82 automobiles reported by Heavenrich, Murrell, and Hellman (1991). Use the loess method and the kernel method to obtain a plot of MPG (x) versus SP (y).

MPG	SP	MPG	SP	MPG	SP	MPG	SP
65.4	96	40.7	101	32.2	106	25.3	114
56.0	97	40.0	111	32.2	109	23.9	114
55.9	97	39.3	105	32.2	106	23.6	117
49.0	105	38.8	111	31.5	105	23.6	122
46.5	96	38.4	110	31.5	108	23.6	122
46.2	105	38.4	110	31.4	108	23.6	122
45.4	97	38.4	110	31.4	107	23.6	122
59.2	98	29.5	109	31.2	120	23.5	148
53.3	98	46.9	90	33.7	109	23.4	160
43.4	107	36.3	112	32.6	109	23.4	121
41.1	103	36.1	103	31.3	109	23.1	121
40.9	113	36.1	103	31.3	109	22.9	110
40.9	113	35.4	111	30.4	133	22.9	110
40.4	103	35.3	111	28.9	125	19.5	121
39.6	100	35.1	102	28.0	115	18.1	165
39.3	103	35.1	106	28.0	102	17.2	140
38.9	106	35.0	106	28.0	109	17.0	147
38.8	113	33.2	109	28.0	104	16.7	157
38.2	106	32.9	109	28.0	105	13.2	130
42.2	109	32.3	120	27.7	120		
40.9	110	32.2	106	25.6	107		

4   For the daughter's height (DH), mother's height (MH), and father's height (FH) data in Exercise 11 of Chapter 8, use the optimal M-estimation method in S-Plus and the JHM method of rank-based regression to obtain estimates of the coefficients for the model DH = β_0 + β_1(MH) + β_2(FH). Compare with the ordinary least squares estimates.

5   Fit the quadratic model SP = β_0 + β_1(MPG) + β_2(MPG)^2 to the data in Exercise 3 using the optimal M-estimation method in S-Plus and the JHM method of rank-based regression.

Theory and Complements

6   **a**   Discuss how the JHM method may be applied to analysis of variance problems.

   b   Apply the JHM procedure for rank-based two-way analysis of variance to the data in Exercise 1 of Chapter 9. Compare the results with the standard two-way analysis of variance. See McKean's web site at http://www.stat.wmich.edu/slab/RGLM for a program for doing rank-based two-way analysis of variance.

Appendix

Binomial Probabilities

		p								
n	k	.10	.15	.20	.25	.30	.35	.40	.45	.50
5	0	.5905	.4437	.3277	.2373	.1681	.1160	.0778	.0503	.0313
	1	.3281	.3915	.4096	.3955	.3602	.3124	.2592	.2059	.1563
	2	.0729	.1382	.2048	.2637	.3087	.3364	.3456	.3369	.3125
	3	.0081	.0244	.0512	.0879	.1323	.1811	.2304	.2757	.3125
	4	.0005	.0022	.0064	.0146	.0284	.0488	.0768	.1128	.1563
	5	.0000	.0001	.0003	.0010	.0024	.0053	.0102	.0185	.0313
10	0	.3487	.1969	.1074	.0563	.0282	.0135	.0060	.0025	.0010
	1	.3874	.3474	.2684	.1877	.1211	.0725	.0403	.0207	.0098
	2	.1937	.2759	.3020	.2816	.2335	.1757	.1209	.0763	.0439
	3	.0574	.1298	.2013	.2503	.2668	.2522	.2150	.1665	.1172
	4	.0112	.0401	.0881	.1460	.2001	.2377	.2508	.2384	.2051
	5	.0015	.0085	.0264	.0584	.1029	.1536	.2007	.2340	.2461
	6	.0001	.0012	.0055	.0162	.0368	.0689	.1115	.1596	.2051
	7	.0000	.0001	.0008	.0031	.0090	.0212	.0425	.0746	.1172
	8	.0000	.0000	.0001	.0004	.0014	.0043	.0106	.0229	.0439
	9	.0000	.0000	.0000	.0000	.0001	.0005	.0016	.0042	.0098
	10	.0000	.0000	.0000	.0000	.0000	.0000	.0001	.0003	.0010
15	0	.2059	.0874	.0352	.0134	.0047	.0016	.0005	.0001	.0000
	1	.3432	.2312	.1319	.0668	.0305	.0126	.0047	.0016	.0005
	2	.2669	.2856	.2309	.1559	.0916	.0476	.0219	.0090	.0032
	3	.1285	.2184	.2501	.2252	.1700	.1110	.0634	.0318	.0139
	4	.0428	.1156	.1876	.2252	.2186	.1792	.1268	.0780	.0417
	5	.0105	.0449	.1032	.1651	.2061	.2123	.1859	.1404	.0916

(continued on next page)

TABLE A1

Binomial Probabilities *(continued)*

						p				
n	*k*	.10	.15	.20	.25	.30	.35	.40	.45	.50
15	6	.0019	.0132	.0430	.0917	.1472	.1906	.2066	.1914	.1527
	7	.0003	.0030	.0138	.0393	.0811	.1319	.1771	.2013	.1964
	8	.0000	.0005	.0035	.0131	.0348	.0710	.1181	.1647	.1964
	9	.0000	.0001	.0007	.0034	.0116	.0298	.0612	.1048	.1527
	10	.0000	.0000	.0001	.0007	.0030	.0096	.0245	.0515	.0916
	11	.0000	.0000	.0000	.0001	.0006	.0024	.0074	.0191	.0417
	12	.0000	.0000	.0000	.0000	.0001	.0004	.0016	.0052	.0139
	13	.0000	.0000	.0000	.0000	.0000	.0001	.0003	.0010	.0032
	14	.0000	.0000	.0000	.0000	.0000	.0000	.0000	.0001	.0005
	15	.0000	.0000	.0000	.0000	.0000	.0000	.0000	.0000	.0000
20	0	.1216	.0388	.0115	.0032	.0008	.0002	.0000	.0000	.0000
	1	.2702	.1368	.0576	.0211	.0068	.0020	.0005	.0001	.0000
	2	.2852	.2293	.1369	.0669	.0278	.0100	.0031	.0008	.0002
	3	.1901	.2428	.2054	.1339	.0716	.0323	.0123	.0040	.0011
	4	.0898	.1821	.2182	.1897	.1304	.0738	.0350	.0139	.0046
	5	.0319	.1028	.1746	.2023	.1789	.1272	.0746	.0365	.0148
	6	.0089	.0454	.1091	.1686	.1916	.1712	.1244	.0746	.0370
	7	.0020	.0160	.0545	.1124	.1643	.1844	.1659	.1221	.0739
	8	.0004	.0046	.0222	.0609	.1144	.1614	.1797	.1623	.1201
	9	.0001	.0011	.0074	.0271	.0654	.1158	.1597	.1771	.1602
	10	.0000	.0002	.0020	.0099	.0308	.0686	.1171	.1593	.1762
	11	.0000	.0000	.0005	.0030	.0120	.0336	.0710	.1185	.1602
	12	.0000	.0000	.0001	.0008	.0039	.0136	.0355	.0727	.1201
	13	.0000	.0000	.0000	.0002	.0010	.0045	.0146	.0366	.0739
	14	.0000	.0000	.0000	.0000	.0002	.0012	.0049	.0150	.0370
	15	.0000	.0000	.0000	.0000	.0000	.0003	.0013	.0049	.0148
	16	.0000	.0000	.0000	.0000	.0000	.0000	.0003	.0013	.0046
	17	.0000	.0000	.0000	.0000	.0000	.0000	.0000	.0002	.0011
	18	.0000	.0000	.0000	.0000	.0000	.0000	.0000	.0000	.0002
	19	.0000	.0000	.0000	.0000	.0000	.0000	.0000	.0000	.0000
	20	.0000	.0000	.0000	.0000	.0000	.0000	.0000	.0000	.0000

TABLE A2
Standard Normal Cumulative Probabilities

z	0.00	0.01	0.02	0.03	0.04	0.05	0.06	0.07	0.08	0.09
0.0	.5000	.5040	.5080	.5120	.5160	.5199	.5239	.5279	.5319	.5359
0.1	.5398	.5438	.5478	.5517	.5557	.5596	.5636	.5675	.5714	.5753
0.2	.5793	.5832	.5871	.5910	.5948	.5987	.6026	.6064	.6103	.6141
0.3	.6179	.6217	.6255	.6293	.6331	.6368	.6406	.6443	.6480	.6517
0.4	.6554	.6591	.6628	.6664	.6700	.6736	.6772	.6808	.6844	.6879
0.5	.6915	.6950	.6985	.7019	.7054	.7088	.7123	.7157	.7190	.7224
0.6	.7257	.7291	.7324	.7357	.7389	.7422	.7454	.7486	.7517	.7549
0.7	.7580	.7611	.7642	.7673	.7704	.7734	.7764	.7794	.7823	.7852
0.8	.7881	.7910	.7939	.7967	.7995	.8023	.8051	.8078	.8106	.8133
0.9	.8159	.8186	.8212	.8238	.8264	.8289	.8315	.8340	.8365	.8389
1.0	.8413	.8438	.8461	.8485	.8508	.8531	.8554	.8577	.8599	.8621
1.1	.8643	.8665	.8686	.8708	.8729	.8749	.8770	.8790	.8810	.8830
1.2	.8849	.8869	.8888	.8907	.8925	.8944	.8962	.8980	.8997	.9015
1.3	.9032	.9049	.9066	.9082	.9099	.9115	.9131	.9147	.9162	.9177
1.4	.9192	.9207	.9222	.9236	.9251	.9265	.9279	.9292	.9306	.9319
1.5	.9332	.9345	.9357	.9370	.9382	.9394	.9406	.9418	.9429	.9441
1.6	.9452	.9463	.9474	.9484	.9495	.9505	.9515	.9525	.9535	.9545
1.7	.9554	.9564	.9573	.9582	.9591	.9599	.9608	.9616	.9625	.9633
1.8	.9641	.9649	.9656	.9664	.9671	.9678	.9686	.9693	.9699	.9706
1.9	.9713	.9719	.9726	.9732	.9738	.9744	.9750	.9756	.9761	.9767
2.0	.9772	.9778	.9783	.9788	.9793	.9798	.9803	.9808	.9812	.9817
2.1	.9821	.9826	.9830	.9834	.9838	.9842	.9846	.9850	.9854	.9857
2.2	.9861	.9864	.9868	.9871	.9875	.9878	.9881	.9884	.9887	.9890
2.3	.9893	.9896	.9898	.9901	.9904	.9906	.9909	.9911	.9913	.9916
2.4	.9918	.9920	.9922	.9925	.9927	.9929	.9931	.9932	.9934	.9936
2.5	.9938	.9940	.9941	.9943	.9945	.9946	.9948	.9949	.9951	.9952
2.6	.9953	.9955	.9956	.9957	.9959	.9960	.9961	.9962	.9963	.9964
2.7	.9965	.9966	.9967	.9968	.9969	.9970	.9971	.9972	.9973	.9974
2.8	.9974	.9975	.9976	.9977	.9977	.9978	.9979	.9979	.9980	.9981
2.9	.9981	.9982	.9982	.9983	.9984	.9984	.9985	.9985	.9986	.9986
3.0	.9987	.9987	.9987	.9988	.9988	.9989	.9989	.9989	.9990	.9990

z	.842	1.036	1.282	1.645	1.960	2.326
prob $\leq z$	.800	.850	.900	.950	.975	.990

TABLE A3
Critical Values for Wilcoxon Rank-Sum Statistic: Sum Is Taken for Treatment with n Observations

5%

$n \rightarrow$	4		5		6		7		8		9		10	
$m \downarrow$	Lower	Upper	Lower	Upper	Lower	Upper	Lower	Upper	Lower	Upper	Lower	Upper	Lower	Upper
4	11	25	17	33	24	42	32	52	41	63	51	75	62	88
5	12	28	19	36	26	46	34	57	44	68	54	81	66	94
6	13	31	20	40	28	50	36	62	46	74	57	87	69	101
7	14	34	21	44	29	55	39	66	49	79	60	93	72	108
8	15	37	23	47	31	59	41	71	51	85	63	99	75	115
9	16	40	24	51	33	63	43	76	54	90	66	105	79	121
10	17	43	26	54	35	67	45	81	56	96	69	111	82	128

2.5%

$n \rightarrow$	4		5		6		7		8		9		10	
$m \downarrow$	Lower	Upper	Lower	Upper	Lower	Upper	Lower	Upper	Lower	Upper	Lower	Upper	Lower	Upper
4	10	26	16	34	23	43	31	53	40	64	49	77	60	90
5	11	29	17	38	24	48	33	58	42	70	52	83	63	97
6	12	32	18	42	26	52	34	64	44	76	55	89	66	104
7	13	35	20	45	27	57	36	69	46	82	57	96	69	111
8	14	38	21	49	29	61	38	74	49	87	60	102	72	118
9	14	42	22	53	31	65	40	79	51	93	62	109	75	125
10	15	45	23	57	32	70	42	84	53	99	65	115	78	132

1%

$n \rightarrow$	4		5		6		7		8		9		10	
$m \downarrow$	Lower	Upper	Lower	Upper	Lower	Upper	Lower	Upper	Lower	Upper	Lower	Upper	Lower	Upper
4			15	35	22	44	29	55	38	66	48	78	58	92
5	10	30	16	39	23	49	31	60	40	72	50	85	61	99
6	11	33	17	43	24	54	32	66	42	78	52	92	63	107
7	11	37	18	47	25	59	34	71	43	85	54	99	66	114
8	12	40	19	51	27	63	35	77	45	91	56	106	68	122
9	13	43	20	55	28	68	37	82	47	97	59	112	71	129
10	13	47	21	59	29	73	39	87	49	103	61	119	74	136

TABLE A4
Lower and Upper Critical Values for Mann–Whitney Statistic

5%

$n \rightarrow$	4		5		6		7		8		9		10	
m	Lower	Upper	Lower	Upper	Lower	Upper	Lower	Upper	Lower	Upper	Lower	Upper	Lower	Upper
↓ 4	1	15	2	18	3	21	4	24	5	27	6	30	7	33
5	2	18	4	21	5	25	6	29	8	32	9	36	11	39
6	3	21	5	25	7	29	8	34	10	38	12	42	14	46
7	4	24	6	29	8	34	11	38	13	43	15	48	17	53
8	5	27	8	32	10	38	13	43	15	49	18	54	20	60
9	6	30	9	36	12	42	15	48	18	54	21	60	24	66
10	7	33	11	39	14	46	17	53	20	60	24	66	27	73

2.5%

$n \rightarrow$	4		5		6		7		8		9		10	
m	Lower	Upper	Lower	Upper	Lower	Upper	Lower	Upper	Lower	Upper	Lower	Upper	Lower	Upper
↓ 4	0	16	1	19	2	22	3	25	4	28	4	32	5	35
5	1	19	2	23	3	27	5	30	6	34	7	38	8	42
6	2	22	3	27	5	31	6	36	8	40	10	44	11	49
7	3	25	5	30	6	36	8	41	10	46	12	51	14	56
8	4	28	6	34	8	40	10	46	13	51	15	57	17	63
9	4	32	7	38	10	44	12	51	15	57	17	64	20	70
10	5	35	8	42	11	49	14	56	17	63	20	70	23	77

1%

$n \rightarrow$	4		5		6		7		8		9		10	
m	Lower	Upper	Lower	Upper	Lower	Upper	Lower	Upper	Lower	Upper	Lower	Upper	Lower	Upper
↓ 4			0	20	1	23	1	27	2	30	3	33	3	37
5	0	20	1	24	2	28	3	32	4	36	5	40	6	44
6	1	23	2	28	3	33	4	38	6	42	7	47	8	52
7	1	27	3	32	4	38	6	43	7	49	9	54	11	59
8	2	30	4	36	6	42	7	49	9	55	11	61	13	67
9	3	33	5	40	7	47	9	54	11	61	14	67	16	74
10	3	37	6	44	8	52	11	59	13	67	16	74	19	81

TABLE A5
Van der Waerden Scores

N = 10	Scores
1	−1.335
2	−0.908
3	−0.605
4	−0.349
5	−0.114
6	0.114
7	0.349
8	0.605
9	0.908
10	1.335

N = 11	Scores
1	−1.383
2	−0.967
3	−0.674
4	−0.431
5	−0.210
6	0.000
7	0.210
8	0.431
9	0.674
10	0.967
11	1.383

N = 12	Scores
1	−1.426
2	−1.020
3	−0.736
4	−0.502
5	−0.293
6	−0.097
7	0.097
8	0.293
9	0.502
10	0.736
11	1.020
12	1.426

N = 13	Scores
1	−1.465
2	−1.068
3	−0.792
4	−0.566
5	−0.366
6	−0.180
7	0.000
8	0.180
9	0.366
10	0.566
11	0.792
12	1.068
13	1.465

N = 14	Scores
1	−1.501
2	−1.111
3	−0.842
4	−0.623
5	−0.431
6	−0.253
7	−0.084
8	0.084
9	0.253
10	0.431
11	0.623
12	0.842
13	1.111
14	1.501

N = 15	Scores
1	−1.534
2	−1.150
3	−0.887
4	−0.674
5	−0.489
6	−0.319
7	−0.157
8	0.000
9	0.157
10	0.319
11	0.489
12	0.674
13	0.887
14	1.150
15	1.534

N = 16	Scores
1	−1.565
2	−1.187
3	−0.929
4	−0.722
5	−0.541
6	−0.377
7	−0.223
8	−0.074
9	0.074
10	0.223
11	0.377
12	0.541
13	0.722
14	0.929
15	1.187
16	1.565

N = 17	Scores
1	−1.593
2	−1.221
3	−0.967
4	−0.765
5	−0.589
6	−0.431
7	−0.282
8	−0.140
9	0.000
10	0.140
11	0.282
12	0.431
13	0.589
14	0.765
15	0.967
16	1.221
17	1.593

N = 18	Scores
1	−1.620
2	−1.252
3	−1.003
4	−0.805
5	−0.634
6	−0.480
7	−0.336
8	−0.199
9	−0.066
10	0.066
11	0.199
12	0.336
13	0.480
14	0.634
15	0.805
16	1.003
17	1.252
18	1.620

N = 19	Scores
1	−1.645
2	−1.282
3	−1.036
4	−0.842
5	−0.674
6	−0.524
7	−0.385
8	−0.253
9	−0.126
10	0.000
11	0.126
12	0.253
13	0.385
14	0.524
15	0.674
16	0.842
17	1.036
18	1.282
19	1.645

N = 20	Scores
1	−1.668
2	−1.309
3	−1.068
4	−0.876
5	−0.712
6	−0.566
7	−0.431
8	−0.303
9	−0.180
10	−0.060
11	0.060
12	0.180
13	0.303
14	0.431
15	0.566
16	0.712
17	0.876
18	1.068
19	1.309
20	1.668

TABLE A6
Critical Values of Kruskal–Wallis Statistic

		Three Treatments		
n_1, n_2, n_3	*10%*	*5%*	*2.5%*	*1%*
2, 2, 2	4.57	*	*	*
3, 3, 3	4.62	5.60	5.96	7.20
3, 3, 2	4.56	5.36	5.56	*
3, 2, 2	4.50	4.71	*	*
4, 4, 4	4.65	5.69	6.62	7.65
4, 4, 3	4.55	5.60	6.39	7.14
4, 4, 2	4.55	5.45	6.33	7.04
4, 3, 3	4.71	5.79	6.15	6.75
4, 3, 2	4.51	5.44	6.00	6.55
4, 2, 2	4.46	5.33	5.50	*
5, 5, 5	4.56	5.78	6.74	8.00
5, 5, 4	4.52	5.64	6.76	7.90
5, 5, 3	4.55	5.71	6.55	7.58
5, 5, 2	4.62	5.34	6.35	7.34
5, 4, 4	4.67	5.66	6.67	7.76
5, 4, 3	4.59	5.66	6.41	7.44
5, 4, 2	4.54	5.27	6.07	7.20
5, 3, 3	4.53	5.65	6.32	7.08
5, 3, 2	4.65	5.25	6.00	6.91
5, 2, 2	4.37	5.16	6.00	6.53

		Four Treatments		
n_1, n_2, n_3, n_4	*10%*	*5%*	*2.5%*	*1%*
2, 2, 2, 2	5.67	6.17	6.67	6.67
3, 3, 3, 3	6.03	7.00	7.67	8.54
3, 3, 3, 2	5.88	6.73	7.52	8.02
3, 3, 2, 2	5.75	6.53	7.05	7.64
3, 2, 2, 2	5.64	6.33	6.98	7.13

TABLE A7
Upper Critical Values of the Chi-Square Distribution

df	Upper-Tail Probability				df	Upper-Tail Probability			
	.1	.05	.025	.01		.1	.05	.025	.01
1	2.71	3.84	5.02	6.63	31	41.4	45.0	48.2	52.2
2	4.61	5.99	7.38	9.21	32	42.6	46.2	49.5	53.5
3	6.25	7.81	9.35	11.3	33	43.7	47.4	50.7	54.8
4	7.78	9.49	11.1	13.3	34	44.9	48.6	52.0	56.1
5	9.24	11.1	12.8	15.1	35	46.1	49.8	53.2	57.3
6	10.6	12.6	14.4	16.8	36	47.2	51.0	54.4	58.6
7	12.0	14.1	16.0	18.5	37	48.4	52.2	55.7	59.9
8	13.4	15.5	17.5	20.1	38	49.5	53.4	56.9	61.2
9	14.7	16.9	19.0	21.7	39	50.7	54.6	58.1	62.4
10	16.0	18.3	20.5	23.2	40	51.8	55.8	59.3	63.7
11	17.3	19.7	21.9	24.7	41	52.9	56.9	60.6	64.9
12	18.5	21.0	23.3	26.2	42	54.1	58.1	61.8	66.2
13	19.8	22.4	24.7	27.7	43	55.2	59.3	63.0	67.5
14	21.1	23.7	26.1	29.1	44	56.4	60.5	64.2	68.7
15	22.3	25.0	27.5	30.6	45	57.5	61.7	65.4	70.0
16	23.5	26.3	28.8	32.0	46	58.6	62.8	66.6	71.2
17	24.8	27.6	30.2	33.4	47	59.8	64.0	67.8	72.4
18	26.0	28.9	31.5	34.8	48	60.9	65.2	69.0	73.7
19	27.2	30.1	32.9	36.2	49	62.0	66.3	70.2	74.9
20	28.4	31.4	34.2	37.6	50	63.2	67.5	71.4	76.2
21	29.6	32.7	35.5	38.9	51	64.3	68.7	72.6	77.4
22	30.8	33.9	36.8	40.3	52	65.4	69.8	73.8	78.6
23	32.0	35.2	38.1	41.6	53	66.5	71.0	75.0	79.8
24	33.2	36.4	39.4	43.0	54	67.7	72.2	76.2	81.1
25	34.4	37.7	40.6	44.3	55	68.8	73.3	77.4	82.3
26	35.6	38.9	41.9	45.6	56	69.9	74.5	78.6	83.5
27	36.7	40.1	43.2	47.0	57	71.0	75.6	79.8	84.7
28	37.9	41.3	44.5	48.3	58	72.2	76.8	80.9	86.0
29	39.1	42.6	45.7	49.6	59	73.3	77.9	82.1	87.2
30	40.3	43.8	47.0	50.9	60	74.4	79.1	83.3	88.4

TABLE A8

Tukey's HSD 5% Critical Values, $q(.05, k, df)$

$k = number\ of\ treatments$

df	2	3	4	5	6	7	8	9	10
2	6.08	8.33	9.80	10.88	11.73	12.43	13.03	13.54	13.99
3	4.50	5.91	6.82	7.50	8.04	8.48	8.85	9.18	9.46
4	3.93	5.04	5.76	6.29	6.71	7.05	7.35	7.60	7.83
5	3.64	4.60	5.22	5.67	6.03	6.33	6.58	6.80	6.99
6	3.46	4.34	4.90	5.30	5.63	5.90	6.12	6.32	6.49
7	3.34	4.16	4.68	5.06	5.36	5.61	5.82	6.00	6.16
8	3.26	4.04	4.53	4.89	5.17	5.40	5.60	5.77	5.92
9	3.20	3.95	4.41	4.76	5.02	5.24	5.43	5.59	5.74
10	3.15	3.88	4.33	4.65	4.91	5.12	5.30	5.46	5.60
11	3.11	3.82	4.26	4.57	4.82	5.03	5.20	5.35	5.49
12	3.08	3.77	4.20	4.51	4.75	4.95	5.12	5.26	5.39
13	3.06	3.73	4.15	4.45	4.69	4.88	5.05	5.19	5.32
14	3.03	3.70	4.11	4.41	4.64	4.83	4.99	5.13	5.25
15	3.01	3.67	4.08	4.37	4.59	4.78	4.94	5.08	5.20
16	3.00	3.65	4.05	4.33	4.56	4.74	4.90	5.03	5.15
17	2.98	3.63	4.02	4.30	4.52	4.70	4.86	4.99	5.11
18	2.97	3.61	4.00	4.28	4.49	4.67	4.82	4.96	5.07
19	2.96	3.59	3.98	4.25	4.47	4.65	4.79	4.92	5.04
20	2.95	3.58	3.96	4.23	4.45	4.62	4.77	4.90	5.01
21	2.94	3.56	3.94	4.21	4.42	4.60	4.74	4.87	4.98
22	2.93	3.55	3.93	4.20	4.41	4.58	4.72	4.85	4.96
23	2.93	3.54	3.91	4.18	4.39	4.56	4.70	4.83	4.94
24	2.92	3.53	3.90	4.17	4.37	4.54	4.68	4.81	4.92
25	2.91	3.52	3.89	4.15	4.36	4.53	4.67	4.79	4.90
26	2.91	3.51	3.88	4.14	4.35	4.51	4.65	4.77	4.88
27	2.90	3.51	3.87	4.13	4.33	4.50	4.64	4.76	4.86
28	2.90	3.50	3.86	4.12	4.32	4.49	4.62	4.74	4.85
29	2.89	3.49	3.85	4.11	4.31	4.47	4.61	4.73	4.84
30	2.89	3.49	3.85	4.10	4.30	4.46	4.60	4.72	4.82
40	2.86	3.44	3.79	4.04	4.23	4.39	4.52	4.63	4.73
50	2.84	3.42	3.76	4.00	4.19	4.34	4.47	4.58	4.68
60	2.83	3.40	3.74	3.98	4.16	4.31	4.44	4.55	4.65
70	2.82	3.39	3.72	3.96	4.14	4.29	4.42	4.53	4.62
80	2.81	3.38	3.71	3.95	4.13	4.28	4.40	4.51	4.60
90	2.81	3.37	3.70	3.94	4.12	4.27	4.39	4.50	4.59
100	2.81	3.36	3.70	3.93	4.11	4.26	4.38	4.48	4.58
200	2.79	3.34	3.66	3.89	4.07	4.21	4.33	4.44	4.53
∞	2.77	3.31	3.63	3.86	4.03	4.17	4.29	4.39	4.47

A critical value may be obtained in SAS® using the following commands. The illustration is for df = 10 and k = 3.

```
data;
p = probmc("RANGE",.,.95,10,3);
proc print;
```

TABLE A9
Signed-Rank Tail Probabilities, $P(SR_+ \geq c)$

c	$n = 4$	c	$n = 8$	c	$n = 10$	c	$n = 11$	c	$n = 12$
7	0.313	23	0.273	34	0.278	41	0.260	48	0.259
8	0.188	24	0.230	35	0.246	42	0.232	49	0.235
9	0.125	25	0.191	36	0.216	43	0.207	50	0.212
10	0.063	26	0.156	37	0.188	44	0.183	51	0.190
		27	0.125	38	0.161	45	0.160	52	0.170
c	$n = 5$	28	0.098	39	0.138	46	0.139	53	0.151
10	0.313	29	0.074	40	0.116	47	0.120	54	0.133
11	0.219	30	0.055	41	0.097	48	0.103	55	0.117
12	0.156	31	0.039	42	0.080	49	0.087	56	0.102
13	0.094	32	0.027	43	0.065	50	0.074	57	0.088
14	0.063	33	0.020	44	0.053	51	0.062	58	0.076
15	0.031	34	0.012	45	0.042	52	0.051	59	0.065
		35	0.008	46	0.032	53	0.042	60	0.055
c	$n = 6$	36	0.004	47	0.024	54	0.034	61	0.046
14	0.281			48	0.019	55	0.027	62	0.039
15	0.219	c	$n = 9$	49	0.014	56	0.021	63	0.032
16	0.156	28	0.285	50	0.010	57	0.016	64	0.026
17	0.109	29	0.248	51	0.007	58	0.012	65	0.021
18	0.078	30	0.213	52	0.005	59	0.009	66	0.017
19	0.047	31	0.180	53	0.003	60	0.007	67	0.013
20	0.031	32	0.150	54	0.002	61	0.005	68	0.010
21	0.016	33	0.125	55	0.001	62	0.003	69	0.008
		34	0.102			63	0.002	70	0.006
c	$n = 7$	35	0.082			64	0.001	71	0.005
18	0.289	36	0.064			65	0.001	72	0.003
19	0.234	37	0.049			66	0.000	73	0.002
20	0.188	38	0.037					74	0.002
21	0.148	39	0.027					75	0.001
22	0.109	40	0.020					76	0.001
23	0.078	41	0.014					77	0.000
24	0.055	42	0.010					78	0.000
25	0.039	43	0.006						
26	0.023	44	0.004						
27	0.016	45	0.002						
28	0.008								

Lower-tail probabilities may be obtained as $P(SR_+ \leq c) = P(SR_+ \geq n(n + 1)/2 - c)$.

TABLE A10
Upper-Tail Probabilities for Sign Test, $P(SN_+ \geq k)$

n	k	Prob	n	k	Prob	n	k	Prob	n	k	Prob	n	k	Prob	n	k	Prob
5	0	1.000	10	0	1.000	13	0	1.000	15	0	1.000	17	0	1.000	19	0	1.000
	1	.969		1	.999		1	1.000		1	1.000		1	1.000		1	1.000
	2	.813		2	.989		2	.998		2	1.000		2	1.000		2	1.000
	3	.500		3	.945		3	.989		3	.996		3	.999		3	1.000
	4	.188		4	.828		4	.954		4	.982		4	.994		4	.998
	5	.031		5	.623		5	.867		5	.941		5	.975		5	.990
				6	.377		6	.709		6	.849		6	.928		6	.968
6	0	1.000		7	.172		7	.500		7	.696		7	.834		7	.916
	1	.984		8	.055		8	.291		8	.500		8	.685		8	.820
	2	.891		9	.011		9	.133		9	.304		9	.500		9	.676
	3	.656		10	.001		10	.046		10	.151		10	.315		10	.500
	4	.344					11	.011		11	.059		11	.166		11	.324
	5	.109	11	0	1.000		12	.002		12	.018		12	.072		12	.180
	6	.016		1	1.000		13	.000		13	.004		13	.025		13	.084
				2	.994					14	.000		14	.006		14	.032
7	0	1.000		3	.967	14	0	1.000		15	.000		15	.001		15	.010
	1	.992		4	.887		1	1.000					16	.000		16	.002
	2	.938		5	.726		2	.999	16	0	1.000		17	.000		17	.000
	3	.773		6	.500		3	.994		1	1.000					18	.000
	4	.500		7	.274		4	.971		2	1.000	18	0	1.000		19	.000
	5	.227		8	.113		5	.910		3	.998		1	1.000			
	6	.062		9	.033		6	.788		4	.989		2	1.000	20	0	1.000
	7	.008		10	.006		7	.605		5	.962		3	.999		1	1.000
				11	.000		8	.395		6	.895		4	.996		2	1.000
8	0	1.000					9	.212		7	.773		5	.985		3	1.000
	1	.996	12	0	1.000		10	.090		8	.598		6	.952		4	.999
	2	.965		1	1.000		11	.029		9	.402		7	.881		5	.994
	3	.855		2	.997		12	.006		10	.227		8	.760		6	.979
	4	.637		3	.981		13	.001		11	.105		9	.593		7	.942
	5	.363		4	.927		14	.000		12	.038		10	.407		8	.868
	6	.145		5	.806					13	.011		11	.240		9	.748
	7	.035		6	.613					14	.002		12	.119		10	.588
	8	.004		7	.387					15	.000		13	.048		11	.412
				8	.194					16	.000		14	.015		12	.252
9	0	1.000		9	.073								15	.004		13	.132
	1	.998		0	.019								16	.001		14	.058
	2	.980		11	1.000								17	.000		15	.021
	3	.910		12	.000								18	.000		16	.006
	4	.746														17	.001
	5	.500														18	.000
	6	.254														19	.000
	7	.090														20	.000
	8	.020															
	9	.002															

Lower-tail probablitities may be obtained as $P(SN_+ \leq k) = P(SN_+ \geq n - k)$.

TABLE A11
Upper-Tail Probabilities for Friedman's Test, $P(\text{FM} \geq c)$

k = 3, b = 2		k = 4, b = 4		k = 5, b = 2		k = 5, b = 4		k = 5, b = 5	
c	Prob	c	Prob	c	Prob	c	Prob	c	Prob
3.00	.500	4.80	.200	6.00	.225	6.00	.205	5.92	.218
4.00	.167	5.10	.190	6.40	.175	6.20	.197	6.08	.195
		5.40	.158	6.80	.117	6.40	.178	6.24	.183
k = 3, b = 3		5.70	.141	7.20	.067	6.60	.161	6.40	.174
c	Prob	6.00	.105	7.60	.042	6.80	.143	6.56	.164
2.67	.361	6.30	.094	8.00	.008	7.00	.136	6.72	.151
4.67	.194	6.60	.077			7.20	.121	6.88	.146
6.00	.028	6.90	.068	**k = 5, b = 3**		7.40	.113	7.04	.130
		7.20	.054	c	Prob	7.60	.095	7.20	.121
k = 3, b = 4		7.50	.052	6.13	.213	7.80	.086	7.36	.112
c	Prob	7.80	.036	6.40	.172	8.00	.080	7.52	.107
3.50	.273	8.10	.033	6.67	.163	8.20	.072	7.68	.094
4.50	.125	8.40	.019	6.93	.127	8.40	.063	7.84	.089
6.00	.069	8.70	.014	7.20	.117	8.60	.060	8.00	.082
6.50	.042	9.30	.012	7.47	.096	8.80	.049	8.16	.077
8.00	.005	Greater	< .01	7.73	.080	9.00	.043	8.32	.073
				8.00	.063	9.20	.038	8.48	.066
k = 3, b = 5		**k = 4, b = 5**		8.27	.056	9.40	.035	8.64	.058
c	Prob	c	Prob	8.53	.045	9.60	.028	8.80	.056
				8.80	.038	9.80	.025	8.96	.049
2.80	.367	4.92	.210	9.07	.028	10.00	.021	9.12	.046
3.60	.182	5.16	.162	9.33	.026	10.20	.019	9.28	.042
4.80	.124	5.40	.151	9.60	.017	10.40	.017	9.44	.038
5.20	.093	5.88	.123	9.87	.015	10.60	.014	9.60	.035
6.40	.039	6.12	.107	Greater	< .01	10.80	.011	9.76	.032
7.60	.024	6.36	.093			11.00	.010	9.92	.029
Greater	< .01	6.84	.075			Greater	< .01	10.08	.026
		7.08	.067					10.24	.024
k = 4, b = 2		7.32	.055					10.40	.022
c	Prob	7.80	.044					10.56	.019
		8.04	.034					10.72	.018
4.80	.208	8.28	.031					10.88	.015
5.40	.167	8.76	.023					11.04	.013
6.00	.042	9.00	.020					11.20	.012
		9.24	.017					11.36	.012
k = 4, b = 3		9.72	.012					11.52	.010
c	Prob	Greater	< .01					Greater	< .01
5.00	.207								
5.40	.175								
5.80	.148								
6.60	.075								
7.00	.054								
7.40	.033								
8.20	.017								
Greater	< .01								

TABLE A12
Upper-Tail Probabilities for Spearman Rank Correlation, $P(r_s \geq c)$

$n = 4$		$n = 7$		$n = 8$		$n = 9$		$n = 10$		$n = 10$	
c	Prob	c	Prob	c	Prob	c	Prob	c	Prob	c	Prob
.00	.542	.00	.518	.00	.512	.00	.509	.01	.500	.62	.030
.20	.458	.04	.482	.02	.488	.02	.491	.02	.486	.64	.027
.40	.375	.07	.453	.05	.467	.03	.474	.03	.473	.65	.024
.60	.208	.11	.420	.07	.441	.05	.456	.04	.459	.66	.022
.80	.167	.14	.391	.10	.420	.07	.440	.05	.446	.67	.019
1.00	.042	.18	.357	.12	.397	.08	.422	.07	.433	.68	.017
		.21	.331	.14	.376	.10	.405	.08	.419	.70	.015
$n = 5$		.25	.297	.17	.352	.12	.388	.09	.406	.71	.013
c	Prob	.29	.278	.19	.332	.13	.372	.10	.393	.72	.012
		.32	.249	.21	.310	.15	.354	.12	.379	.73	.010
.00	.525	.36	.222	.24	.291	.17	.339	.13	.367	.75	.009
.10	.475	.39	.198	.26	.268	.18	.322	.14	.354	.76	.007
.20	.392	.43	.177	.29	.250	.20	.307	.15	.341	.77	.006
.30	.342	.46	.151	.31	.231	.22	.290	.16	.328	.78	.005
.40	.258	.50	.133	.33	.214	.23	.276	.18	.316	Greater	< .005
.50	.225	.54	.118	.36	.195	.25	.260	.19	.304		
.60	.175	.57	.100	.38	.180	.27	.247	.20	.292		
.70	.117	.61	.083	.40	.163	.28	.231	.21	.280		
.80	.067	.64	.069	.43	.150	.30	.218	.22	.268		
.90	.042	.68	.055	.45	.134	.32	.205	.24	.257		
1.00	.008	.71	.044	.48	.122	.33	.193	.25	.246		
		.75	.033	.50	.108	.35	.179	.26	.235		
$n = 6$		.79	.024	.52	.098	.37	.168	.27	.224		
c	Prob	.82	.017	.55	.085	.38	.156	.28	.214		
		.86	.012	.57	.076	.40	.146	.30	.203		
.03	.500	.89	.006	.60	.066	.42	.135	.31	.193		
.09	.460	.93	.003	.62	.057	.43	.125	.32	.184		
.14	.401	.96	.001	.64	.048	.45	.115	.33	.174		
.20	.357	1.00	.000	.67	.042	.47	.106	.35	.165		
.26	.329			.69	.035	.48	.097	.36	.156		
.31	.282			.71	.029	.50	.089	.37	.148		
.37	.249			.74	.023	.52	.081	.38	.139		
.43	.210			.76	.018	.53	.074	.39	.132		
.49	.178			.79	.014	.55	.066	.41	.124		
.54	.149			.81	.011	.57	.060	.42	.116		
.60	.121			.83	.008	.58	.054	.43	.109		
.66	.088			.86	.005	.60	.048	.44	.102		
.71	.068			.88	.004	.62	.043	.45	.096		
.77	.051			.90	.002	.63	.038	.47	.089		
.83	.029			.93	.001	.65	.033	.48	.083		
.89	.017			.95	.001	.67	.029	.49	.077		
.94	.008			.98	.000	.68	.025	.50	.072		
1.00	.001			1.00	.000	.70	.022	.52	.067		
						.72	.018	.53	.062		
						.73	.016	.54	.057		
						.75	.013	.55	.052		
						.77	.011	.56	.048		
						.78	.009	.58	.044		
						.80	.007	.59	.040		
						.82	.005	.60	.037		
						Greater	< .005	.61	.033		

For negative values of c, $P(r_s \leq c) = P(r_s \geq -c)$.

TABLE A13
Upper-Tail Probabilities for Kendall's Tau, $P(r_\tau \geq c)$

n = 4 c	Prob	n = 7 c	Prob	n = 8 c	Prob	n = 9 c	Prob	n = 10 c	Prob
.00	.625	.05	.500	.00	.548	.00	.540	.02	.500
.33	.375	.14	.386	.07	.452	.06	.460	.07	.431
.67	.167	.24	.281	.14	.360	.11	.381	.11	.364
1.00	.042	.33	.191	.21	.274	.17	.306	.16	.300
		.43	.119	.29	.199	.22	.238	.20	.242
n = 5 c	Prob	.52	.068	.36	.138	.28	.179	.24	.190
		.62	.035	.43	.089	.33	.130	.29	.146
.00	.592	.71	.015	.50	.054	.39	.090	.33	.108
.20	.408	.81	.005	.57	.031	.44	.060	.38	.078
.40	.242	.90	.001	.64	.016	.50	.038	.42	.054
.60	.117	1.00	.000	.71	.007	.56	.022	.47	.036
.80	.042			.79	.003	.61	.012	.51	.023
1.00	.008			.86	.001	.67	.006	.56	.014
				.93	.000	.72	.003	.60	.008
n = 6 c	Prob			1.00	.000	.78	.001	.64	.005
						.83	.000	.69	.002
						.89	.000	.73	.001
.07	.500					.94	.000	.78	.001
.20	.360					1.00	.000	.82	.000
.33	.235							.87	.000
.47	.136							.91	.000
.60	.068							.96	.000
.73	.028							1.00	.000
.87	.008								
1.00	.001								

For negative values of c, $P(r_\tau \leq c) = P(r_\tau \geq -c)$.

References

Agar, J. W., Jr., and Brent, S. (1978). An Index of Agreement Between a Hypothesized Partial Order and an Empirical Rank Order. *Journal of the American Statistical Association,* 73, 827–830.

Agresti, A. (1990). *Categorical Data Analysis.* John Wiley & Sons, New York.

Agresti, A., and Liu, I.-M. (1999). Modeling a Categorical Variable Allowing Arbitrary Many Category Choices. *Biometrics*, 55, 935–943.

Akritas, M. G. (1990). The Rank Transform Method in Some Two-Factor Designs. *Journal of the American Statistical Association,* 85, 73–78.

Ansari, A. R., and Bradley, R. A. (1960). Rank Sum Tests for Dispersion. *Annals of Mathematical Statistics*, 31, 1174–1189.

Bailer, A. J. (1988). Testing Variance Equality with Randomization Tests. *Journal of Statistical Computation and Simulation*, 31, 1–8.

Bain, L. J., and Engelhardt, M. (1992). *Introduction to Probability and Mathematical Statistics*, 2nd ed. PWS-Kent, Boston.

Bean, S. J. and Tsokos, C. P. (1980). Developments in Nonparametric Density Estimation. *International Statistical Review*, 48, 267–287.

Beyer, W. H., ed. (1968). *CRC Handbook of Tables for Probability and Statistics,* 2nd ed. Chemical Rubber Co., Cleveland.

Bilder, C. R. (2000). *Testing for Marginal Independence with Pick Any/C Variables*. Ph.D. Dissertation, Department of Statistics, Kansas State University, Manhattan.

Bilder, C. R., Loughin, T. M., and Nettleton, D. (2000). Multiple Marginal Independence Testing for Pick Any/c Variables. *Communications in Statistics: Simulation and Computation,* 29(4).

Blair, R. C., and Higgins, J. J. (1978). Tests of Hypotheses for Unbalanced Factorial Designs Under Various Regression/Coding Method Combinations. *Educational and Psychological Measurement*, 38, 621–631.

Blair, R. C., and Higgins, J. J. (1980). A Comparison of the Power of Wilcoxon's Rank-Sum Test to That of Student's *T* Statistic Under Various Non-Normal Distributions. *Journal of Educational Statistics*, 5(4), 309–335.

Blair, R. C., and Higgins, J. J. (1985). A Comparison of the Power of the Paired Sample *T*-Test to That of Wilcoxon's Signed-Ranks Test Under Various Population Shapes. *Psychological Bulletin*, 97, 119–128.

Blair, R. C., Higgins, J. J., Karniski, W., and Kromrey, J. D. (1994). A Study of Multivariate Permutation Tests Which May Replace Hotelling's *T*-Square Test in Prescribed Circumstances. *Multivariate Behavioral Research*, 29, 141–163.

Blair, R. C., Higgins, J. J., and Sawilowsky, S. S. (1987). Limitations of the Rank Transform Statistic in Tests for Interaction. *Communications in Statistics: Simulation and Computation,* 16, 1133–1145.

Blank, S., Seiter, C., and Bruce, P. (1999). *Resampling Stats in Excel.* Resampling Stats, Inc., Arlington, VA.

Booth, J. G., and Sarkar, S. (1998). Monte Carlo Approximation of Bootstrap Variances. *The American Statistician*, 52, 354–357.

Box, G. E. P. (1954). I. Some Theorems on Quadratic Forms Applied in the Study of Analysis of Variance Problems. *Annals of Mathematical Statistics*, 25, 290–302.

Boyett, J. M., and Shuster, J. J. (1977). Nonparametric One-Sided Tests in Multivariate Analysis with Medical Applications. *Journal of the American Statistical Association*, 72, 665–668.

Cleveland, W. S., and Devlin, S. J. (1988). Locally-Weighted Regression: An Approach to Regression Analysis by Local Fitting. *Journal of the American Statistical Association,* 83, 596–610.

Cochran, W. G. (1963). *Sampling Techniques,* 2nd ed. John Wiley & Sons, New York.

Conover, W. J. (1999). *Practical Nonparametric Statistics,* 3rd ed. John Wiley & Sons, New York.

Conover, W. J., Johnson, M. E., and Johnson, M. M. (1981). Comparative Study of Tests for Homogeneity of Variances, with Applications to the Outer Continental Shelf Bidding Data. *Technometrics*, 23, 351–361.

Cressie, N. A. C. (1982). Playing Safe with Misweighted Means. *Journal of the American Statistical Association*, 77, 754–759.

Davison, A. C., and Hinkley, D. V. (1997). *Bootstrap Methods and Their Application.* Cambridge University Press, Cambridge, UK.

Efron, B. (1979). Bootstrap Methods: Another Look at the Jackknife. *Annals of Statistics*, 7, 1–26.

Efron, B., and Tibshirani, R. (1993). *An Introduction to the Bootstrap.* Chapman & Hall, London.

Fisher, R. A. (1935). *The Design of Experiments.* Oliver & Boyd, Edinburgh.

Fleming, T. R., O'Fallon, J. R., and O'Brien, P. C. (1980). Modified Kolmogorov–Smirnov Test Procedures with Application to Arbitrarily Right-Censored Data. *Biometrics*, 36, 607–625.

Freeman, G. H., and Halton, J. H. (1951). Note on an Exact Treatment of Contingency Tables, Goodness of Fit, and Other Problems of Significance. *Biometricka*, 38, 141–149.

Freireich, E. J., et al. (1963). The Effect of 6-Mercaptopurine on the Duration of Steroid-Induced Remissions in Acute Leukemia. *Blood*, 21, 699–716.

Gehan, E. A. (1965). A Generalized Wilcoxon Test for Comparing Arbitrarily Singly-Censored Samples. *Biometrika*, 52, 203–224.

Good, P. (2000). *Permutation Tests. A Practical Guide to Resampling Methods for Testing Hypotheses,* 2nd ed. Springer, New York.

Graybill, F. A. (1976). *Theory and Applications of the Linear Model*, Duxbury, North Scituate, MA.

Hardle, W. (1991). *Smoothing Techniques with Implementation in S.* Springer-Verlag, New York.

Hart, J. D. (1997). *Nonparametric Smoothing and Lack-of-Fit Tests.* Springer, New York.

Heavenrich, R. M., Murrell, J. D., and Hellman, K. H. (1991). *Light Duty Automotive Technology and Fuel Economy Trends Through 1991.* U.S. Environmental Protection Agency, EPA/AA/CTAB/91-02.

Hettmansperger, T. P. (1984). *Statistical Inference Based on Ranks.* John Wiley & Sons, New York.

Hettmansperger, T. P., and McKean, J. W. (1998). *Robust Nonparametric Statistical Methods*. Arnold, London.

Higgins, J. J., and Bain, P. T. (1999). Nonparametric Tests for Lattice-Ordered Alternatives in Unreplicated Two-Factor Experiments. *Journal of Nonparametric Statistics*, 11, 307–318.

Higgins, J. J., Blair, R. C., and Tashtoush, S. (1990). The Aligned Rank Transform Procedure. *Proceedings of the 1990 Kansas State University Conference on Applied Statistics in Agriculture*, 185–195.

Higgins, J. J., and Keller-McNulty, S. (1995). *Concepts in Probability and Stochastic Modeling*. Duxbury, Belmont, CA.

Higgins, J. J., and Tashtoush, S. (1994). An Aligned Rank Transform Test for Interaction. *Nonlinear World*, 1, 201–211.

Hodges J. L., and Lehmann, E. L. (1956). The Efficiency of Some Nonparametric Competitors of the *t*-Test. *Annals of Mathematical Statistics,* 27, 324–335.

Hoeffding, W. (1952). The Large-Sample Power of Tests Based on Permutations of Observations. *Annals of Mathematical Statistics,* 23, 169–192.

Hogg, R. V., and Craig, A. T. (1995). *Introduction to Mathematical Statistics*, 5th ed. Prentice-Hall, Englewood Cliffs, NJ.

Hollander, M., and Wolfe, D. (1999). *Nonparametric Statistical Methods*, 2nd ed. John Wiley & Sons, New York.

Huber, P. J. (1964). Robust Estimation of a Location Parameter, *Annals of Mathematical Statistics*, 35, 73–101.

Iman, R., and Conover, J. (1981). Rank Transform as a Bridge Between Parametric and Nonparametric Statistics. *American Statistician*, 35, 124–133.

Jaeckel, L. A. (1972). Estimating Regression Coefficients by Minimizing the Dispersion of the Residuals. *Annals of Mathematical Statistics*, 43, 1449–1458.

Keller-McNulty, S., and Higgins, J. J. (1987). Effect of Tail Weight and Outliers on Power and Type-I Error of Robust Permutation Tests for Location. *Communications in Statistics: Simulation and Computation*, 16(1), 17–36.

Keller-McNulty, S., and McNulty, M. (1987). The Independent Pairs Assumption in Hypothesis Tests Based on Rank Correlation Coefficients. *The American Statistician*, 41, 40–41.

Lawless, J. F. (1982). *Statistical Models and Methods for Lifetime Data*. John Wiley & Sons, New York.

Lee, E. T., Desu, M. M., and Gehan, E. A. (1975). A Monte Carlo Study of the Power of Some Two-Sample Tests. *Biometrika*, 62, 425–432.

Lehmann, E. L. (1975). *Nonparametrics: Statistical Methods Based on Ranks*. Holden-Day, San Francisco.

Lehmann, E. L. (1986). *Testing Statistical Hypotheses*, 2nd ed. John Wiley & Sons, New York.

Lingenfelser, J. (2001). *Comparison of Genotype and Cultural Practices to Control Iron Deficiency Chlorosis in Soybeans*. M.S. Thesis, Department of Agronomy, Kansas State University, Manhattan.

Lunneborg, C. E. (2000). *Data Analysis by Resampling: Concepts and Applications*. Duxbury, Pacific Grove, CA.

Manly, B. F. J. (1997). *Randomization, Bootstrap and Monte Carlo Methods in Biology.* 2nd ed. Chapman & Hall, London.

Mansouri, H. (1999). Multifactor Analysis of Variance Based on the Aligned Rank Transform Technique. *Computational Statistics & Data Analysis,* 29, 177–189.

Matlack, R. S. (2001). *The Ecology of Shrews in Tallgrass Prairie*. Ph.D. Dissertation, Division of Biology, Kansas State University, Manhattan.

McCullough, B. D., and Wilson, B. (1999). On the Accuracy of Statistical Procedures in Microsoft Excel 97. *Computational Statistics & Data Analysis*, 31, 27–37.

McCullough, B. D., and Wilson, B. (2002). On the Accuracy of Statistical Procedures in Microsoft Excel 2000 and Excel XP. *Computational Statistics & Data Analysis*, 40(4), 685–711.

McKean, J. W. (2002). RANOVA-RGLM's ANOVA. Web site address http://www.stat.wmich.edu/slab/RGLM.

McKean, J. W., and Ryan, T. A. (1977). An Algorithm for Obtaining Confidence Intervals and Point Estimates Based on Ranks in the Two Sample Location Problem. *Transactions on Mathematical Software*, 183–185.

Mehta, C. R., and Patel, N. R. (1983). A Network Algorithm for Performing Fisher's Exact Test in $r \times c$ Contingency Tables. *Journal of the American Statistical Association,* 78, 427–434.

Mehta, C. R., Patel, N. R., and Senchaudhuri, P. (1988). Importance Sampling for Estimating Exact Probabilities in Permutational Inference. *Journal of the American Statistical Association,* 83, 999–1005.

Mehta, C. R., Patel, N. R., and Wei, L. J. (1988). Constructing Exact Significance Tests with Restricted Randomization Rules. *Biometrika*, 75, 295–302.

Milliken, G. A., and Johnson, D. E. (1984). *Analysis of Messy Data, Volume I: Designed Experiments*. Chapman & Hall, London.

Moore, D. S. and McCabe, G. P. (2002). *Introduction to the Practice of Statistics,* 4th ed. W. H. Freeman, New York.

Nadaraya, E. A. (1964). On Estimating Regression. *Theory of Probability and Applications*, 9, 141–142.

Nandi, C., and Nelson, M. (1992). Maternal Pregravid Weight, Age, and Smoking Status as Risk Factors for Low Birth Weight Births. *Public Health Reports*, 107, 658–662.

Nelson, W. (1982). *Applied Life Data Analysis*. John Wiley & Sons, New York.

Ott, R. L., and Longnecker, M. (2001). *An Introduction to Statistical Methods and Data Analysis,* 5th ed. Duxbury, Pacific Grove, CA.

Parzen, E. (1962). On Estimation of a Probability Density Function and Mode. *Annals of Mathematical Statistics*, 33, 1065–1076.

Peto, R. (1972). Rank Tests of Maximal Power Against Lehmann-Type Alternatives. *Biometrika*, 59, 472–475.

Pitman, E. J. G. (1937a). Significance Tests Which May be Applied to Samples from Any Population. *Journal of the Royal Statistical Society*, Series B, 4, 119–130.

Pitman, E. J. G. (1937b). Significance Tests Which May Be Applied to Samples from Any Population. II. The Correlation Coefficient Test. *Journal of the Royal Statistical Society*, Series B, 4, 225–232.

Pitman, E. J. G. (1938). Significance Tests Which May Be Applied to Samples from Any Population. III. The Analysis of Variance Test. *Biometrika,* 29, 322–335.

Pratt, J. W. (1959). Remarks on Zeros and Ties in the Wilcoxon Signed Rank Procedures. *Journal of the American Statistical Association*, 54, 655–667.

Prentice, R. L. (1978). Linear Rank Tests with Right Censored Data. *Biometrika*, 65, 167–180.

Prentice, R. L., and Marek, P. (1979). A Quantitative Discrepancy Between Censored Data Rank Tests. *Biometrics*, 35, 861–886.

Puri, M. L., and Sen, P. K. (1985). *Nonparametric Methods in General Linear Models.* John Wiley & Sons, New York.

Randles, R. H., and Wolfe, D. A. (1979). *Introduction to the Theory of Nonparametric Statistics*. John Wiley & Sons, New York.

Resampling Stats User's Guide. (1999). Resampling Stats, Inc., Arlington, VA.

Robins, J., Breslow, N., and Greenland, S. (1986). Estimators of the Mantel–Haenszel Variance Consistent in Both Sparse Data and Large-Strata Limiting Models. *Biometrics*, 42, 311–323.

Rousseeuw, P. J., and Leroy, A. M. (1987). *Robust Regression and Outlier Detection*. John Wiley and Sons, New York.

S-Plus 2000 Guide to Statistics, Vol. 1. (1999). Data Analysis Products Division, MathSoft, Seattle.

S-Plus 2000 Guide to Statistics, Vol. 2. (1999). Data Analysis Products Division, MathSoft, Seattle.

Salter, K. C., and Fawcett, R. F. (1993). The ART Test of Interaction: A Robust and Powerful Rank Test of Interaction in Factorial Models. *Communications in Statistics: Theory and Methods,* 22, 137–153.

Sawilowsky, S. S., Blair, R. C., and Higgins, J. J. (1989). An Investigation of the Type I Error and Power Properties of the Rank Transform Procedure in Factorial ANOVA. *Journal of Educational Statistics,* 14, 225–268.

Schenker, N. (1985). Qualms About Bootstrap Confidence Intervals. *Journal of the American Statistical Association*, 80, 360–361.

Scott, D. W. (1992). *Multivariate Density Estimation*. John Wiley & Sons, New York.

StatXact 4 for Windows. (2000). Cytel Software Corporation, Cambridge, MA.

Strand, M. (2000). A Generalized Nonparametric Test for Lattice-Ordered Means. *Biometrics*, 56, 1222–1226.

ter Braak, C. J. F. (1992). Permutation versus Bootstrap Significance Tests in Multiple Regression and ANOVA. In *Bootstrapping and Related Techniques*, edited by K. H. Jöckel, 79–86. Springer-Verlag, Berlin.

Theil, H. (1950). A Rank-Invariant Method of Linear and Polynomial Regression. *Koninkl. Ned. Akad. Wetenschap*, 53, 386–392, 521–525, 1397–1412.

Thompson, G. L. (1991). A Unified Approach to Rank Tests for Multivariate Repeated Measures Designs. *Journal of the American Statistical Association,* 86, 410–419.

Watson, G. S. (1964). Smooth Regression Analysis. *Sankhya Series A*, 26, 359–372.

Welch, B. L. (1951). On the Comparison of Several Mean Values: An Alternative Approach. *Biometrika*, 38, 330–336.

Westfall, P., Tobias, R., Rom, D., Wolfinger, R., and Hochberg, Y. (1999). *Multiple Comparisons and Multiple Tests Using the SAS® System*. SAS Institute Inc., Cary, NC.

Woreck, C. L., (1997). *A Comparison of Parametric and Nonparametric Procedures for Testing Equality of Scale Parameters in a Two-Sample Case*. M.S. Report, Kansas State University, Manhattan.

Yohai, V. J., and Zamar, R. H. (1998). Optimal Locally Robust *M*-Estimates of Regression. *Journal of Statistical Planning and Inference,* 64, 309–323.

Index